Structural Optimization

Related Titles:

H. H. E. Leipholz, Structural Control. 1985.
ISBN 90–247–3429–0

J. D. Achenbach and Y. Rajapakse, Solid Mechanics Research for Quantitative Non-Destructive Evaluation. 1985.
ISBN 90–247–3428–2

Structural Optimization

Proceedings of the IUTAM Symposium on Structural Optimization, Melbourne, Australia, 9–13 February 1988

Edited by

G. I. N. Rozvany
Department of Structural Design, Essen University, Essen, F.R.G.

and

B. L. Karihaloo
School of Civil and Mining Engineering, The University of Sydney, Sydney, Australia

KLUWER ACADEMIC PUBLISHERS
DORDRECHT / BOSTON / LONDON

Library of Congress Cataloging in Publication Data

Structural Optimization

ISBN-13:978-94-010-7132-1 e-ISBN-13:978-94-009-1413-1
DOI: 10.1007/978-94-009-1413-1

Published by Kluwer Academic Publishers,
P.O. Box 17, 3300 AA Dordrecht, The Netherlands.

Kluwer Academic Publishers incorporates
the publishing programmes of
D. Reidel, Martinus Nijhoff, Dr W. Junk and MTP Press.

Sold and distributed in the U.S.A. and Canada
by Kluwer Academic Publishers,
101 Philip Drive, Norwell, MA 02061, U.S.A.

In all other countries, sold and distributed
by Kluwer Academic Publishers Group,
P.O. Box 322, 3300 AH Dordrecht, The Netherlands.

Softcover reprint of the hardcover 1st edition 1988

CONTENTS

Abstracts of papers presented at the Symposium which are likely to appear in an early issue of the Journal "Structural Optimization", (Springer-Verlag).

SCIENTIFIC COMMITTEE

V.V Bolotin (USSR)
D.C. Drucker (USA)
H. Eschenauer (FRG)
E. Haug (USA)
B.L. Karihaloo (Australia, Joint Chairman)
G. Maier (Italy)
Y. Murotsu (Japan)
N. Olhoff (Denmark)
G.I.N. Rozvany (FRG, Joint Chairman)
M. Save (Belgium)
R.T. Shield (USA)
C.A. Mota Soares (Portugal)
M. Życzkowski (Poland)

LIST OF PARTICIPANTS

M.S. Anderson
Dept. of Mechanical Engrg. and Mechanics, Old Dominion University, Norfolk, VA. 23508, USA

J. Bäcklund
Dept. of Aeronautical Struct. and Materials, The Royal Institute of Technology, S-10044 Stockholm, SWEDEN

M. Balachandran
Dept. of Architectural Science, University of Sydney, Sydney, NSW 2006, AUSTRALIA

N.V. Banichuk
Institute for Problems in Mechanics, USSR Academy of Sciences, Vernadskogo 101, 117526 Moscow V–526, USSR

M.P. Bendsøe
Mathematical Institute, The Technical University of Denmark, Building 303, DK-2800 Lyngby, DENMARK

L. Berke
NASA Lewis Research Center, Cleveland, Ohio 44136, USA

R. Bogacz
Institute for Fundamental, Technological Research, Swietokrzyska 21, 00-049 Warsaw, POLAND

V.V. Bolotin*
Institute of Machine Design, Griboedov Street 4, 10 18 30 Moscow, USSR

---

* Participants who did not contribute a paper are marked with an asterisk.

G. Boully*
Road Construction Authority, 60 Denmark Street, Kew, VIC 3101, AUSTRALIA

S. Brown*
Road Construction Authority, 60 Denmark Street, Kew, VIC. 3101, AUSTRALIA

G. Cheng
Dalian Institute of Technology, Dalian 116024, P.R. CHINA

C. Cinquini
Dept. of Struct. Mechanics, University of Pavia, Via Abbiategrasso 211, 27100 Pavia, ITALY

R. Contro
Dept. of Engineering, University of Trento, Mesiano di Povo, 38050 Trento, ITALY

V. De Angelis
Dept. of Statistics & Probability, University "La Sapienza", Rome, ITALY

R. de Boer
Fachbereich Bauwesen, Universität–Gesamthochschule–Essen, Universitätstraße 15, 4300 Essen 1, FRG

F. Ellyin
Dept. of Mechanical Engineering, University of Alberta, 4-9 Mechanical Engineering Building, Edmonton, Alberta T6G 2G8, CANADA

H.A. Eschenauer
University of Siegen, Research Laboratory for Applied, Structural Optimization, 5900 Siegen 21, FRG

B.J.D. Esping
Dept. of Aeronautical Struct. and Materials, The Royal Institute of Technology, S-10044 Stockholm, SWEDEN

J.F. Gero
Dept. of Architectural Science, University of Sydney, Sydney, NSW 2006, AUSTRALIA

P. Grundy*
Dept. of Civil Engineering, Monash University, Clayton, VIC. 3168, AUSTRALIA

C.S. Gurujee
Department of Civil Engineering, Indian Institute of Technology, P.O. I.I.T. Powai, Bombay 400076, INDIA

P. Hajela
Department of Engineering Sciences, University of Florida, 231 Aerospace Engineering Building, Gainsville, Florida 32611, USA

W.S. Hemp
Duffryn House, Horton-cum-Studley, Oxford OX9 1AW, ENGLAND

D. Holm
Dept. of Aeronautical Struct. and Materials, The Royal Institute of Technology, S-10044 Stockholm, SWEDEN

M. Hýča
Nat. Res. Inst. of Machine Design (SVÚSS), Department of Loading Capacity, Husova 8, CS - 110 00 Praha 1, CSSR

R. Isby
Dept. of Aeronautical Struct. and Materials The Royal Institute of Technology, S-10044 Stockholm, SWEDEN

L.F. Jansen
University of Twente, P.O. Box 217, 7500 AE Enschede, THE NETHERLANDS

S. Kanagasundaram
School of Civil and Mining Engineering, The University of Sydney, Sydney, NSW 2006, AUSTRALIA

B.L. Karihaloo
School of Civil and Mining Engineering, The University of Sydney, Sydney, NSW 2006, AUSTRALIA

N.S. Khot
Air Force Wright Aeronaut. Lab., (AFWAL/FIBR), Wright–Patterson Air Force Base, Ohio 45 433 – 6553, USA

J. Koski
University of Oulu and Tampere University of Technology, Linnanmaa, SF-90570 Oulu, FINLAND

M. Lawo
Kernforschungszentrum Karlsruhe GmbH, Inst. f. Datenverarb. i.d. Technik, Postfach 3640, 7500 Karlsruhe 1, FRG

P.G. Lowe
Dept. of Civil Engineering, University of Auckland, Private Bag, Auckland, NEW ZEALAND

O. Mahrenholtz
Arbeitsgebiet Meerestechnik II, TU Hamburg – Harburg, Eissendorfer Strasse 38, 2100 Hamburg 90, FRG

R.E. Melchers
Dept. of Civil Engrg. and Surveying, The University of Newcastle, Newcastle, NSW 2308, AUSTRALIA

R. Morgan*
BHP Engineering, BHP House, 152 Wharf St., Brisbane, QLD 4000, AUSTRALIA

C.A. Mota Soares
CEMUL, Instituto Superior Técnico, Avenida Rovisco Pais, 1096 Lisboa Codex, PORTUGAL

Y. Murotsu
Dept. of Aeronautical Engineering, University of Osaka Prefecture, Sakai, Osaka 591, JAPAN

J. Oda
Dept. of Mechanical Engineering, Kanazawa University, 2-40-20 Kodatsuno, Kanazawa, 920, JAPAN

N. Olhoff
University of Aalborg, Mechanical Engineering and Energy Techn., Pontoppidanstraede 101, DK – 9220 Aalborg, DENMARK

R.D. Parbery
Dept. of Mechanical Engineering, University of Newcastle, Newcastle, NSW 2308, AUSTRALIA

Z.A. Parszewski
Mechanical Engineering, The University of Melbourne, Parkville, Victoria 3052, AUSTRALIA

P. Pedersen
Department of Solid Mechanics, The Technical University of Denmark, Building 404, DK-2800 Lyngby, DENMARK

A. Pičuga
Faculty of Mechanical Engineering, University Džemal Bijedić, B. Parovića bb, 88000 Mostar, YUGOSLAVIA

R.H. Plaut
Department of Civil Engineering, Virginia Polytechnic Inst. and State Univ., Blacksburg, VA 24061, USA

E. Polak
Dept. of Electrical Engineering and Computer Sciences, University of California, Berkeley, CA 94720, USA

P.U. Post
University of Siegen, Research Laboratory for Applied, Structural Optimization, 5900 Siegen 21, FRG

U.T. Ringertz
Dept. of Aeronautical Struct. and Materials, The Royal Institute of Technology, S-10044 Stockholm, SWEDEN

G.I.N. Rozvany
Fachbereich Bauwesen, Universität–Gesamthochschule–Essen, Universitätstraße 15, 4300 Essen 1, FRG

E. Sandgren
School of Mechanical Engineering, Room 302, Purdue University, West Lafayette, Indiana 47906, USA

W.O. Schiehlen
Institute B of Mechanics, University of Stuttgart, Pfaffenwaldring 9, 7000 Stuttgart 80, FRG

K. Schittkowski
Mathematical Institute, University of Bayreuth, P.O. Box 3008, 8580 Bayreuth, FRG

E. Schnack
Institute of Solid Mechanics, University of Karlsruhe, Kaiserstrasse 12, 7500 Karlsruhe 1, FRG

A.J.G. Schoofs
Eindhoven University of Technology, P.O.Box 513, 5600 MB Eindhoven, THE NETHERLANDS

Y. Seguchi
Department of Mechanical Engineering, Osaka University, Toyonaka, Osaka 560, JAPAN

L.M.C. Simões
Departamento de Engenharia Civil, Faculdade de Ciências e Tecnologia, Universidade de Coimbra, 3000 Coimbra, PORTUGAL

J. Sobieski
Interdisciplinary Research Office, MS 246, Structures Directorate, NASA Langley Research Center, Hampton, VA 23669-5225, USA

L.J. Sparke
Holden Motor Company, GPO Box 1714, Melbourne, VIC 3001, AUSTRALIA

W. Stadler
Division of Engineering, San Francisco State University, 1600 Holloway Avenue, San Francisco, CA 94132, USA

Y. Tada
Dept. of Systems Engineering, Kobe University, Rokkodai, Nada, Kobe, 657, JAPAN

D.P. Thambiratnam
Department of Civil Engineering, National University of Singapore, Kent Ridge, Singapore 0511, SINGAPORE

C.M. Wang
Department of Civil Engineering, National University of Singapore, Kent Ridge, Singapore 0511, SINGAPORE

H. Yamakawa
Department of Mechanical Engineering, Waseda University, 3-4-1 Okubo, Shinjuku-ku, Tokyo 160, JAPAN

K. Yamazaki
Department of Mechanical Systems Engrg., Kanazawa University, Kodatsuno 2-40-20, Kanazawa, 920 JAPAN

K.-Y. Yeh
Department of Mechanics, Lanzhou University, Lanzhou, Gansu 730 001, P.R. CHINA

M. Yoshimura
Dept. of Precision Engineering, Kyoto University, Sakyo-ku, Kyoto 606, JAPAN

M. Życzkowski
Politechnika Krakowska, ul. Warszawska 24, Krakow, POLAND

SCIENTIFIC PROGRAMME

9 February, Tuesday

Opening Session:

Welcoming speeches by Prof. B.L. Karihaloo, Joint Symposium Chairman, and Prof. W.O. Schiehlen, Secretary General of IUTAM. Opening address by Prof. G.I.N. Rozvany, Joint Symposium Chairman: *"Historical Review of Layout Optimization"*

Session A: Layout Optimization in Structural Design

Chairman: N. Olhoff, Co-Chairman: R. de Boer

G. Rozvany:
Optimality Criteria and Layout Theory in Structural Design: Recent Developments and Applications

W.S. Hemp:
A Michell Type Criterion for Shells

P.G. Lowe:
Optimization of Systems in Bending

M.P. Bendsøe:
Composite Materials as a Basis for Generating Topologies in Optimal Shape Design

Session B: Shape Optimization I

Chairman: E. Polak, Co-Chairman: N.S. Khot

J. Sobieski:
Structural Shape Optimization in Multidisciplinary Engineering System Synthesis

E. Sandgren & M. El Sayed:
Shape Optimization: Creating a Useful Design Tool

P. Hajela & J. Jih:
Boundary Element Methods in Optimal Shape Design — An Integrated Approach

W.O. Schiehlen:
Optimal Shape of Pendulum Links

P. Pedersen:
Shape Design for Minimum Stress Concentration

E. Schnack:
Gradientless Shape Optimization with FEM

10 February, Wednesday

Session C: Numerical Methods

Chairman: J. Sobieski, Co-Chairman: Y. Murotsu

B.L. Karihaloo & S. Kanagasundaram:
Comparative Study of NLP Techniques in Optimum Structural Frame Design

E. Polak:
Minimax Algorithms for Structural Optimization

N. Olhoff:
Solution of Max-Min Problems via Bound Formulation and Mathematical Programming

Y. Tada & H. Taketani:
Boundary Element Approach to Optimal Structural Design Based on the Inverse Variational Principle

U.T. Ringertz:
Discrete Continuous Structural Optimization

H.A. Eschenauer & P.U. Post:
Optimization Procedure SAPOP Applied to Optimal Layouts of Complex Structures

H.A. Eschenauer:
On Shape Optimization of Sattelite Tanks

Session D: Expert Systems and Multicriterion Design

Chairman: O. Mahrenholtz, Co-Chairman: Y. Tada

N.S. Khot:
Structural and Control Optimization for Space Structures

M. Balachandran & J.S. Gero:
Development of a Knowledge-Based System for Structural Optimization

K. Schittkowski:
Integrated Problem Solving Systems in Structural Optimization

W. Stadler & V. Krishnan:
Natural Structural Shapes of Membrane Shells

J. Koski, M. Lawo & R. Silvennoinen:
Multicriterion Plate Optimization

A. Pičuga:
Optimality Conditions for Multiple Loaded Structures: Integration of Control and Finite Element Methods

11 February, Thursday

Session E: Safety, Dynamic & Stability Problems

Chairman: M. Życzkowski, Co-Chairman: R.H. Plaut

L. Berke & N.S. Khot:
Performance Characteristics of Optimality Criteria Methods

R. Bogacz & O. Mahrenholtz:
On the Optimal Design of Columns Subjected to Circulatory Loads

R. de Boer:
Optimization of Vibrating Thin-Walled Structures

A.N. Kounadis & O. Mahrenholtz:
Divergence Instability Conditions in the Optimum Design of Nonlinear Elastic Systems under Follower Loads

Y. Seguchi & S. Kojima:
On the Shape Determination of Non-Conservative System: A Case of Column under Follower Force

H. Yamakawa:
Optimum Designs of Rotating Shaft Systems for Dynamic Nonlinear Responses

J. Oda:
On Optimality in Structural and Material Composition of Bamboo

Session F: Plates, Shells, Panels and Machine Systems

Chairman: J. Bäcklund, Co-Chairman: K. Schittkowski

M.S. Anderson:
Application of Optimization Methods to the Practical Design of Shear and Compression Loaded Stiffened Panels

D.T. Young & R.H. Plaut:
A Variational Principle for Optimizing Rectangular-Base Shallow Shells

Z.A. Parszweski:
Dynamic Optimization of Machine Systems Configuration

K. Yamazaki:
Minimum Compliance Design of the Stiffener Layout of Thin Plates

D.P. Thambiratnam:
Optimization of Conical Shells for Static and Dynamic Loads

12 February, Friday

Session G: Stochastic Methods and Miscellaneous Topics

Chairman: H.A. Eschenauer, Co-Chairman: E. Sandgren

M. Życzkowski:
Optimal Design of Structures under Creep Conditions

R.E. Melchers:
Structural Optimization in a Non-Deterministic Setting

L.M.C. Simões:
A Mathematical Programming Approach for Finding the Stochastically More Relevant Failure Mechanism

Y. Murotsu, M. Yonezawa & M. Kishi:
Optimization of Frame Structures Based on Reliability Concept

V. De Angelis:
Linear Complementarity Problems: A Cutting Plane Method Based on the Convex Hull of Polyhedra

K.L. Teo & C.-M. Wang:
Optimal Shape of Least Weight Arches

C.S. Gurujee & S.N. Agashe:
Optimal Design of R.C. Frames Based on Improved Inelastic Analysis Method

<u>Session H: Shape Optimization II</u>

Chairman: P. Pedersen, Co-Chairman: P. Hajela

J. Bäcklund:
Shape Optimization of Holes in Shear Panels of Composite Materials

B.J.D. Esping & D. Holm:
Structural Shape Optimization Using OASIS

F. Ellyin:
Shape Optimization of Intersecting Pressure Vessels

M. Cannarozzi C. Cinquini & R. Contro:
Optimal Shape of Cable Structures

M. Yoshimura:
Detailed Machine Structural Shapes Generated from Simplified Models

R.D. Parbery:
Optimization of a Hollow Beam-Shaft with Prescribed Inner Contour

13 February, Saturday

<u>Session J: Miscellaneous Topics</u>

Chairman: W. Hemp, Co-Chairman: C. Cinquini

A.J.G. Schoofs:
Experimental Design and Structural Optimization

K.-T. Cheng & F. Biao:
Shape Optimization of Continuum with Cracks

N.V. Banichuk:
Modern Trends in Elastic-Plastic Design and Internal Structure Optimization

M. Hýča:
Optimization of the Cross-Section of Short Thin-Walled Beams Subject to Bending and Shear with Allowance for the Shear-Lag Effect

L.F. Jansen:
Optimization of Structures Using the Finite Element Method

K.-Y. Yeh:
New Approach to Structural Optimization by Analytical Methods

<u>Closing Remarks by the Secretary-General of IUTAM and the Joint Chairmen</u>

OPENING ADDRESS:
HISTORICAL REVIEW OF LAYOUT OPTIMIZATION

G.I.N. ROZVANY
Joint Symposium Chairman

Mr. Secretary-General, Mr. Chairman, Ladies and Gentlemen. It is a great pleasure, indeed a rare privilege, to present an opening address to such a distinguished audience. I have been asked to say a few words about the history of layout optimization for two reasons: first, layout optimization is intellectually the most challenging and economically the most rewarding task in structural design; and second, the original aim of this meeting, as proposed by myself at the IUTAM General Assembly in August 1986 in London, was to discuss primarily layout and shape optimization of structures. The geographical location of the symposium was selected accordingly, for it was in this city of Melbourne where the theory of layout optimization was pioneered by A.G.M Michell at the turn of the century and extended considerably by my research group in the seventies and early eighties. The latter development was carried out in close collaboration with the late Prof. William Prager, to whose memory this meeting is dedicated. Two of my former research associates, Prof. Melchers and Dr. Wang, are with us in the audience.

Returning to Michell's contribution, his pioneering efforts are all the more remarkable, if we consider that he was a most versatile inventor (crankless engine, thrust-bearing), consulting engineer (hydraulics, lubrication) and author (mathematics, mechanics) who studied architecture, mining and civil engineering in Melbourne. He presented, at the age of 34, his revolutionary contribution to structural optimization in a paper entitled "The Limits of Economy of Material in Frame-Structures" (Philosophical Magazine, 1904) in which he introduced two fundamental concepts of far-reaching implications, namely static-kinematic optimality criteria and optimal layout theory.

Optimality criteria are necessary (and sometimes sufficient) conditions of cost minimality which are derived, for distributed parameter problems, from energy theorems or variational and control-theoretic principles. These conditions can be expressed as fictitious strain-stress relations which, together with static/kinematic admissibility, ensure optimality of a structural design. This means that *static-kinematic* optimality criteria convert, in effect, a problem of structural optimization into a problem of analysis.

The first *continuum-type* static-kinematic optimality condition was introduced by Michell for trusses and could be described in the current notation as:

$$(\text{for } N^S \neq 0) \quad \epsilon^K = k \operatorname{sgn} N^S, \qquad (\text{for } N^S = 0) \quad |\epsilon^K| \leq k \tag{1}$$

where ϵ is the longitudinal strain, k is a given constant, N is the member force and the superscripts K and S denote kinematic and static admissibility.

Further continuum-type optimality conditions were proposed for segment-wise prismatic beams by Foulkes (1954), for non-prismatic frames (with freely varying cross-section) by Heyman (1959) and for plastic design in general by Prager and Shield (1967). The latter can also be derived from theorems of Mróz (1963), Masur (1970) or Save (1972).

Quite independently from the above developments, optimality criteria for *discretized systems* were derived by Khot, Venkayya, Berke and others from 1968 onwards.

The Prager-Shield optimality condition, derived from energy theorems, was extended by my research groups in Melbourne and Essen to a comprehensive set of design requirements in both plastic and elastic design. I shall discuss these conditions in my main lecture, and in greater detail in a forthcoming book (Structural Design via Optimality Criteria – Kluwer Academic Publishers). This new generation of optimality criteria were derived from extended variational principles, which I originally learned at Oxford from Prof. W.S. Hemp who is fortunately, also with us here.

Applications of optimality criteria may include

(1) *idealised systems*, with a view to determining (a) fundamental features of optimal solutions, (b) the range of validity and applicability of various numerical methods and (c) the relative economy of more realistic designs (basis of comparison); and

(2) *large real systems*, where they can be used (a) for checking the optimality of solutions determined by other methods or (b) in highly efficient iterative re-sizing strategies.

Structural optimization problems fall into three broad categories: (a) In *cross-sectional parameter optimization*, the cross-sectional geometry is partially prescribed and can be fully defined by a finite number of parameters whose variation along the centroidal axes (middle-surfaces) is to be optimized. (b) In *shape optimization*, the unspecified boundary of a continuum or the shape of a middle surface (e.g. shells) is to be selected optimally. Whilst in the above two problems we seek one or several unknown scalar functions, (c) *layout optimization* is much more difficult because we simultaneously optimize the topology, certain layout parameters within a given topology and the cross-sectional dimensions. These problems involve an infinite number of potential topologies or, alternatively, an infinite number of potential member directions and sizes at each point of the feasible space. Moreover, it will be shown in my main lecture that layout optimization can result in much more substantial savings than cross-section or shape optimization.

Partial layout optimization may be achieved by employing *numerical methods* (e.g. mathematical programming) used in improving the location of the gridpoints for a given topology or eliminating non-optimal members from a preassigned grid. This work is of considerable practical importance, as was shown by Eschenauer in designing e.g. radio-telescopes.

In *exact layout theory*, developed in the late seventies by William Prager and myself, the optimal topology is not prescribed and is to be determined directly. The two underlying concepts of this method are (a) static-kinematic optimality criteria and (b) the *structural universe*, consisting of all potential members (elements). As optimality criteria also give strain requirements for vanishing members. their fulfilment (in convex problems) for the entire structural universe constitutes a sufficient condition of layout optimality.

We further distinguish between *classical layout theory*, dealing with "low density" systems where the members are relatively slender and the effect of intersections on cost, strength and stiffness is neglected, as in Michell frames; and *advanced layout theory*, concerning "high density" systems in which a high proportion of the feasible space is occupied by material. In the latter, the microstructure must be optimized locally and a specific cost function depending on local strength and/or stiffness parameters is to be derived, before layout optimization of the macrostructure is carried out. Applications of these two branches of layout theory will be discussed in my main lecture (see p. 265 in these proceedings).

In conclusion, I would like to wish all participants a very pleasant and fruitful stay here in Melbourne.

PRACTICAL DESIGN OF SHEAR AND COMPRESSION LOADED STIFFENED PANELS

Melvin S. Anderson
Research Professor, Old Dominion University
Norfolk, Virginia 23508 USA

ABSTRACT. Many applications of optimization methods to structural design problems have been based on material strength. When buckling is considered, the problem is generally more complicated and difficult. In this paper, several of these difficulties and means to handle them for practical design will be described in the context of the development of the computer program PASCO (Panel Analysis and Sizing COde).

INTRODUCTION

The goal of the analysis is to provide a designer with a tool to determine dimensions of composite stiffened panel structure subject to all the practical requirements and limitations that may occur. The approach that is recognized as one being able to do such a task is to express the requirements and limitations in the form of mathematical constraints which having a certain value, say less than zero, indicate that the design is satisfactory with respect to this condition. The sensitivity of these constraints to the chosen design variables is determined and a standard nonlinear optimizing program can be used to minimize the mass of the panel and insure that all constraints are satisfied. This paper will discuss the special requirements in setting up the constraints when buckling is a design condition as well as indicate many of the other types of constraints necessary to achieve a practical design. The details of the optimization process will not be discussed, but results will be given to illustrate some of the important features of the total analysis-optimization procedure.

METHOD OF APPROACH

Description of Problem

The hat stiffened panel in figure 1 is typical of the type of configuration which may be considered. The structure is modeled as a simply supported prismatic panel of length L. The cross-sectional shape is completely arbitrary and is defined by an assembly of laminated plates having symmetric walls that may contain several different materials. The most frequent application and the one that has the greatest computational efficiency is a cross-section having several repeated identical skin-stiffener combinations requiring the input of only the repeating portion shown on the right. The ply thicknesses have been exaggerated to show the detailed modeling capability including offsets of one plate relative to another. Besides the loadings shown on the figure, variable temperatures and an overall bow imperfection may be considered. Multiple load cases are included. Any plate width, ply thickness and ply angle may be a design variable. A means of linking the variable through a set of linear equations is available. Bounds may

G. I. N. Rozvany and B. L. Karihaloo (eds.), Structural Optimization, 1–8.

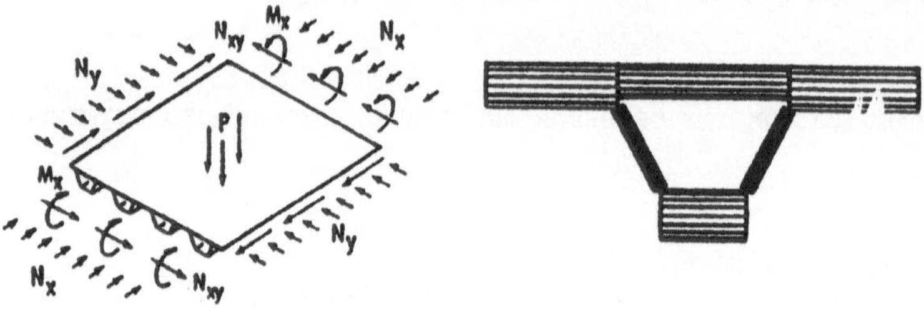

Figure 1. Typical panel and design loadings.

be placed on design variables. Besides buckling, material strength and stiffness constraints may also be imposed. Stiffness constraints are determined from inplane and bending stiffnesses calculated on the basis of average or smeared orthotropic stiffnesses. The above capabilities are believed to represent a minimum required for practical design.

Sizing Strategy

The program can be thought of as having an analysis module, a Taylor series module, and an optimization module. In the analysis module, critical buckling modes are identified and calculated, stresses from all the applied loadings including the effects of imperfections are determined. Constraints that are violated or those that have the potential of being violated are identified and calculated. The sensitivity of these constraints to changes in design variables are determined by making a small change in turn of each of the design variables and repeating the analysis to obtain a numerical derivative. These constraints and their derivatives are used to generate a first order Taylor series for each constraint. This information is all that is passed to the optimization module which for the PASCO program is the CONMIN (ref. 1) optimizer. Because of move limits on the design variables, a satisfactory design is not necessarily achieved in one cycle. Even if one was found, the approximations involved in the Taylor series expansion would not make the result accurate and even more important, new modes of failure might have arisen during the redesign which were not originally identified. The basic approach is then to return to the analysis module after each optimization cycle to check the new design, and if more design cycles are called for, determine new constraints and derivatives and repeat the process. In the design process the move limits are actually reduced with each major design cycle to prevent large excursions when final design is approached.

Buckling Analysis

One of the most important requirements for structural optimization is an efficient and accurate analysis method. Efficient, because many analyses of trial designs and their derivatives to changes in design variable must be made to arrive at a minimum weight design satisfying all design constraints. Accurate (or conservative) analysis is required, otherwise the optimization process will surely lead to a design where the analysis is in error on the unsafe side. The PASCO program (refs. 2, 3 and 4) uses the VIPASA analysis (ref. 5) which is an exact solution for buckling of arbitrary cross-section prismatic panels subject to inplane direct and shear loadings. The VIPASA analysis assumes a sinusoidal mode shape along the panel length which corresponds to simply supported ends for orthotropic panels under direct stress and yields conservative results for simply supported shear loaded anisotropic panels. The analysis has a multilevel substructuring capability that makes analysis of panels having several identical

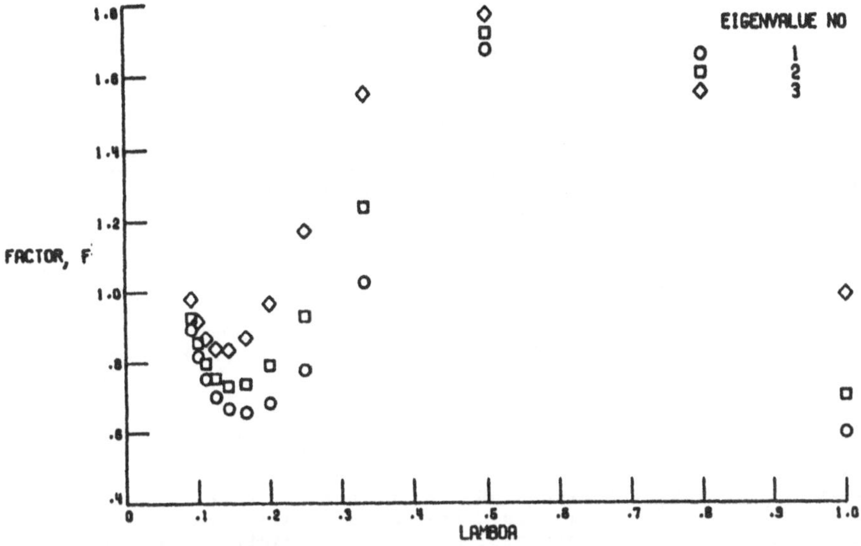

Figure 2. Buckling load factor as a function of axial half wavelength.

stiffeners very rapid. Even though the program is among the fastest for solving a given problem to which it applies, typically 90% of the total time is spent in analysis indicating the importance of striving for the most efficient analysis possible.

The analysis must be able to distinguish and identify all of the potentially critical modes of buckling so that, in the optimization process, cycling between various buckling modes does not occur. This is easily accomplished with the VIPASA analysis because of the nature of the solution method. The assumption of sinusoidal response results in separate buckling calculations being made for half-wavelengths λ. If a sequence of λ's is chosen L, L/2, L/3, ... till the smallest λ is sufficiently less than the smallest plate dimension, one can be assured of covering all possible buckling modes. Such a calculation is shown in figure 2 for a blade stiffened panel. The first three eigenvalues are shown for each λ. The ordinate is the factor by which the design load must be multiplied to determine the state at buckling; a factor less than one means the panel would buckle at less than the design load. The lowest value of the factor corresponds to the buckling load of the panel and might be the only load of interest for an analysis. However in a design situation, other values of λ and factor may be just as important in ensuring an adequate panel when the design is changed. The classification of buckling loads at different λ's is an ideal way of keeping track of critical and nearly critical buckling loads. For the example in figure 2, values of λ/L of .2 and 1. are seen to produce nearly identical buckling loads. The program always selects $\lambda/L = 1$ because the potential of column buckling is nearly always present. It is sufficient to select only the λ corresponding to the minimum buckling load for the short wavelengths, which the program automatically does. (For cross-sections having complex geometries with different plate widths and thicknesses, more than one relative minimum for local buckling may occur). As the design progresses the location of this minimum may change and new critical values of λ are selected. However, the old ones are retained to prevent a design from going back to a previous state. (If the list of λ's becomes greater than some reasonable number, the least critical are discarded). For λ = L, the possibility of a stiffener twisting or rolling mode exists as well as the column mode. The second eigenvalue at $\lambda = L$ is such a case and at the user's option, can be retained. Experience has shown that in many cases a design will oscillate between a column and a twist mode unless two or more eigenvalues are retained at the long wavelengths. This would not be easy to do in a general purpose analysis such as a finite element code. For the case of figure 2, the second eigenvalue at $\lambda = L$ is the sixth eigenvalue of the structure considering all values of λ. If one were using a general finite element analysis method, it might be very difficult to identify a few representative critical modes from the large number of similar modes that can occur for typical panels (especially if eigenvalues were closely spaced). For example, it would be necessary for the panel of figure 2 to determine that the first, second and sixth eigenvalue need to be considered in determining the constraints and their derivatives and that the other eigenvalues could be ignored. There are other methods that would do this, such as expanding the solution in terms of a few modes using a reduced basis technique, however there must be a means of assuring that no critical or potentially critical modes are neglected. The VIPASA solution used in PASCO is exact and the solution algorithm guarantees no eigenvalues are missed. This fact is especially important for composite panels since a large backlog of experience such as exists for metal panels is not available to guide designers. For metal panels, practical guidelines for limits on various proportions, such as thickness ratios, width-thickness ratios, and ratios between plate breadths (in particular if one is an outstanding flange) have prevented unexpected buckling modes from occuring and a few simple local buckling and column buckling formulas are adequate for analysis. To take full advantage of composite construction, it is desirable not to place severe restrictions on design, but a completely accurate analysis must be done to avoid buckling failures at lower than expected loads.

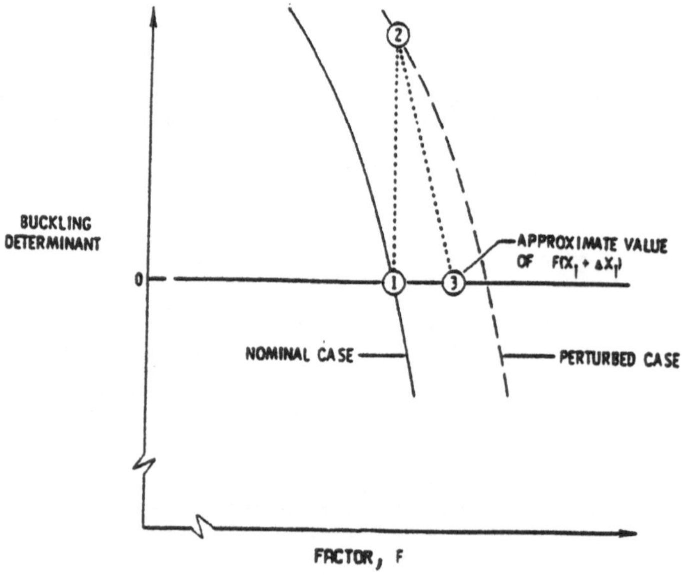

Figure 3. Determinant plot for nominal and perturbed cases.

Sensitivity Analysis for Taylor Series Expansion

The VIPASA stiffness matrix is a transcendental matrix that is an implicit function of the eigenvalue and λ.
This of course allows the exact solution to be obtained without subdividing any plate into a finer grid and substructures are treated exactly as well. Derivatives with respect to design variables can not be obtained by techniques used for problems governed by linear stiffness matrices. Use of substructuring to reduce problem size has made the analysis very efficient but a complete reanalysis of a perturbed design in order to obtain a derivative greatly increases the computational effort, especially if there are a large number of design variables. There is a procedure for obtaining the solution for a small perturbation of a design variable from information obtained in the solution of the nominal case. The VIPASA solution involves repeated reduction of the stiffness matrix at different values of the eigenvalue till convergence is obtained. The determinant of the stiffness matrix, which must be zero at an eigenvalue, is available after each iteration. A schematic plot of the determinant is shown in figure 3 by the solid line with the value of factor for buckling indicated by point 1. The dashed line is a plot of the determinant where one of the design variables has been perturbed a small amount. If one iteration is performed for the perturbed case at a value of factor, F, equal to the eigenvalue of the nominal case a value of the determinant depicted by point 2 will be obtained. Projecting from point 2 along a slope equal to that determined for the nominal case, point 3 is obtained as the estimate of F for the perturbed case. Thus the derivative can be determined in one iteration compared to the 10 to 15 required for a full analysis.

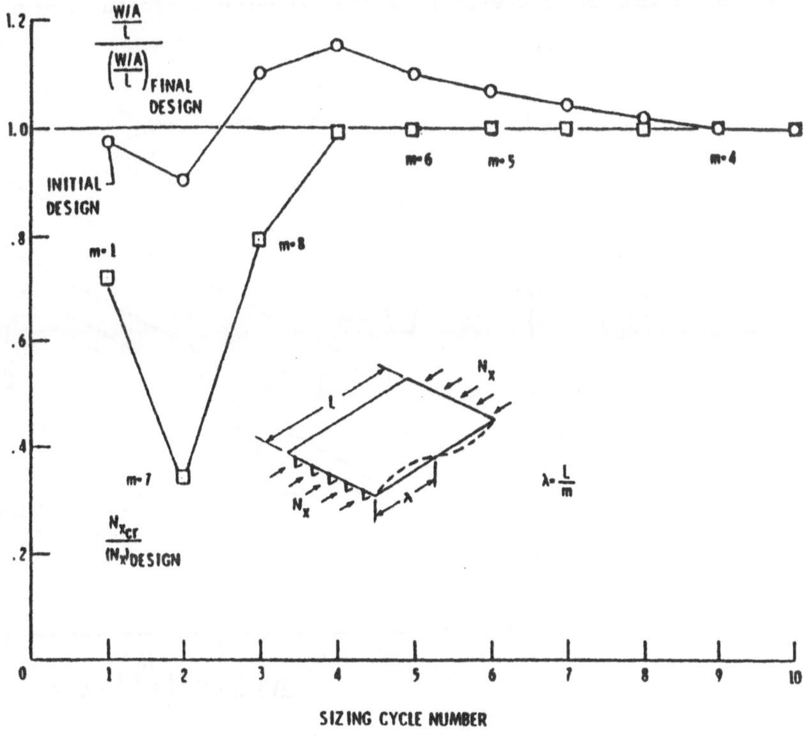

Figure 4 . Mass and load history for a typical design problem.

OPTIMIZATION RESULTS

A typical history for the design cycles just described is illustrated in figure 4 in terms of buckling load and mass as a function of design cycle. For the first cycle, only the buckling load for $\lambda = L$ was used as a constraint. The design was changed so that this mode was brought up to the design value with even a reduction in mass. However, these changes were such that local buckling occured at a very low load. The key in figure 4 shows the new wave numbers m identified in each design cycle. Any m number identified is retained in subsequent design cycles. The next two design cycles were spent in raising the local buckling load to the design value while maintaining the column mode at the design value. The move limits in the problem prevented recovery in just one cycle. These two cycles were accompanied by a significant increase in mass. At this point where all constraints were satisfied, the program was able to rearrange the material to bring the mass down to its final value, just slightly greater than the original design.

Design Studies

One application of PASCO is to make structural efficiency comparisons of various panel configurations with different materials. Such a result is shown in figure 5. A mass index (mass per unit area divided by L) is plotted against the structural index for wide columns for a variety of configurations fabricated from aluminum and graphite epoxy. The relative mass of the different composite configurations can be seen. The significant reduction in mass of the hat stiffened composite panel over its aluminum counterpart is also apparent. There were no other

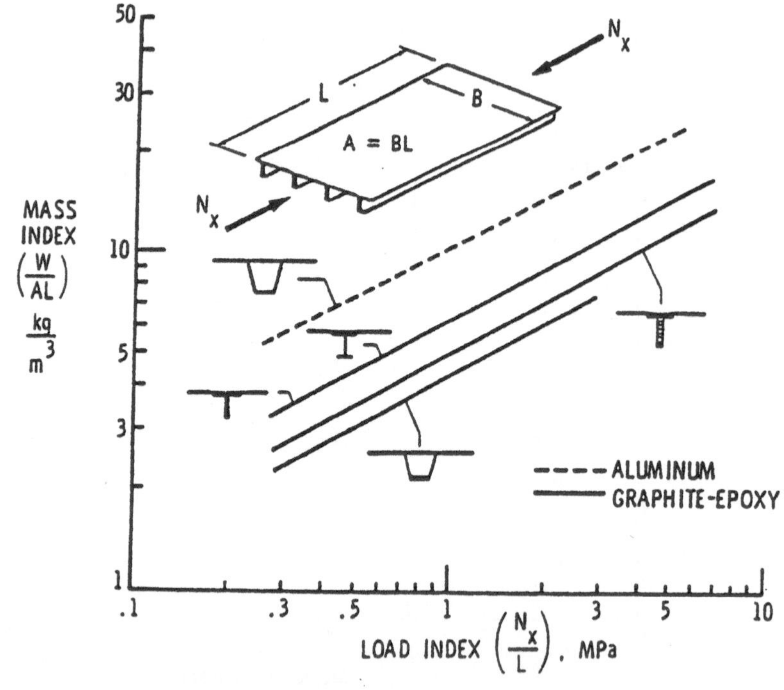

Figure 5. Comparison of structural efficiency for various panels.

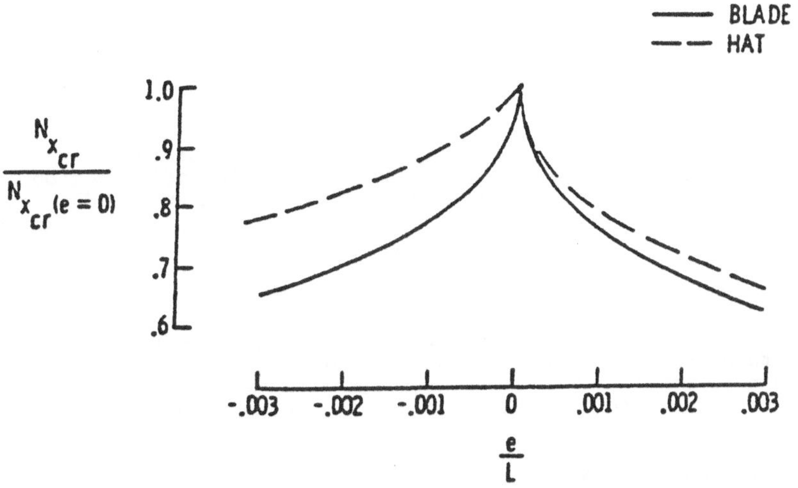

Figure 6. Effect of imperfection on buckling of optimized panels.

constraints except buckling for these results. At higher values of the load index, material strength would limit the design. Even more serious is not accounting for imperfections. Panels designed on the basis of buckling may have several coincident buckling loads that make them sensitve to imperfections as shown in figure 6. The buckling load of panels designed on the basis of buckling with no imperfection is shown

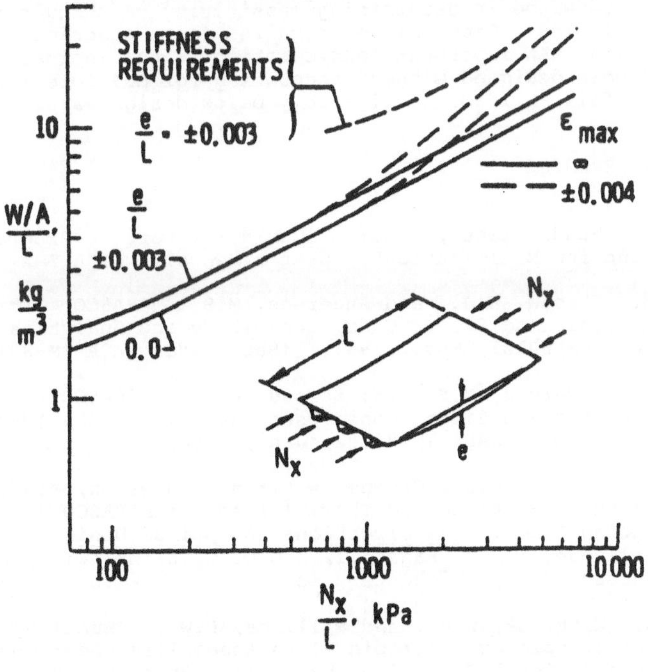

Figure 7. Effect of practical considerations on panel mass.

as a function of the amplitude e of an overall bow imperfection (see sketch in figure 7). Results for both blade stiffened and hat stiffened are shown. Drastic reductions in buckling load are seen for rather small imperfections. Depending on the direction of the imperfection, the results are different because in one direction the skin is subject to increased compression and in the other direction the outstanding portion of the stiffener is subject to the increased compression. Based on the results of figure 6, it is necessary to design for both positive and negative imperfections which are treated as two load cases.

The PASCO code has been used to study the effect of imperfections on panel design. The results of such a design study for a composite hat stiffened panel are shown in figure 7. The increase in mass over a panel designed with no imperfection is seen to be modest compared to the drastic reduction in buckling load of a panel designed for no imperfection but actually having $e/L = \pm.003$ (see fig. 6).The effect of introducing a material strain limit of $\varepsilon = .004$ is also shown. Typical shear stiffness requirements for aircraft wing construction might further increase the mass as shown by the upper dashed curve.

CONCLUDING REMARKS

The special problems of structural optimization for buckling constraints are discussed as applied to a program for optimization of stiffened composite panels. The critical requirement in the analysis is that a means of identifying critical and potentially critical buckling modes must be possible. For a simply supported stiffened panel, the analysis assumed a sinusoidal buckling mode along the panel length which results in separate buckling calculations being made at different possible values of the axial half wavelength. Thus overall and local modes can be retained in determining sensitivity to changes in the design variables. Capabilities required for a practical design application are given. An important consideration is the inclusion of imperfections. Panels designed without accounting for possible imperfections were shown to fail at loads considerably below design values.

REFERENCES

1. Vanderplaats, G.N.: 'CONMIN - A Fortran Program for Constrained Function Minimization'. User's Manual. NASA TMX-62,282, 1973.

2. Stroud, W.J.; and Anderson, M.S.: 'PASCO: Structural Panel Analysis and Sizing Code, Capability and Analytical Foundations'. NASA TM-80181, Nov., 1981. (Supersedes NASA TM-80181, January, 1980)

3. Anderson, M.S.; and Stroud, W.J. : 'PASCO: Structural Panel Analysis and Sizing Code, Users Manual'. NASA TM-80182, November, 1981. (Supersedes NASA TM-80182, Jan., 1980)

4. Stroud, W.J.; Greene, W.H.; and Anderson, M.S.: 'Current Research on Shear Buckling and Thermal Loads with PASCO - Panel Analysis and Sizing Code'. New Directions in Optimum Design, E. Atrek, R. H. Gallagher, K. M. Ragsdell, and O. C. Zienkiewicz, Eds. Wiley-Interscience, Chichester, England and New York, 1984.

5. Wittrick, W.H.; and Williams, F.W.: 'Buckling and Vibration of Anisotropic or Isotropic Plate Assemblies Under Combined Loadings'. International Journal of Mechanical Sciences, Vol. 16, 1974, pp.209-239.

SHAPE OPTIMIZATION OF HOLES IN COMPOSITE SHEAR PANELS

Jan Bäcklund and Rikard Isby
The Royal Institute of Technology
Department of Aeronautical Structures and Materials
S-100 44 Stockholm
Sweden

ABSTRACT. The aim of this work is to find optimum shapes of holes in shear panels made of carbon/epoxy composite materials. The study is restricted to non-buckling shear panels, and a simplified fracture criterion is applied. Starting with a baseline square shear panel with a circular hole, the optimization results in a larger non-circular hole that gives the same maximum stress level in the laminate. Optimum shaped holes are determined for four different laminate configurations, the differences between the shapes being small.

1. INTRODUCTION

Shear panels of composite materials are important components in load-carrying structures, e g in aircraft beams. Holes are frequent in such panels due to demands of carrying ducts through the panel. These holes are often circular or elliptical. With the availability of efficient optimization computer codes such as OASIS [1], it is now possible by using shape optimization to determine hole shapes other than circular or elliptic that gives a lower panel weight without increasing the stress levels in the laminate.

A typical composite material aircraft beam structure from the new Swedish fighter aircraft JAS 39 Gripen is shown in Fig 1. With this type of structure as a background, the present work was focused on simple square non-buckling shear panels with a central hole. An additional load case of uniaxial tension was also investigated.

2. COMPOSITE MATERIAL LAMINATES

The data used in this investigation are typical of carbon fiber/epoxy resin T300/914C prepreg [2]. Shear panels of four different lay-ups were studied, see Table I. Even though a +/-45 degree laminate is most efficient as an unnotched shear panel, this might not be the case with a hole present. It should also be interesting to find the difference between optimum hole shapes for various lay-ups.

G. I. N. Rozvany and B. L. Karihaloo (eds.), Structural Optimization, 9–16.

TABLE I. Laminate lay-ups

Laminate type	Percentage of fibers in 0, 90 and +/- 45 degrees: 0	90	+45/-45
A	25%	25%	25% each
B	18%	6%	38% each
C	6%	6%	44% each
D	0%	0%	50% each

Based on the ply data, the stiffnesses and the tensile, compressive and shear strengths of the different laminates were calculated using classical laminate theory, the Tsai-Hill fracture criterion [3] and the assumption of first-ply-failure. The resulting strengths are given in Table II.

TABLE II. Laminate strengths in MPa. The x-direction coincides with the 0 degree direction

Laminate type	x-direction tens	x-direction comp	y-direction tens	y-direction comp	shear xy
A	290	570	290	570	207
B	222	300	161	267	295
C	134	175	134	175	338
D	140	158	140	158	382

3. DESIGN CRITERIA

The Tsai-Hill fracture criterion [3] was used for determining the strengths of the different laminates according to Table II. It was also applied to the shear panels, with the strengths in Table II as input, to estimate the external shear load that corresponds to first-ply-failure around the hole. According to Tsai-Hill, fracture is assumed to occur when the Tsai-Hill Number TH in Eq (1) equals unity.

$$TH = \sqrt{\left(\frac{\sigma_x}{\sigma_{xf}}\right)^2 - \frac{\sigma_x \sigma_y}{\sigma_{xf}\sigma_{yf}} + \left(\frac{\sigma_y}{\sigma_{yf}}\right)^2 + \left(\frac{\tau_{xy}}{\tau_{xyf}}\right)^2} \qquad (1)$$

In Eq (1), σ_x denotes laminate stress in x-direction and σ_{xf} denotes laminate strength in x-direction (according to Table II), etc.

In the optimization process, a simplified notched strength criterion known as the Point Stress Criterion (PSC) [4] was used. Although more accurate and physically tractable criteria exist [5], PSC was considered sufficiently accurate for the design process, and it was simple to include in the shape optimization. To illustrate PSC, a composite laminate with a circular hole under tension is considered, see the Appendix. Due to damage in the laminate, the net section laminate stress is reduced in the vicinity of the hole, as compared with the linear elastic solution. To compensate for this in the design process, the stress (or rather the Tsai-Hill Number TH) is evaluated at a distance "d" from the edge of the hole. The critical distance d varies with material system but also with laminating sequence. For simplicity, however, d was assumed to be constant and equal to 1 mm throughout this study.

4. STRATEGY OF SHAPE OPTIMIZATION

For each laminate type A through D, a baseline shear panel was first analysed with a central circular hole with the diameter equal to 20% of the panel side, Fig 2a. The external load τ_B was adjusted such that the maximum TH value along the PSC line "L" attained unity. Then the shape of the hole was allowed to change, using the CAD based shape optimization technique in OASIS [6], so as to give minimum weight of the panel without increasing the maximum TH value (=1), Fig 2b. Four Ferguson splines controlled by 16 design variables were used for describing the boundary curve, Fig 3. The boundary curve was controlled by OASIS to be smooth, hence avoiding sharp notches that would be detrimental for the fatigue life of the shear panel. This was obtained by enforcing the splines to have common tangents at their junctions (points 1, 5, 9 and 13 in Fig 3).

Two different cases were investigated. In the first (#1), the laminate strengths in Table II were used, and in the second (#2), the laminate strengths were modified such that the compressive strengths were set equal to the tensile strengths. The rationale for this is that case #1 then corresponds to optimization with one load case, and case #2 corresponds to optimization with two load cases; shear load as shown in Fig 2, and shear load with the reverse direction.

5. RESULTS OF SHAPE OPTIMIZATION

Tsai-Hill contour plots for laminate A, case #1 are shown in Fig 4 for the circular hole, and in Fig 5 for the optimized hole. It can be seen from Fig 5 and Table III that the area of the hole could be increased considerably without increasing the maximum TH value. The optimum shape is elongated due to the nature of the load and the difference in tensile and compressive strengths of the laminate. The optimized hole shapes for all laminates A through D are shown in Fig 6a. Despite the drastic difference in stiffnesses and strengths between the laminates, the optimized holes are surprisingly similar.

For case #2 with equal tensile and compressive strengths, "square" holes were obtained, Fig 6b. In this case, the hole shapes were even closer. It is interesting to note the diamond shape of the optimized holes as compared with the holes in the lower part of the JAS 39 beam, Fig 1.

All computations were performed on a VAX 11/750 with an FPS 164 attached processor. One optimization run took typically 75 minutes CPU time on the computer.

6. ADDITIONAL LOAD CASE

To investigate whether small differences in the optimized hole shapes were likely to occur also for other loads than shear, a case with uniaxial tension was run through the optimizer. Clearly, the results are quite similar for the different laminates also in this case, Fig 7, even though the deviations are greater than for the shear load cases, see Table III. Like in metals, elliptic or "super-elliptic" holes appear to be the best design for this load case.

7. SUMMARY

Shape optimization is a powerful tool for designing composite shear panels with holes for minimum weight. The optimum hole in a shear panel is of smoothed diamond shape, whereas panels subjected to uniaxial tension should have "super-elliptic" holes. The optimum shapes appear to be rather insensitive to fiber lay-up in the panel.

8. REFERENCES

[1] B. Esping: 'The OASIS Structural Optimization System', Computers & Structures 23 (1986), pp 365-377

[2] L. A. Carlsson & R. B. Pipes: Experimental Characterization of Advanced Composite Materials, Prentice-Hall 1987

[3] R. M. Jones: Mechanics of Composite Materials, Scripta 1975

[4] J. M. Whitney & R. J. Nuismer: 'Stress Fracture Criteria for Laminated Composites Containing Stress Concentrations', Journal of Composite Materials 8 (1974), pp 253-265

[5] J. Bäcklund & C. G. Aronsson: 'Tensile Fracture of Laminates with Holes', Journal of Composite Materials 20 (1986), pp 259-286

[6] B. Esping & D. Holm: 'A CAD Approach to Shape Optimization', in Computer Aided Optimal Design: Structural and Mechanical Systems (ed C A Mota), Springer 1987

TABLE III. Areas of the optimized holes normalized with respect to the circular hole (area=1)

Laminate type	Shear #1 (unequal)	Shear #2 (equal)	Tension
A	2.5	2.4	4.7
B	2.6	2.5	4.5
C	2.5	2.5	4.3
D	2.3	2.2	4.0

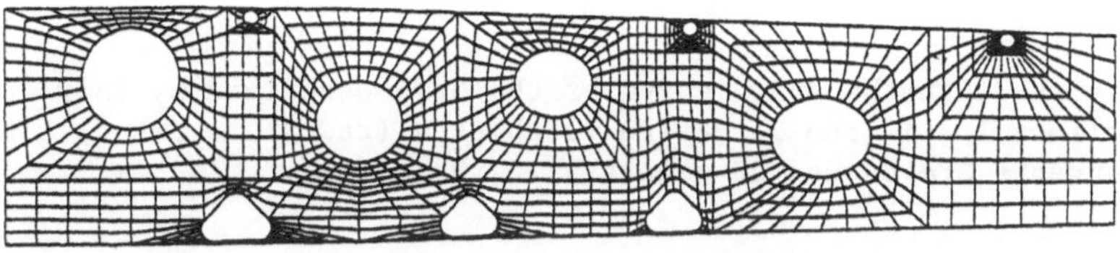

Figure 1. Web of typical carbon epoxy beam in the new Swedish fighter aircraft JAS 39 Gripen.

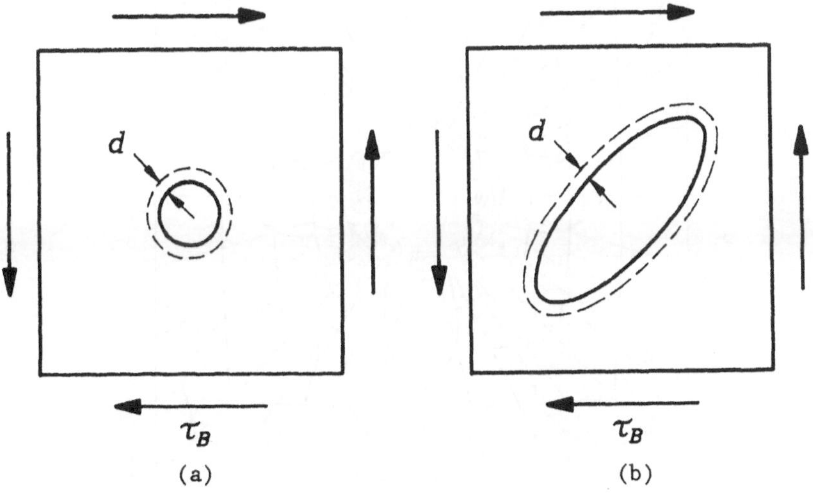

Figure 2. Shear panel with hole. The fracture criterion is evaluated along the dashed curve.
(a) baseline circular hole
(b) schematic optimized hole

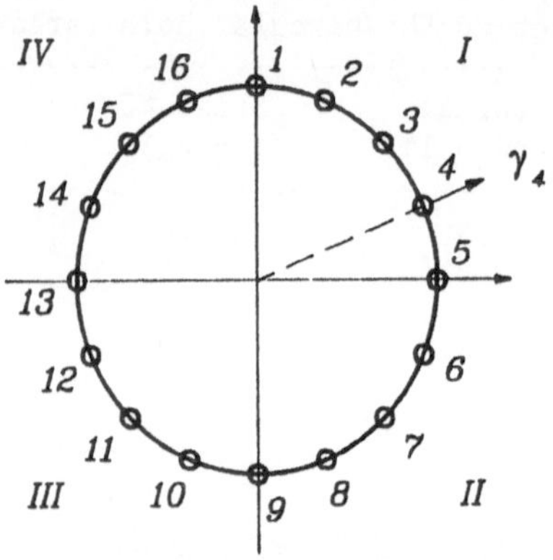

Figure 3. The boundary curve of the hole described by four Ferguson splines I - IV and 16 design varables γ (radial movement). Spline tangents are common at points 1, 5, 9, and 13.

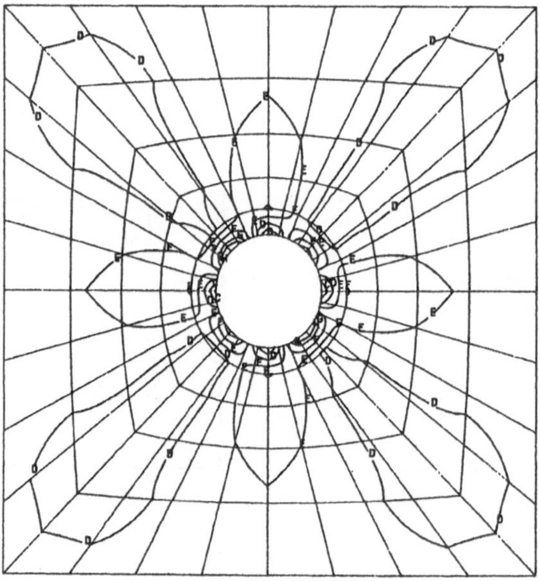

Figure 4. Contour plots of the Tsai-Hill Number for shear load on laminate A (quasi-isotropic) with baseline circular hole and unequal tensile and compressive strengths (case #1).

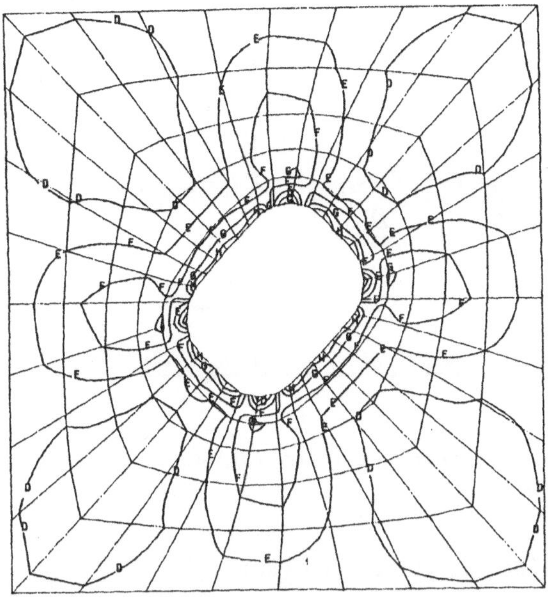

Figure 5. Contour plots of the Tsai-Hill Number for shear load on laminate A (quasi-isotropic) with optimized hole and unequal tensile and compressive strengths (case #1).

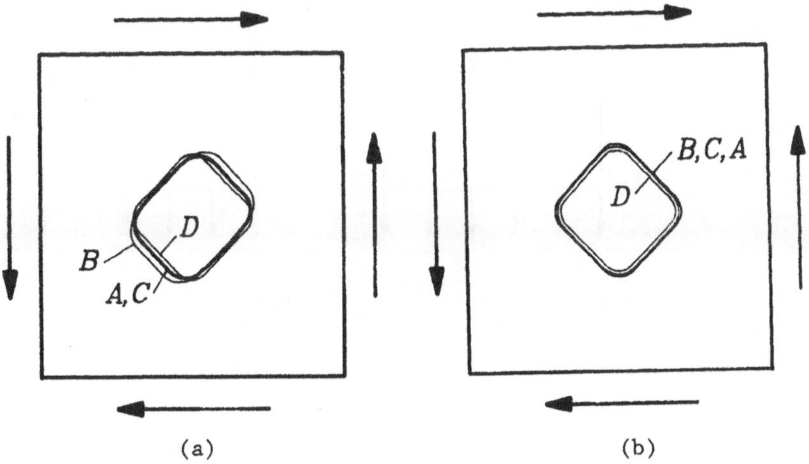

Figure 6. Optimzed hole shapes for the various laminates for shear loading.

(a) Laminates with unequal tensile and compressive strengths
(b) Laminates with equal tensile and compressive strengths

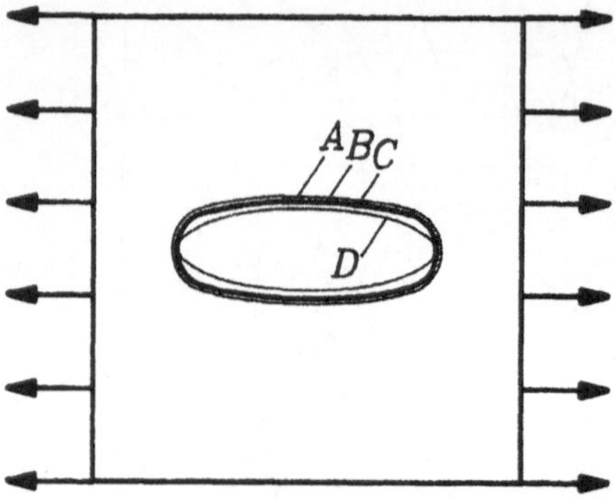

Figure 7. Optimized hole shapes for the various laminates for unidirectional tensile load.

APPENDIX. The Point Stress Criterion (PSC).

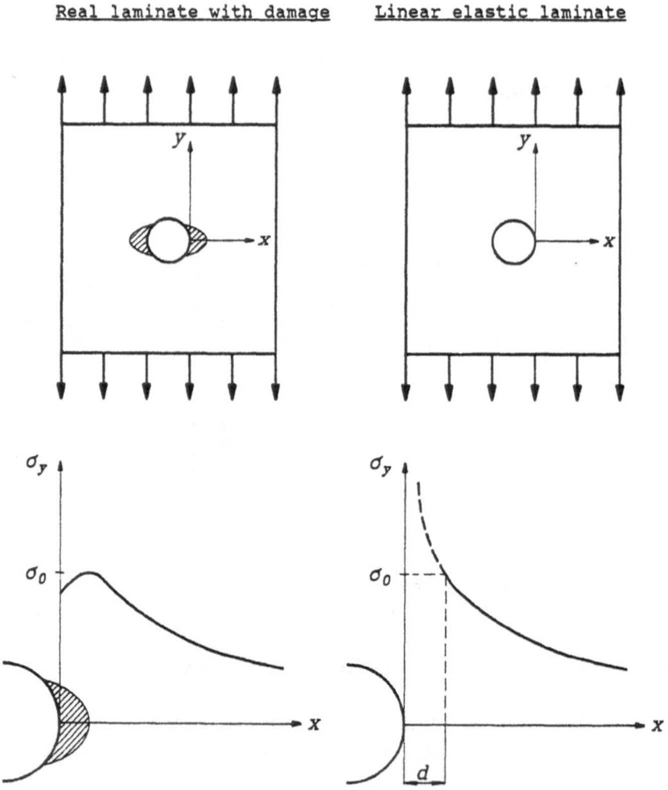

DEVELOPMENT OF A KNOWLEDGE-BASED SYSTEM FOR STRUCTURAL OPTIMIZATION

M. BALACHANDRAN AND J. S. GERO
Architectural Computing Unit
Department of Architectural Science
The University of Sydney
NSW 2006 Australia

ABSTRACT. In structural optimization, computer-based systems have been used to assist in the numerical aspects of the optimization process. However, structural optimization involves a number of tasks which require human expertise and are traditionally assisted by human designers. These include design optimization formulation, problem recognition and the selection of appropriate algorithm(s). This paper presents a framework for developing an integrated knowledge-based system for structural optimization. The operation of a prototype system, called OPTIMA, implemented in a combination of Lisp, Prolog and C on SUN workstations is described and demonstrated through an example problem. The system clearly demonstrates the potential of applying knowledge-based systems to structural optimization.

1. Introduction

Optimization is a useful and challenging activity in structural design. It provides designers with tools for producing better designs while saving time in the design process. Traditional programs have been used to automate analysis and numeric computation involved in the optimization process. Some major difficulties associated with conventional optimization systems are discussed. It is shown that knowledge-based system methodologies provide some means to alleviate those difficulties and aim to make optimization an easier tool for designers to use. Such systems can assist designers who do not have a detailed knowledge of optimization methodology in formulating their problems and provide assistance in selecting appropriate algorithms. In this paper we present a computer system, called OPTIMA, which addresses these issues. The role of such knowledge-based optimization systems in structural design is demonstrated through an example.

2. Conventional Optimization Systems

Many algorithms for finding the solution to structural optimization problems have been developed and implemented during the last two decades. In spite of the rapid development of the numerical optimization techniques, their application to structural design practice has not been widespread. A structural designer may have many reasons to avoid optimization techniques and resort to conventional trail and error methods based on his judgment, intuition and experience. Some major

G. I. N. Rozvany and B. L. Karihaloo (eds.), Structural Optimization, 17–24.

difficulties faced by a designer who may not be an expert in optimization techniques in using conventional optimization systems are discussed below.

Problem Formulation

Most optimization algorithms require the complete formulation of the optimization problem; that is, the objective function (s), the constraint equations and the bounds on the design variables should be clearly defined in mathematical form. This task is normally beyond the expertise of many designers as they do not usually thind in terms of such mathematical models. Furthermore, in many practical design situations, the designer may want to introduce changes to his design and reformulate the problem during the design process. The conventional optimization systems do not provide sufficient flexibility to allow the designer to modify his design easily.

Choosing Algorithms

The designer has to select from different methods available for optimization to suit his problem domain. In general a number of aspects have to be considered in selecting the appropriate algorithm for the problem at hand, which include :

- type of design variables (discrete or continuous);
- number of objectives (single objective or multiple objectives);
- nature of objective function(s) and constraints (linear or nonlinear);
- availability of gradients of objective function(s) and constraint functions (differentiable or nondifferentiable).

Thus, considerable experience and mathematical background is expected from the designer in selecting the best possible optimization algorithm.

Computational Effort

Most of the available techniques to solve optimization problems, particularly nonlinear and multiple criteria optimization algorithms are usually implemented in a black box environment. Thus, they do not utilize either the qualitative value of the information generated during the solution process or the advantage of previous design results. Hence, such programs may take excessive computational effort resulting in poor performance.

Selecting Initial Design (Initial Starting Point)

Most nonlinear optimization techniques require an initial solution for the problem. This requires previous experience and judgment of the designer. One major problem is that the initial solution given by the designer may be either highly infeasible or far away from the optimum design resulting in poor performance of the algorithm.

Redundant and Unrealistic Constraints

Many available optimization algorithms do not recognize unrealistic or redundant constraints specified by the designer. The unrealistic constraints result in infeasible solution to the problem, whereas redundant constraints increase the computational effort. Conventional optimization programs are often not capable of reformulating the problem to exclude such infeasible or surplus constraints.

3. Knowledge-Based Systems for Structural Optimization

Today's optimization systems are sufficiently powerful to handle complex numerical computation but they do not emulate the problem solving capabilities of a human expert. Knowledge-based system methodologies now allow us to incorporate different forms of knowledge required in optimization tasks which are non-numeric (Agogino and Almgren, 1987). In recent years there has been some interest in applying knowledge-based methodologies to structural optimization. Arora and Baenziger (1985) discuss the use of artificial intelligence in structural optimization. Jozwiak (1986) describes ideas for improving structural optimization programs by taking advantage of previous design results in the selection of the initial values of decision variables.

Knowledge-based approaches attempt to alleviate manyof the difficulties associated with conventional optimization systems. Knowledge-based systems bring the automation of reasoning and with it the ability to perform knowledge intensive tasks. In the following section, a prototype knowledge-based system, called OPTIMA, designed to assist in structural optimization processes is presented.

4. The OPTIMA System

The major issues considered in the development of the system are the following. The system should:

- accept and represent a designer's description of his problem in an effective and manipulatible form;
- formulate his problem into a canonical form of optimization model providing functional relationships for objectives and constraints;
- recognize the types of design variables and functional types of objectives and constraints;
- select an appropriate algorithm and carry out the solution procedure; and
- provide a simple and effective interface with flexible modelling features.

4.1 THE ARCHITECTURE OF OPTIMA

One of the major concepts utilized in the development of the system is a clear separation between the generic optimization module, which performs the numeric computations, and knowledge-based modules, which perform symbolic and knowledge processing. The kernel of the whole system is the communication controller which allows the four major components, user interface, problem formulator, problem recognizer and the problem solver to communicate with each other. The architecture of the system is shown in Figure 1. A brief description of the major modules of the system are given below. Further details can be found in Balachandran and Gero (1987a, 1987b).

4.1.1. *The User/System Interface.* The OPTIMA system provides the user with a simple and efficient command language to interact with the system in the form of a highly restricted subset of English. This language utilizes a set of reserved words with specific meanings. The user-interface translates the information provided by the user into a set of *frames* (Fikes and Kehler, 1985). For

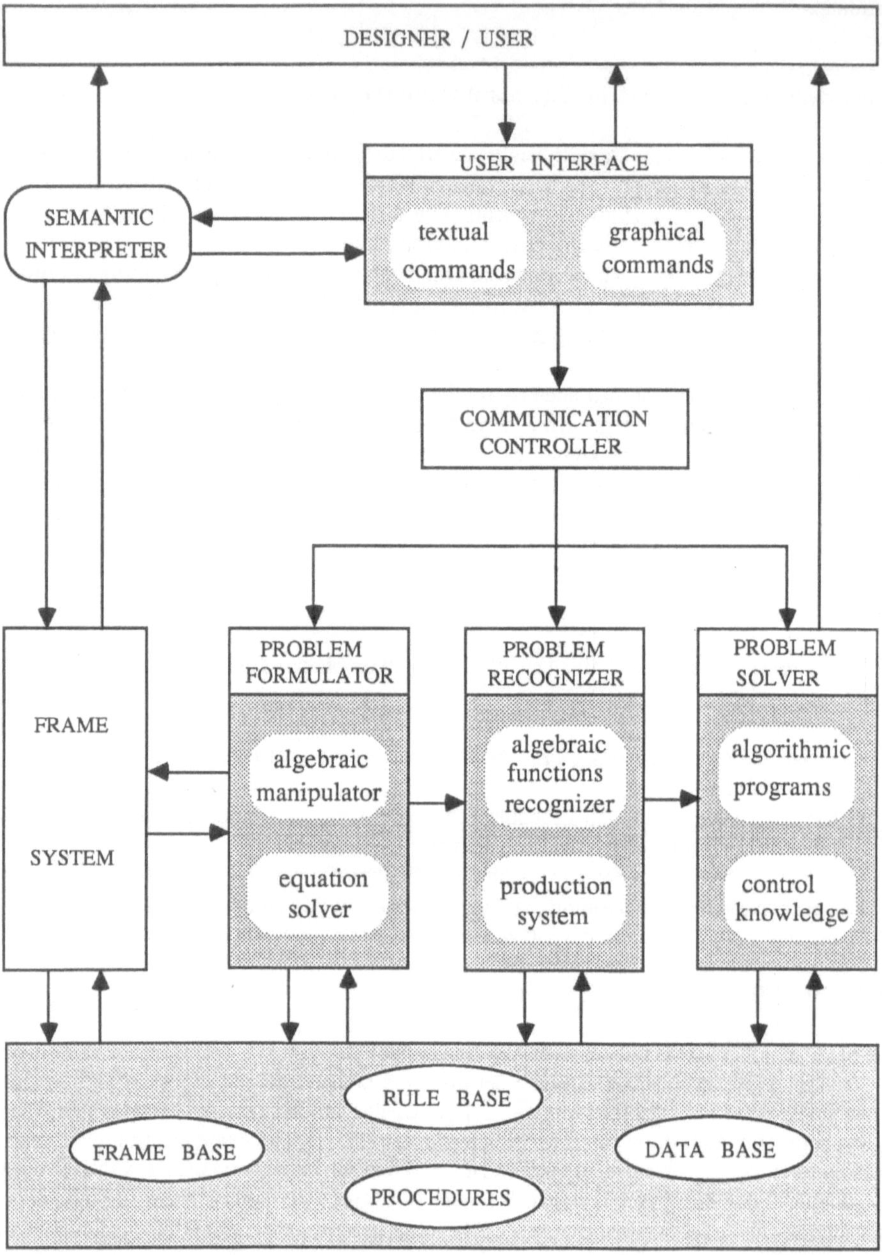

Figure 1. The data flows in the OPTIMA system

many design applications the use of visual interfaces are more appropriate as people understand visually represented information much better than purely textual information. Furthermore, certain design problems can be described more easily by graphics than using text. Domain specific but problem independent graphical interface have been developed for structural design. An important task of such interfaces is the semantic interpretation of the graphics (Balachandran and Gero, 1987c).

4.1.2. *The Optimization Problem Formulator.* The problem formulator module carries out the task of formulating a design problem as a canonical optimization model. This module incorporates an algebraic package capable of carrying out algebraic simplification, equation solving and inequality processing. The major functions of the formulator module include identification of the parameters and decision variables involved in the design problem and provision of functional relationships in terms of variables for the constraints and objectives described declaratively. It is also capable of identifying and eliminating redundant constraints. The designer can interactively modify models, adding new design parameters and relationships or deleting existing parameters and relationships, and the problem can be reformulated.

4.1.3. *The Optimization Model Recognizer.* This module selects an optimization algorithm which is suitable to the problem. The first step in the selection of a method is to identify various components of the model. This is carried out by a pattern matching program which has the ability to recognize the various functions and variable types involved in optimization models. Appropriate algorithms are selected by a *rule-based system* which contains knowledge about various optimization models and their algorithms. An example rule to identify a multiple objective linear programming model is given below.

If all the variables are of continuous type
and number of objectives is greater than 1
and all the constraints are linear
and all the objective functions are linear
then conclude that the model is multiple objective linear programming
and execute MOLP algorithm

4.1.4. *The Optimization Problem Solver.* The inclusion of a wide range of numerical optimization methods is a desirable feature in a system which is designed to solve a broad class of structural optimization problems. A variety of routines that contain the optimization algorithms have been developed as a package. These algorithms range from simple linear programming to complex nonlinear programming to multiple criteria algorithms. This module prepares data for the algorithm selected and runs the algorithm and sends the results back to the top-level controller. Problem solving processes can also be controlled by a knowledge-based controller. (Balachandran and Gero, 1987b).

4.2 IMPLEMENTATION OF THE OPTIMA SYSTEM

The prototypical knowledge-based optimal design system OPTIMA has been implemented on a network of SUN 3 workstations under the UNIX operating system. The domain specific knowledge is all represented as frames or production rules and encoded in Lisp (Wilensky, 1984). The pattern matching knowledge used to recognize variables and their relationships and the graphical interface are implemented in Prolog (Clocksin and Mellish, 1981). The optimization algorithms are all encoded in C. A top level controller for the entire system to allow the major components, namely problem formulator, problem recognizer and problem solver, to communicate with each other is written in Lisp.

5. Illustrative Example

Here a simple beam design problem will be used to illustrate the performance of the system. This problem involves determining sectional dimensions to a simply supported beam which carries a uniformly distributed load. The problem is described using both graphics and text. The beam configuration, span and loads are described through graphics (see Figure 2) and the beam design specifications are presented in textual form as follows. The words and symbols in bold type are some of the key words of the system.

beam is_a_kind_of I_section
beam material is steel
beam deflection = 5 * w * L^4 / {384 * E * I}
L = beam span
w = beam uniformly_distributed_load
E = beam elastic_modulus
I = beam second_moment_of_area
beam bending_stress = w * L^2 * D / {16 * I}
beam shear_stress = w * L / {2 * d * t}
d = beam web_depth
t = beam web_thickness
maximum beam bending_stress = 0.660 times beam yield_stress
maximum beam shear_stress = 0.370 times beam yield_stress
maximum beam deflection = 1/360 times beam span
sum_of beam web_depth and 2 times beam beam flange_thickness = beam overall_depth
quotient_of beam web_depth and beam web_thickness <= 180
quotient_of beam web_depth and beam web_thickness <= 816/{0.37*Fy)^0.5
Fy = beam yield_stress
1 / 15 times beam span <= beam overall_depth <= 1/8 times beam span
1 / 5 times beam overall_depth <= beam flange_thickness <= 1 / 3 times beam overall_depth
0.003 m <= beam flange_thickness <= 0.100 m
0.003 m <= beam web_thickness <= 0.100 m
minimize beam weight

Figures 2 and 3 show screen dumps of the session during which the above problem was solved by the system. Figure 2 shows the graphical description of the problem and the mathematical model constructed by the system. Figure 3 shows the interpretation derived from the mathematical description and the solution to the problem obtained using the sequential linear programming method.

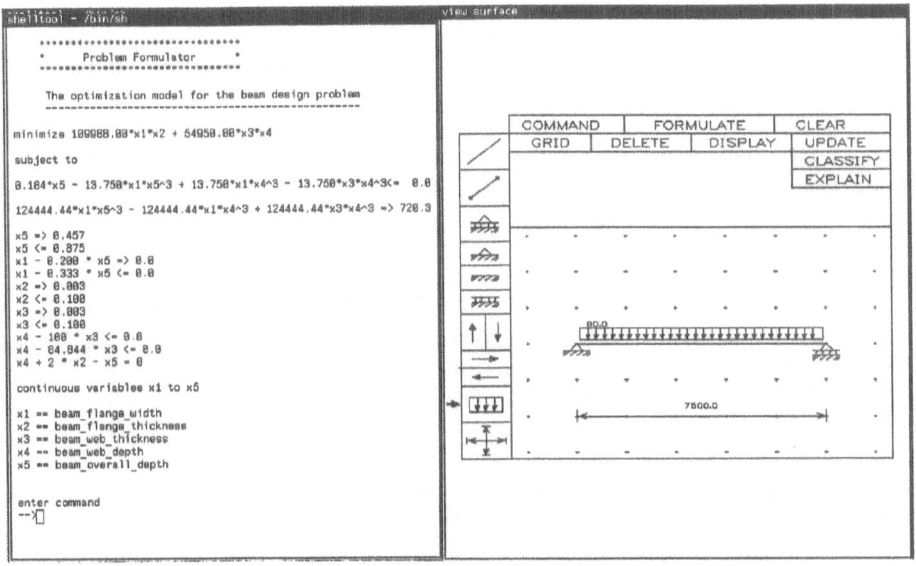

Figure 2. The right window shows the graphical input of the problem and left window shows the mathematical model constructed by the system.

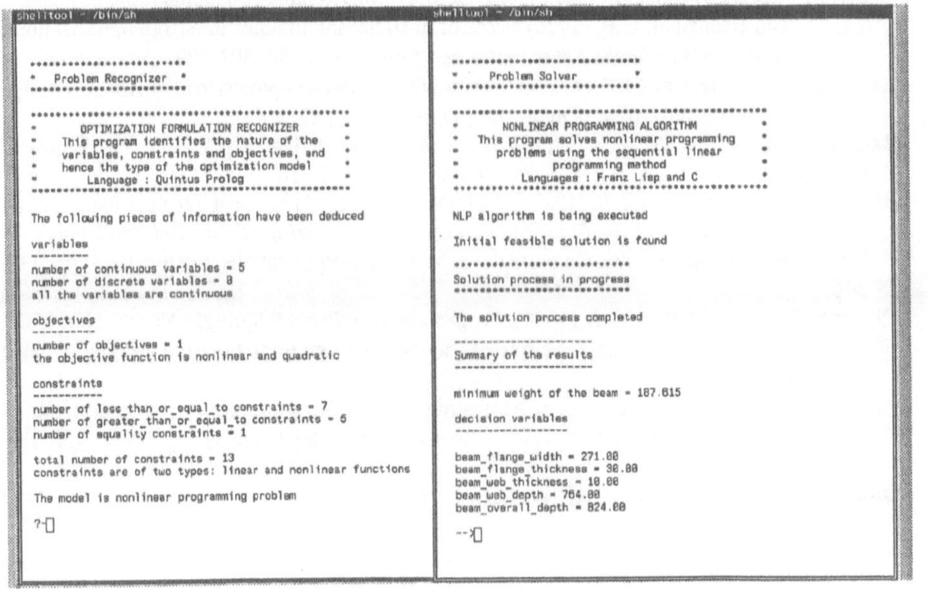

Figure 3. The left window shows the interpretation derived from the canonical model constructed by the system. The right window shows the final solution obtained by solving the model by a sequential linear programming method.

6. Discussion

The structure and operation of a knowledge-based system, used in conjunction with numerical optimization techniques for structural optimization, has been presented. The introduction of knowledge-based systems in this area is significant. Structural optimization processes involve several types of knowledge. The OPTIMA system which was developed incorporating those distinct types of knowledge is more versatile and useful than conventional systems. The example illustrated in the last section obviously shows the potential of this approach to structural optimization.

Optimization has a useful and valid place in structural design decision making, however, knowledge-based systems such as OPTIMA allow designers to handle their problems more easily and provide better interaction compared to most of the conventional systems. It is hoped that the methodology presented in this paper can be applied to a wide variety of structural optimization problems employing mathematical programming techniques.

Acknowledgment. The work described here is supported by continuing grants from the Australian Research Grants Scheme and by a Sydney University Postgraduate Research Studentship.

References

Agogino, A. and Almgren, A. (1987). 'Symbolic Computation in computer-aided optimal design', *Expert Systems in CAD*, ed. J.S. Gero, North-Holland, Amsterdam, pp. 267-284.

Arora, J.S. and Baenziger, G.P. (1986) 'Uses of artificial intelligence in design optimization', *Computer Methods in Applied Mechanics and Engineering*, **54**: 303-323.

Balachandran, M. and Gero, J. S. (1987a). 'A Knowledge-based approach to mathematical design modelling and optimization', *Engineering Optimization*, **12**:2: 99-115 .

Balachandran, M. and Gero, J. S. (1987b). 'Use of knowledge in selection and control of optimization algorithms', *Engineering Optimization*, **12**:2: 163-173.

Balachandran, M. and Gero, J. S. (1987c): 'A knowledge-based graphical interface for structural design', *Applications of Artificial Intelligence in Engineering: Tools and Techniques*, D. Sriram and R.A. Adey, (eds), Computational Mechanics Publications, Boston, USA, pp. 335-346.

Clocksin, W. F. and Mellish, C. S. (1981). *Programming in Prolog*, Springer-Verlag, New York.

Fikes, R. and Kehler, T. (1985). 'The role of frame-based representation in reasoning', *Comm ACM*, **28**:9: 904-920.

Jozwiak, S.F. (1986). 'Applications of AI in structural optimization', *Applications of Artificial Intelligence in Engineering Problems*, D. Sriram and R. Adey (eds), Springer-Verlag, Berlin, pp. 823-831.

Wilensky, R. (1984). *LISPcraft*, W. W. Norton, New York.

MODERN TREND IN ELASTIC-PLASTIC DESIGN. SHAPE AND INTERNAL STRUCTURE OPTIMIZATION

N.V. Banichuk
Institute for Problems in Mechanics
USSR Academy of Sciences, Moscow

ABSTRACT. The paper deals with optimal design of anisotropic elastic-plastic structures. General formulations of the optimization problems are presented. Some results concerning solutions of the problems with strength and load capacity constraints are discussed.

1. SHAPE OPTIMIZATION FOR ELASTIC BODIES

Let us consider the equilibrium state of an elastic body, occupying domain Ω. The body is under the action of volume forces $\underline{q}$ and external loads $\underline{T}$ applied to the part Γ_σ of the body surface Γ. The rigid clamping conditions are supposed to be satisfied for the other part Γ_u of the body surface ($\Gamma_u + \Gamma_\sigma = \Gamma$). The part of the boundary Γ_v of body is taken as a design variable. It is required to minimize the functional depending on the strain-stress state of the body.

$$J[\underline{\sigma},\underline{u}] = \int_\Omega F(\underline{\sigma},\underline{u})d\Omega \tag{1}$$

subject to constant mass isoperimetric condition. Stresses and displacements are defined with the help of the solution of the problem of the theory of elasticity

$$\underline{\nabla}\cdot\underline{\sigma} + \underline{q}=0,\ \underline{\sigma}=\underline{C}\cdot\cdot\underline{\varepsilon} \tag{2}$$

$$\underline{\varepsilon}=1/2(\underline{\nabla}\underline{u} + \underline{\nabla}\overset{*}{\underline{u}}) \tag{3}$$

$$(\underline{\sigma}\cdot\underline{n})_{\Gamma_\sigma} = \underline{T},\ (\underline{u})_{\Gamma_u} = \underline{U} \tag{4}$$

where $\underline{C}$-elastic constant tensor of rank four, F-given function. Here and below the dot between the symbols means scalar product and two dots denote the double scalar product.

Next, an expression for the variation of augmented functional δJ may be sought by varying the surface Γ_v and taking into account (1) - (4)

G. I. N. Rozvany and B. L. Karihaloo (eds.), Structural Optimization, 25–30.

$$\delta J = \int_{\Gamma_u \cap \Gamma_v} [F + \lambda - \underline{\xi}..\underline{\nabla}(\underline{U}-\underline{u})]\, \delta t d\Gamma +$$

$$+ \int_{\Gamma_\sigma \cap \Gamma_v} [F + \lambda + (\underline{n}.\underline{\nabla})(\underline{T}.\underline{\psi}) - \underline{\nabla}.(\underline{\sigma}.\underline{\psi}) + 2H(\underline{T}.\underline{\psi})]\delta t d\Gamma \tag{5}$$

where scalar δt- the value of variation of Γ_v in the n-direction (n-unit vector normal to Γ_v), scalar λ - Lagrange multiplier, H-mean curvature of the body surface; $\underline{\psi}$ - vector of adjoint displacements; $\underline{\xi}$ - tensor of adjoint stresses. The adjoint variables $\underline{\psi}$, $\underline{\xi}$ are determined as functions satisfying the partial differential equations and boundary conditions

$$\underline{\nabla}.\underline{\xi} + \frac{\delta F}{\delta \underline{u}} = 0, \quad \underline{\xi} = \underline{C}..\underline{\phi} \tag{6}$$

$$\underline{\phi} = \frac{1}{2}(\nabla\underline{\psi} + \nabla^*\underline{\psi}) + \frac{\partial F}{\partial \underline{\sigma}} \tag{7}$$

$$(\underline{\xi}.\underline{n})_{\Gamma_\sigma} = 0, \quad (\underline{\psi})_{\Gamma_u} = 0 \tag{8}$$

To exclude the dependence of δJ on $\delta\underline{u}$ and $\delta\underline{\sigma}$ and to obtain explicit relation between δJ and δt the governing equations (6)-(8) for adjoint variables were introduced. Formula (5) determines the sensitivity of the optimized functional to the variations of shape and gives us the possibility to develop numerical methods of successive optimization. Various methods were used within the successive optimization algorithm, but the gradient method appears to be the most efficient from a practical point of view. A lot of questions of application of design sensitivity analysis to structural design problems were discussed in monographs [1,2]. A numerical approach, based on formula (5) and finite element technique, was developed in [3]. Application of boundary element technique to the shape optimization problems also appears to be efficient [4,5].

2. OPTIMIZATION OF ELASTIC-PLASTIC BODIES

Shape optimization for a body of the minimum volume (mass) under load carrying capacity constraints is one of the most important problems in the theory of optimization of elastic-plastic bodies. The optimization problem is formulated with the help of statical theorem of the limit analysis. It is required to determine the part of the boundary $\Gamma_v \subset \Gamma_\sigma$ which minimizes the volume of the body subject to constraints

$$\underline{\nabla} \cdot \underline{\sigma} + \underline{q} = 0 \tag{9}$$

$$(\underline{\sigma} \cdot \underline{n})_{\Gamma_\sigma} = T \tag{10}$$

$$g(\underline{\sigma}, k_f) \leq 0 \tag{11}$$

Here, g, k_f are a given function and parameter. Yield surfaces g ($\underline{\sigma}$, k_f) = 0 contain the origin of coordinates in stress space and depend on parameter k_f ($k = k_f + \varepsilon$, $\varepsilon > 0$ - small parameter, k - yield constant).

Existence of a safe statically admissible stress field σ, satisfying (9) - (11), ensures that the carrying capacity of the body is not exhausted.

Augmented Lagrange functional has the following form

$$J = \int_\Omega d\Omega + \int_\Omega \underline{\psi}\cdot(\underline{\nabla}\,\underline{\sigma} + \underline{q})d\Omega + \int_\Omega \lambda\,(g + \mu^2)d\Omega$$

In the derivation of the relation between δJ and variation of the surface Γ_v we define the adjoint vector $\underline{\psi}$ as a solution of partial differential equations and boundary conditions

$$\frac{1}{2}(\underline{\nabla}\,\underline{\psi} + \underline{\nabla}^*\underline{\psi}) = \frac{\partial g}{\partial \underline{\sigma}}\,,\quad (\underline{\psi})_{\Gamma_u} = 0 \tag{12}$$

As a result, we have the following expression for δJ

$$\delta J = \int_{\Gamma_v} \{1+\lambda g-2H(\underline{T}.\underline{\psi})-\underline{\nabla}.(\underline{\sigma}.\underline{\psi}) + (\underline{n}.\underline{\nabla})(\underline{T}.\underline{\psi})\}\delta t d\Gamma \tag{13}$$

Sensitivity analysis formula (13) and the stationarity principle $\delta J = 0$ give us the necessary optimality condition. We attain classical optimality condition $\underline{\sigma}..\underline{\varepsilon} = 1$ on Γ_v for the particular case when the volume forces are absent and free boundary is varied. This condition means that the rate of energy dissipation is constant along the unknown part of the boundary Γ_v. Examples of numerical solution obtained with the help of (13) and finite element methods, are given in [2,7].

3. OPTIMAL PLASTIC ANISOTROPY

An elastic-plastic body is clamped along the part of the boundary Γ_u and is loaded by volume and surface forces $\underline{q}=p\underline{q}^\circ$, $\underline{T}=p\underline{T}^\circ$ which are proportional to parameter P($\underline{q}^\circ$; $\underline{T}^\circ$ - given functions of space coordinates). Suppose the body consists of identical infinitesimal elements arbitrarily oriented with respect to each other. The fact that the elements are identical but arbitrarily oriented means that the position of the axes of plastic symmetry with respect to fixed Cartesian reference frame changes with the position within the body, but the values of the plastic moduli measured along the axes of plastic symmetry remain unchanged. The fourth-rank tensor $\underline{b}$ of plasticity constants in the principal axes of symmetry and fourth-rank tensor $\underline{B}$ of plasticity constants in a fixed Cartesian coordinate system (global system) are related by $\underline{B}=\underline{0}^*$. ($\underline{0}^*.\underline{b}.\underline{0}$).$\underline{0}$, $\underline{0}^*.\underline{0}=\underline{E}$, where $\underline{0}$ - orthogonal rotational tensor of rank two.

The problem of determining the optimum orientation of the axes of anisotropy [7,8], i.e. finding the rotational tensor $\underline{0}$ from the condition of maximum critical loading parameter is reduced to

$$p_* = \max_{\underline{0}} p(\underline{0}) \tag{14}$$

$$\underline{\nabla}.\underline{\sigma} + \underline{q}=0, \quad (\underline{\sigma}.\underline{n})_{\Gamma_\sigma} = \underline{T} \tag{15}$$

$$g(\underline{\sigma}, k_f) = \underline{\sigma}..\underline{B}..\underline{\sigma} - k_f \leq 0 \tag{16}$$

Let us derive optimality conditions for the problem (14)-(16). To this end variation of $\underline{B}$ corresponding to the variation $\delta\underline{0}$ is written in the form

$$\delta\underline{B} = 4\underline{B}.\underline{0}^*.\delta\underline{0} \tag{17}$$

Taking into account (17), let us construct and vary the augmented Lagrange functional

$$J = \frac{1}{\text{mes}\Omega} \int_\Omega p d\Omega + \int_\Omega \underline{\psi}.(\underline{\nabla}.\underline{\sigma} + \underline{q})d\Omega + \int_\Omega \lambda(g + \mu^2)d\Omega + \int_\Omega \underline{\eta}..(\underline{0}^*.\underline{0}-\underline{E})d\Omega \tag{18}$$

If the relations

$$\frac{1}{2}(\underline{\nabla}\,\underline{\psi} + \underline{\nabla}^*\underline{\psi}) = \lambda \frac{\partial g}{\partial \sigma} \equiv 2\lambda\underline{B}..\underline{\sigma}; \quad (\underline{\psi})_{\Gamma_u} = 0 \tag{19}$$

for adjoint variable $\underline{\psi}$ and the equality $\lambda\,\mu = 0$ are satisfied then the condition $\delta J=0$ is reduced to the following

$$\int_\Omega (\lambda\,\underline{\sigma}..\underline{B}.\underline{\sigma} + \underline{\eta})\,.\underline{0}^*.\delta\underline{0}d\Omega = 0 \tag{20}$$

Using the property of the symmetry of tensor $\underline{\eta}$ we obtain the optimality condition

$$\underline{\sigma}..\underline{B}.\underline{\sigma} = \underline{\sigma}.\underline{B}..\underline{\sigma} \tag{21}$$

The criterion (21) shows that the plastic anisotropy axes have a special orientation and that the optimal body tensor $\underline{\sigma}..\underline{B}\,.\underline{\sigma}$ is a symmetric one. This criterion is now applied to find the optimal plastic anisotropy for bars in torsion. Let the axis of the bar be parallel to the z axis, in the xyz Cartesian coordinate system. The torques applied to the ends of the bar act along that axis. We denote by S the xy-plane cross section of the bar and by Γ the boundary of S. Material of the bar is assumed to be ideal plastic and orthotropic having at each point a plane of symmetry perpendicular to the z axis. Two non-zero components of the stress tensor are related to the stress function as follows

$$\tau_1 = \sigma_{13} = \psi_{,2}, \quad \tau_2 = \sigma_{23} = -\psi_{,1} \tag{22}$$

The equilibrium equation is automatically satisfied through the introduction of stress function ψ defined by (22) and the boundary condition is reduced to the equality $\psi=0$ on Γ. Yield condition is written in the form

$$\underline{\tau}\cdot\underline{B}^{o}\cdot\underline{\tau} = \underline{\nabla}\psi\cdot\underline{A}\cdot\underline{\nabla}\psi \leq k$$

$$\underline{A} = (\underline{B}^{o})^{-1}\;|\underline{B}^{o}| = |\underline{b}^{o}|\;(\underline{0}\cdot\underline{b}^{o}\cdot\underline{0}^{*})^{-1} \tag{23}$$

$$\underline{0}\cdot\underline{0}^{*} = \underline{E}$$

where $B^{o}_{11} = B_{1313}$, $B^{o}_{12} = B_{1323}$, $B^{o}_{22} = B_{2323}$, $b^{o}_{11} = 1/\tau^{o}_{1}$, $b^{o}_{22} = 1/\tau^{o}_{2}$, $b^{o}_{12} = 0$. Parameters τ^{o}_{1}, τ^{o}_{2} are the yield limits for pure shear with respect to the axes of anisotropy. Tensors $\underline{B}^{o}$ and $\underline{b}^{o}$ are related by the matrix equality $\underline{B}^{o}=\underline{0}\cdot\underline{b}^{o}\cdot\underline{0}^{*}$.

Our problem is one of finding the optimum distribution of the angles of orthotropy throughout the material.

The optimization problem consists of finding the tensor $\underline{0}$ for every point of cross section S that maximizes the torsional stiffness of the bar, subject to the condition (23)

$$M = 2\max_{\psi\in D} \int_S \psi\, ds \rightarrow \max_{\underline{0}} \tag{24}$$

$$D = \{\psi . \underline{\nabla}\psi\cdot\underline{A}\cdot\underline{\nabla}\psi \leq k, (\psi)_{\Gamma} = 0\} \tag{25}$$

For the problem (24), (25) the optimality conditions (21) assume the form

$$(B^{o}_{11}\,\tau_1 + B^{o}_{12}\,\tau_2)\,\tau_2 = (B^{o}_{12}\,\tau_1 + B^{o}_{22}\,\tau_2)\,\tau_1 \tag{26}$$

This equation is identically satisfied, if

$$B^{o}_{11}\,\tau_1 + B^{o}_{12}\,\tau_2 = \Lambda\tau_1,$$
$$B^{o}_{12}\,\tau_1 + B^{o}_{22}\,\tau_2 = \Lambda\tau_2 \tag{27}$$

It is easy to show that the relations (27), which take into account (22), (23), may be rewritten as

$$\underline{A}\cdot\underline{\nabla}\psi = \lambda\,\underline{\nabla}\psi$$

where $\lambda = \Lambda\,|\underline{B}^{o}|$, $|\underline{B}^{o}|$ - determinant of matrix B^{o}. Let us introduce the quantities M^{*} and M^{**}

$$M^{*} = 2\max_{\psi\in D^{*}} \int_S \psi\, ds,\quad M^{**} = 2\max_{\psi\in D^{**}} \int_S \psi\, ds$$

$$D^{*} = \{\psi\colon \lambda_{min}(\underline{b}^{o})\,\underline{\nabla}\psi\cdot\nabla\psi \leq k,\ (\psi)_{\Gamma} = 0\}$$

$$D^{**} = \{\psi: \lambda_{max}(\underline{b}^o)\ \underline{\nabla}\ \psi .\ \underline{\nabla}\ \psi \leq k,\ (\psi)_\Gamma = 0\}$$

Note that the eigenvalues of the matrix $\underline{A}$ are the same as those of $\underline{b}^o$.

Next, it was proved in [8] that

$$M^{**} \leq M \leq M^*$$

The lower bound for M (inf $M=M^{**}$) is attained in the equality

$$\underline{A} . \nabla \psi = \lambda_{max}(\underline{b}^o)\ \underline{\nabla}\ \psi$$

is satisfied on S and the upper bound for M (sup $M=M^*$) is realized for the case

$$\underline{A} . \nabla \psi = \lambda_{min}(\underline{b}^o)\ \underline{\nabla}\ \psi$$

Certain problems in parametric optimization of structures made of composite materials were considered in [9, 10].

REFERENCES

1. Haug, E.J., Choi, K.K., Komkov, V., 'Design sensitivity analysis of structural systems'. Orlando: Academic Press 1986.

2. Banichuk, N.V., 'Introduction to structural optimization (in Russian)'. Moscow: Nauka 1986.

3. Banichuk, N.V., Bel'sky, V.G., Kobelev, V.V., 'Optimization for the theory elasticity problems with unknown boundaries'. MTT (Mech. of Solids), 19 (1984).

4. Mota Soares, G.A., Choi, K.K., 'Boundary elements in shape optimal design of structures'. Proc. Advanced Study Institute Computer-Aided Optimal Design'. Vol. 3, Troia, Portugal (1986), 145-185.

5. Aitaliev, Sh.M., Banichuk, N.V., Kaujpov, M.A., 'Optimal design of underground constructions'. Alma-Ata, Nauka 1986.

6. Kobelev, V.V., 'Numerical method for shape optimization'. Proc. Intern. Conf. on Comput. Engng. Mech., Beijing, China (1987), 354-360.

7. Banichuk, N.V., 'Optimum design of structures made of elastic-plastic materials'. Proc. IUTAM Symposium Inelastic Behaviour of Plates and Shells'. Berlin, Springer-Verlag (1986) 325-343.

8. Banichuk, N.V., Kobelev, V.V., 'On optimal plastic anisotropy'. PMM (Appl. Math. and Mechanics), 51, No. 3, (1987).

9. Eschenauer, H., 'Numerical and experimental investigations on structural optimization of engineering design'. Bonn, Druckerei und Verlag, Siegen 1986.

10. Obraztsov, I.F., Vasiljev, V.V., Bunakov, V.A., 'Optimal reinforcement for composite shells of revolution'. Moscow, Mashinostroenie 1977.

COMPOSITE MATERIALS AS A BASIS FOR GENERATING OPTIMAL TOPOLOGIES IN SHAPE DESIGN.

Martin P. Bendsøe
Mathematical Institute,
The Technical University of Denmark,
DK-2800 Lyngby,
Denmark

ABSTRACT. Optimal shape design of structural elements based on boundary variations results in final designs that are topologically equivalent to the initial choice of design, and general, stable computational schemes for this approach often requires some kind of remeshing of the finite element approximation of the analysis problem. This paper presents a methodology for optimal shape design where both these drawbacks can be avoided. The method is related to modern production techniques and consists of computing the optimal distribution in space of a composite material, with the requirement that the resulting structure can carry the given loads as well as satisfy other design requirements.

1. INTRODUCTION

Shape optimization of linearly elastic structures has been studied for more than fifteen years and has reached a level of maturity that makes it viable to implement the methods in CAE(Computer Aided Engineering) systems for production use.

The development of the boundary movement methodolgy for shape optimization has attracted a great deal of attention and the literature on the subject is quite extensive; we refer to the survey [1] by Haftka and Gandhi. The boundary variation method can be implemented in various ways, for example by employing certain mesh moving schemes to define the shape of a given structure. In this case the design variables are the coordinates of nodal points of a finite element model of the structure. A different approach to representing boundaries in shape optimization is to introduce the boundary segment idea which describes the design boundary by a set of simple segments such as straight lines, circular arcs, elliptic arcs, and splines. The optimum is then sought within this restricted definition of the boundary.

The state of the art today for shape design is that shape optimization is possible under the assumption that the initial topology is fixed during the iterative design optimization. A new method that can yield the optimal topology as well as the optimal shape of a structure

G. I. N. Rozvany and B. L. Karihaloo (eds.), Structural Optimization, 31–37.

would be a useful extension of present methodology. Describing the design boundary using parametric equations means that change of topology cannot be expected in the design process, so it is necessary to represent the shape without using "shape" functions. The method introduced in this paper ia a possible alternative approach to shape optimization. Roughly speaking, shape optimization problems are transformed to material distribution problems, using composite materials. Two material constituants, substance and void, are considered, and the microscopic optimal void distribution is considered, instead of shape optimization by boundary variations in the usual sense. An important feature of the procedure is that the homogenization method is applied to determine macroscopic constitutive equations for the material with microscopic material constituants.

The design method described in this paper is strongly inspired by earlier works on composites in optimal design (Rozvany [2], Kohn and Strang [3]) as well as being related to modern production techniques such as numerically controlled milling and plastic forming with controlled porosity through controlled cooling. We take an approach where a structural element is understood in a broad sense as being defined only by the loads it is supposed to carry, its volume (cost), and design requirements such as stress and strain limitations. The only restrictions on the allowable shapes is that the resulting structure should connect to the given surface tractions. The initial design in the iterative design optimization procedure is a rough block of space in which we fill material in an optimal way (or we have a rough block of material and remove material). The use of a fixed domain of simple geometry simplifies the construction of a finite element approximation and the necessity of remeshing as seen in boundary movement techniques is avoided.

The nature of the method is such that it allows a prediction of the topology of the structural member, but it results in a non-smooth estimate of the exact form of the boundary. The method should thus be the first step in a two step shape design procedure, where the second step consists of a traditional boundary variations optimization, based on the design computed in the first step. This second step can be speeded up considerably, as the first step gives a boundary close to the optimal, smooth boundary. Also, the first step will result in estimates of stresses and strains in the structure that will allow construction of an effective finite element mesh for the boundary optimization.

2. OPTIMAL DESIGN OF LINEARLY ELASTIC STRUCTURES

In the following a general formulation for optimal structural design in linear elasticity is presented. The set-up is well known for sizing problems but covers shape design as well.

Consider a mechanical element as a body occupying a domain Ω in $\mathbb{R}^3$ and suppose that the body is subject to body forces f and boundary tractions t. In optimal design for minimum compliance we seek the optimal choice of elasticity tensor E_{ijkl} in some given set of admissible elasticity tensors, $\mathcal{U}$. The admissible tensors will usually be allowed to vary over the domain of the body, so that E_{ijkl} will be a function of the spatial variable $x \in \Omega$, and we have $\mathcal{U} \subseteq (L^\infty(\Omega))^{21}$, in general, corresponding to the 21 independent elements of E_{ijkl}.

Introducing the energy bilinear form

$$a(u,v) = \int_\Omega E_{ijkl}\epsilon_{kl}(u)\epsilon_{ij}(v)dx \tag{1}$$

with linearized strains $\epsilon_{ij}(u) = \frac{1}{2}\left[\frac{\partial u_i}{\partial x_j} + \frac{\partial u_i}{\partial x_i}\right]$ and the load linear form $L(v) = \int_\Omega f\cdot v\, dx + \int_{\Gamma_T} t\cdot v\, ds$ the minimum compliance problem takes the form

$$\underset{E_{ijkl} \in \mathcal{U}}{\text{minimize}} \qquad L(u) \tag{2a}$$

$$\text{subject to : } a(u,v) = L(v), \text{ all } v \in U \tag{2b}$$

$$\text{design constraints} \tag{2c}$$

Here "design constraints" covers constraints on stresses, strains, displacements etc., while sizing constraints, volume constraints etc. are counted for en the choice of $\mathcal{U}$. The space U is the space of kinematically admissible displacement fields.

In the case of optimal shape design the elements E_{ijkl} of $\mathcal{U}$ take on the form

$$E_{ijkl}(x) = \chi(x)E_{ijkl} \tag{3}$$

where E_{ijkl} is the constant elastic tensor for the material employed for the construction of the mechanical element, and $\chi(\mathbf{x})$ is an indicator function for the part Ω^m of Ω that is occupied by the material:

$$\chi(x) = \begin{cases} 1 & \text{if } x \in \Omega^m \\ 0 & \text{if } x \in \Omega\backslash\Omega^m \end{cases} \tag{4}$$

For sizing problems, like design of variable thickness sheets, the admissible E_{ijkl}'s have the form $E_{ijkl}(x) = h(x)E_{ijkl}$ where again E_{ijkl} is a constant tensor and $h \in L^\infty(\Omega)$ is the sizing function.

In the two examples above, it is natural to impose a volume constraint, and this would take on the form:

$$\int_\Omega \chi(x)dx = Vol \qquad \int_\Omega h(x)dx = Vol \tag{5}$$

Of the two design problems described above, the sizing problem is well-posed, solutions exist (cf. Bendsøe [4]) and for computations this problem is straight forward. However, the shape design problem as posed does not, in general, have a solution (cf. Kohn and Strang [3]) unless the problem is regularized in some way by introducing composite materials. Traditionally shape design problems are treated in a different way by defining shapes as given by mappings into R^3, defined on a given reference domain Ω_0 in R^3. Because of smoothness properties required in this method, the class of shapes that is considered will be diffeomorfic with the reference domain Ω_0 and shape changes will be boundary movements. In the mapping method the mechanical element is thus defined as $\phi\ (\Omega_0)$, where ϕ is a diffeomorfism, $\phi\colon \Omega_0 \to \phi(\Omega_0) \subseteq \mathbf{R}^3$. By giving deformation fields etc. on the body $\phi(\Omega_0)$ in terms of deformation fields on the reference domain Ω_0, shape design by boundary variations can also be stated as in (4) on the reference domain Ω_0 (see for example Rousselet and Haug [5]). In this way the energy bilinear form as well as the load linear form will depend on the derivatives of the design variable $\phi \in C^1(\Omega_0, R^3)$. The moving boundary technique leads to a complicated functional dependence on the derivatives of the design function ϕ, and this property has to be taken into account in any discretized, numerical method for solving problems of this type. In order to generate a shape design method which in principle is as general as using a shape design statement as in Eq. (5) and which has the computational attractive simplicity of the sizing problem we propose to use composites of *a priori* simple form which allows a description of the body by a density function. This density function takes on values in the interval [0,1] instead of only the values 0 and 1 as for the indicator function statement above. The structure still consists of material or holes (voids), but on a microlevel, and on a macrolevel the structure is described by a sizing variable, the density of material.

The method consists of the following steps:

- choose a suitable reference domain that allows you to define surface tractions, fixed boundaries, etc.
- choose a composite, constructed by periodic repetition of a unit cell consisting of the given material with one or more holes.
- compute the effective, i.e. homogenized material properties of the composite, using homogenization theory (Sanchez-Palencia [6]). This gives a functional relationship between the density of material in the composite (i.e. sizes of holes) and the effective material properties.
- compute the optimal distribution of this composite material in the reference domain, treating the problem as a sizing problem with the density as the sizing variable.
- interpret the optimal distribution of material as defining a shape, in the sense of the general shape design formulation given above.

3. COMPUTATIONAL RESULTS

The optimization method described above has been tested in various ways on the much studied problem of optimal shape design of a fillet.

Several different settings of the fillet problem was used, allowing for different possibilities for connecting the surface tractions of the problem. Here we show results for a case where only part of the structure is free to be redesigned. The example problem is treated as a plane elasticity problem. The optimization problem that was solved for the examples was the case of minimization of compliance for a given volume of material. In what follows, results for different volume constraints will be illustrated, the volume constraint being given as the percentage of the full design area that is available for the construction of the fillet. For further details on the method and results for other test problems we refer the reader to Bendsøe and Kikuchi [7].

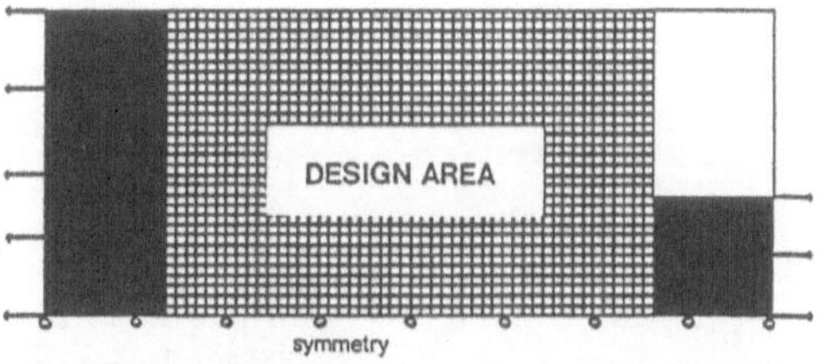

Fig. 1. The example test problem. Black areas indicate material, white areas voids. Figure shows an initial design with the design area filled with a composite with uniform distribution of holes corresponding to a density of 0,64. Note that the voids are at the microlevel.

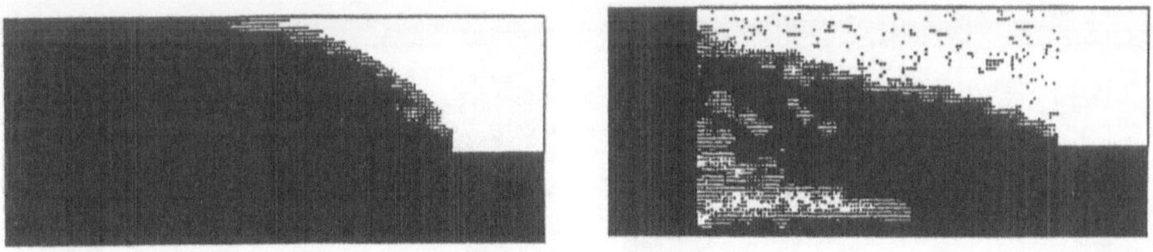

Fig. 2. Optimal distribution of rectangular holes, when cellrotation is a design variable. The volumeconstraint is 91% and 64% respectively.

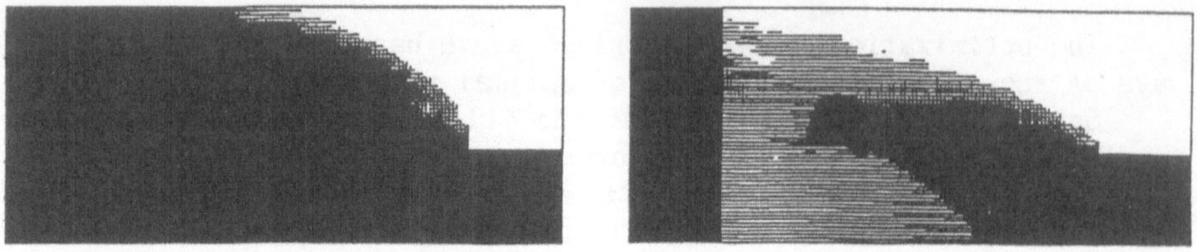

Fig. 3. Optimal distribution of rectangular holes of fixed direction. It is readily seen that the angle of rotation *should* be included as a design variable. Volumeconstraints are 91% and 64%.

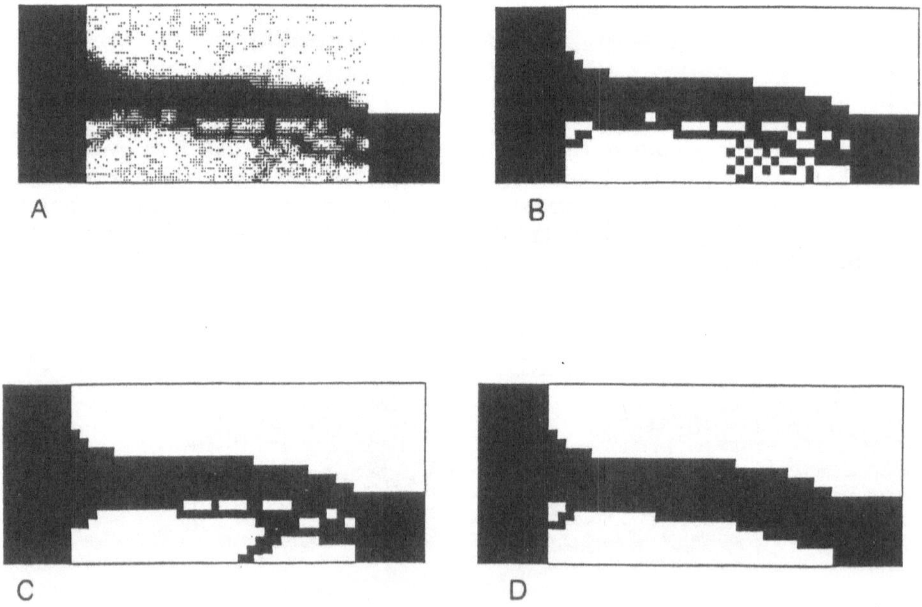

Fig. 4. Optimal shapes for a volume constraint of 36%. A. is obtained with rectangular holes that can rotate, B. is a lumped design that is obtained from A. using a density-cut-off value. C and D show two designs chosen from B. The values of the objective function is: A: 7.55, B: 7.27, C: 7.47, D: 7.28.

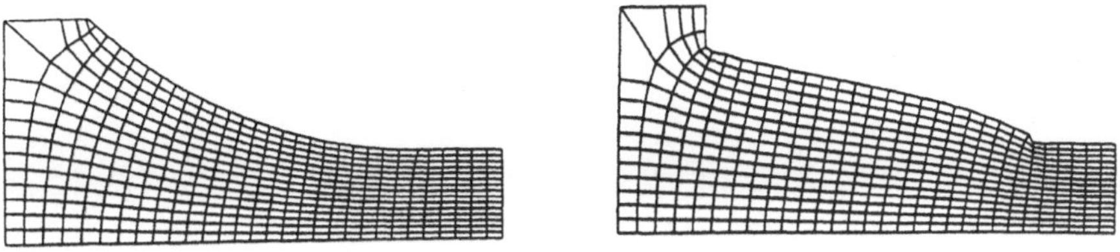

Fig. 5. Optimal shape design obtained by boundary variation,- initial design and final design, respectively (64% volume).A remeshing scheme is used at each iteration and 35 iterations are needed. Using the shape of Fig. 2 as an initial design reduce the number of iterations needed to 5.

References

[1] R.T. Haftka and R.V. Gandhi: *Structural shape optimization - a survey*, Comp Meth. Appl. Mech. Engng., Vol. 57 (1986), pp. 91-106.

[2] G.I.N. Rozvany, T.G. Ong, W.T. Szeto, N. Olhoff, M.P. Bendsøe: *Least-weight design of perforated plates*. Int. J. Solids Struct. Vol. 23 (1987), PP. 521-536 (Part I) and pp. 537-550 (Part II).

[3] R.V. Kohn and G. Strang: *Optimal Design and Relazation of Variational Problems*. Comm. Pure Appl. Math. Vol. 39 (1986), pp. 1-25 (Part I), pp. 139-182 (Part II) and pp. 353-377 (Part III).

[4] M.P. Bendsøe: *On obtaining a solution to optimization problems for solid, elastic plates by restriction of the design space*. J. Struct. Mech., Vol. 11 (1983), pp. 501-521.

[5] B. Rousselet and E.J. Haug, *Design sensitivity analysis in structural mechanics, III. Effects of shape variation*, J. Structural Mech. 10(3) (1983) 273-310.

[6] E. Sanchez-Palencia: *Non-Homogeneous Media and Vibration Theory*. Lecture Notes in Physics, Vol. 127, Springer Verlag 1980.

[7] M.P. Bendsøe and N.Kikuchi: Generating Optimal Topologies in Structural Design Using a Homogenization Method. MAT-Report No. 1988-5, Techn. University of Denmark, 1988.

Performance Characteristics of Optimality Criteria Methods

Laszlo Berke
Chief Scientist for Structures
NASA Lewis Research Center
Cleveland, Ohio 44136

Narendra S. Khot
Aerospace Engineer
Flight Dynamics Laboratory
FIBRA/Wright Patterson AFB, Ohio 45433

ABSTRACT. The formal development of the method is outlined emphasizing the utilization of the separability properties of the objective and constraint functions. Convergence properties are illustrated with examples of increasing numbers of independent size variables ranging from a few hundred to over a thousand.

1. INTRODUCTION

Discretized Optimality Criteria (DOC) methods have a long and somewhat confusing history. Perhaps one should differentiate between the various steps in the design optimization process consisting first of the selection of a structural design concept followed by shape or layout optimization, and finally the sizing of the members in case of discretized structures or models.

In this brief presentation discretized optimality criteria methods will be discussed that can perform efficiently the particular task of element-by-element optimum sizing of structural members, usually the cross sectional properties of finite elements in a large finite element model. The development of the method has been presented in greater detail before (ref. 1-4). The purpose here is to illustrate the weak dependence of convergence on the number of independent size variables.

These derivable optimality criteria statements rest on solid grounds being structures oriented special cases of the more general Khun-Tucker conditions. The iterative procedures to satisfy them are still subject to research. Reference 5. provides a thorough and rigorous treatment of multiplier methods in general.

G. I. N. Rozvany and B. L. Karihaloo (eds.), Structural Optimization, 39–46.

2. DERIVABLE OPTIMALITY CRITERIA

2.1 Discussion

The motivation for the development of the derivable optimality criteria methods was the well known favorable convergence characteristics of the venerable heuristic optimality criteria method of Fully Stressed Design (FSD). Because the mathematical form of stress constraints lacks the important separability properties, FSD cannot be replaced by a derivable optimality criteria, despite a number of attempts (ref. 6). The distinction between derivable and heuristic DOC methods caused considerable confusion relative to the correctness of optimality criteria methods in general. The current state of affairs with displacement formulation for analysis is that for the reason stated above DOC methods for stress constraints are heuristic and for stiffness constraints are derivable. In industry practice the design methods utilize these two kinds of DOC methods sequentially (ref. 7).

2.2 Single Constraints

To introduce the basic ideas of these indirect optimization methods, it is perhaps best to start with the simple, but important, case of a single constraint. In many practical cases there is a single troublesome constraint, or a single composite constraint can be constructed. For example, FSD strength design of the X29-A forward swept wing aircraft was deficient in divergence velocity. The finite element model of the composite wing with thousands of finite element size variables was used to perform the additional stiffening in a few iterations (ref. 7.) employing a method based on ideas presented next.

The essential statement of the optimization problem is:

Optimize $\quad W(X) \quad$ (1)

Subject to $\quad G(X) = C(X) - \bar{C} = 0 \quad$ (2)

where $\quad X = (x_1, \ldots, x_n) \quad x_i > 0 \quad i = 1, \ldots, n$

and
$$W(X) = \sum_{i=1}^{n} w_i(x_i) \qquad C(X) = \sum_{i=1}^{n} C_i(x_i)$$

The function to be optimized (minimized) in case of structures is usually related to weight or cost. G(X) is a constraint function on a behavior variable. Often there are also restrictions on the minimum and maximum acceptable values for the variables x_i. We stipulate that the ith term in W(X) and C(X) is explicitly the function of only x_i. In statically indeterminate structures the $C_i(x_i)$ terms are also the implicit functions of all the variables representing member cross section properties. C is the prescribed value of a constraint, for example, of a displacement limit stated in terms of real or virtual work of a real or virtual load system.

The success of derivable discretized optimality criteria methods for the structural sizing problem, as introduced in references 1-2, hinges on the concept of separability which requires that

$$\frac{\partial^2 w_i(x_i)}{\partial x_i \partial x_j} = \frac{\partial^2 C_i(x_i)}{\partial x_i \partial x_j} = 0 \tag{3}$$

that is

$$\frac{\partial C(X)}{\partial x_i} = \frac{dC_i(x_i)}{dx_i} \tag{4}$$

holds, which in words states that the sum of the implicit derivatives vanishes.

Before going ahead with the derivation, the possibility of passive variables has to be considered. A variable can be passive for a number of reasons. Minimum size constraints, or sizing criteria from earlier considerations can assign a given value to some variables. A practical approach is to include their contribution to the satisfaction of the constraints by appropriately modifying the constraints. To signify this modification we will use the notation $\underline{C}^*$ instead of $\bar{C}$ for the prescribed value of the constraints. Here C^* contains the contribution of the passive variables as a modification to $\bar{C}$.

Employing standard techniques we form the Lagrangian

$$L(X,\lambda) = \sum_{i=1}^{n} w_i(x_i) - \lambda\left(\sum_{i=1}^{n} C_i(x_i) - C^*\right) \tag{5}$$

The optimality criteria for our problem can be stated as

$$\frac{\partial L(X,\lambda)}{\partial x_i} = \frac{dw_i(x_i)}{dx_i} - \lambda\frac{dC_i(x_i)}{dx_i} = 0 \qquad i = 1,\ldots,n \tag{6}$$

with separability expressed by Eq. (4) utilized.

Equation (6) can also be rewritten in the following more useful forms:

$$\frac{\frac{dC_i(x_i)}{dx_i}}{\frac{dw_i(x_i)}{dx_i}} = \frac{1}{\lambda} = constant = \frac{change\ in\ performance}{change\ in\ cost} \tag{7}$$

or

$$1 = \lambda\frac{\frac{dC_i(x_i)}{dx_i}}{\frac{dw_i(x_i)}{dx_i}} = D_i \qquad i = 1,\ldots,n \tag{8}$$

Equation (7) can be interpreted to express the correct optimality criteria statement valid for a general class of optimization problems with diminishing return on investment.

To illustrate how to utilize equation (8) for optimum sizing, a certain degree of specialization of the functions $w_i(x_i)$ and $C_i(x_i)$ is expedient. For simplicity of presentation we assume a linear and a reciprocal dependence, respectively.

$$w_i(x_i) = w_i x_i \tag{9}$$

$$C_i(x_i) = \frac{c_i}{x_i} \tag{10}$$

Further specialization, for example, for trusses and displacement constraints results in the following expressions: $w_i = L_i \rho_i$ and $C_i = S_i^P S_i^V L_i/E_i$ where L_i and ρ_i are the length and specific cost (weight) of the bar respectively. S_i^P and S_i^V are the bar forces due to the actual and the virtual load systems P and V respectively. In this case the displacement constraint is expressed in terms of the (sum of) virtual work $d_k V_k$ of V along the real displacement(s) d due to P.

Substitution of equations (9) and (10) into equation (7) and solving for the size variable x_i yields the simple expression

$$x_i = (\lambda \frac{c_i}{w_i})^{\frac{1}{2}} \tag{11}$$

which gives the value of the size variables at optimum for any value of the constraint. For a given constraint value the Lagrange multiplier λ can be viewed as a simple scaling factor (note the new choice for the sign of the Lagrange multiplier).

For a statically determinate structure the right hand side of equation (11) is not a function of member sizes and it is then a sizing formula for the optimum structure, subject only to scaling. For statically indeterminate structures repeated application of equation (11) is a natural idea similar to the application of repeated FSD iterations. For most practical structural configurations a few iterations should provide sufficient convergence, regardless of the number of the size variables in the structural model. It is this very important property that will be illustrated with examples and serves as the major motivation for considering this indirect approach as a fast algorithm for the optimum sizing of large practical structures. Structural tailoring problems for such constraints as displacement patterns, buckling, dynamic response or aeroelastic requirements have been addressed in the past by optimality criteria methods based essentially on equation (11).

A number of researchers have examined equation (11) and the ideas expressed above and introduced improved algorithms to have a better control over the rate of convergence. Essentially there are three forms that are possible to obtain from equation (8).

If one multiples equation (8) by $(x_i)^q$ on both sides and solves for x_i, the "exponential" iterative form

$$1 = D_i \rightarrow x_i^q = x_i^q D_i \rightarrow x_i^{new} = (x_i D_i^{\frac{1}{q}})^{old} \tag{12}$$

is obtained. After the addition of +1 and -1 to D_i in equation (12) a first term expansion yields the "linearized" form

$$x_i = x_i D_i^{\frac{1}{q}} \rightarrow x_i = x_i(1+D_i-1)^{\frac{1}{q}} \rightarrow x_i^{new} = \left[x_i\left(1+\frac{1}{q}(D_i-1)\right)\right]^{old} \tag{13}$$

by assuming that $(D_i\text{-}1) << 1$ in the neighborhood of the optimum. Finally, if the above steps are performed using reciprocal variables and then reconverting to the original variables, one obtains the "linearized reciprocal" iterative form

$$x_i^{new} = \left[\frac{x_i}{\left(1-\frac{1}{q}(D_i-1)\right)}\right]^{old} \tag{14}$$

The parameter q is called a "step size" parameter. The quantity D_i approaches unity for each variable as the optimum is approached. The deviation from unity is what is utilized to modify the variables. Equation (11), a special case of equation (12) with q=2, can be considered a "natural" form that would assign the correct optimum values to x_i in a single sizing step in case of a statically determinate structure.

Smaller or larger values of q will increase or decrease the modifying effect of the deviation of D_i from unity. A simple adaptive modification is to include another parameter α, slightly different from unity, that would multiply q at each iteration. A good practice appears to be to start with a moderate value for q and increase it slightly near the optimum to compensate for the diminishing deviation of D_i from unity. The best tactics unfortunately are problem dependent.

Equations (11) through (14) all contain the yet undetermined Lagrange multiplier λ which has to be adjusted for the design also to satisfy the prescribed constraint. The following three formulas are recommended choices to modify λ to satisfy the constraints:

simple scaling

$$x_i = x_i'(\frac{C}{C^*}) \tag{15}$$

"first order" iteration

$$\lambda^{new} = [\lambda(1 + pG)]^{old} \qquad G \to 0 \quad as \quad C \to C^* \tag{16}$$

"exponential" iteration

$$\lambda^{new} = [\lambda(C/C^*)^p]^{old} \tag{17}$$

Back substitution of x_i into the constraint equation and solving for λ algebraically is also a possibility. It is beneficial to use scaling even with equations (16) or (17).

2.3 Multiple Constraints

The ideas presented up to this point can be easily extended to multiple constraints and with relatively minor modifications of the algebraic expressions. Following the developments for a single constraint we again form the Lagrangian

$$L(X,\lambda) = \sum_{i=1}^{n} w_i x_i + \sum_{j=1}^{m} \lambda_j \left(\sum_{i=1}^{n} \frac{c_{ij}}{x_i} - C_j^* \right) \tag{18}$$

and obtain the optimally criteria:

$$\frac{\partial L(X,\lambda)}{\partial x_i} = w_i - \sum_{j=1}^{m} \lambda_j \frac{c_{ij}}{x_i^2} = 0 \qquad i = 1,\ldots,n \tag{19}$$

or

$$1 = \sum_{j=1}^{m} \lambda_j \frac{c_{ij}}{w_i x_i} = D_i \tag{20}$$

where the subscript j refers to the jth constraint.

Equations (12), (13) and (14), the three iterative expressions for the size variables, apply with the extended definition of D_i as given in equation (20).

The evaluation of the Lagrange multipliers λ_j becomes somewhat more complicated. The possible approaches are best discussed by first writing the optimality criteria expressed in Eq. (19) in the form

$$\begin{gathered} \lambda_1 \begin{bmatrix} c_{11}/x_1 \\ c_{21}/x_2 \\ \vdots \\ c_{n1}/x_n \end{bmatrix} + \lambda_2 \begin{bmatrix} c_{12}/x_1 \\ c_{22}/x_2 \\ \vdots \\ c_{n2}/x_n \end{bmatrix} + \ldots + \lambda_m \begin{bmatrix} c_{1m}/x_1 \\ c_{2m}/x_2 \\ \vdots \\ c_{nm}/x_n \end{bmatrix} = \begin{bmatrix} w_1 x_1 \\ w_2 x_2 \\ \vdots \\ w_n x_n \end{bmatrix} \\ \lambda_1 C_1 + \lambda_2 C_2 + \ldots + \lambda_m C_m = W \\ \lambda_1 C_1^* + \lambda_2 C_2^* + \ldots + \lambda_m C_m^* = W_{min} \end{gathered} \tag{21}$$

As can be seen the optimum design is a linear combination of designs obtained by considering the constraints one at a time. The way equation (21) is written it resembles a cross-word puzzle. It has to be solved "horizontally" to satisfy the optimality criteria and "vertically" to satisfy the constraints. Furthermore, there is the possibility of some of the constraints to be inactive. If C_j is the current value of the jth constraint the deviation from C_j^* can be used to drive the design to satisfy the constraints using equation (16) or (17) as in case of a single equality constraint. To consider the system of equations as coupled, the simplest approach is to combine the optimality and constraint equations as indicated and obtain a set of linear equations for the λ_j-s:

$$\left[\frac{c_{ij}}{x_i}\right]_{m\times n}\left[\frac{c_{ik}}{w_i x_i^2}\right]_{n\times m}\left[\lambda_j\right]_{m\times 1} = \left[C_j^*\right]_{m\times 1} \Rightarrow \left[E_{jk}\right]\left[\lambda_j\right] = \left[C_j^*\right] \tag{22}$$

where

$$E_{jk} = \frac{c_{ij}c_{ik}}{w_i x_i^3} \tag{23}$$

It is expedient to update the λ_j-s by solving equation (22) with the incremental procedure

$$\left[\lambda_j\right]^{new} = \left[E_{jk}\right]^{-1}\left[(p+1)C_j - pC_j^*\right]^{old} \tag{24}$$

There have been many other approaches proposed in the literature to update the λ_j-s but equations (16) and (17), modified by a subscript j, and equation (24) are effective and are the simplest. Equations (22) and (23) can be derived many other ways but the single line derivation shown here is the simplest. The step size parameter p can also be adoptively upgraded as discussed earlier in connection with the step size parameter q.

3. EXAMPLES

The set of truss-slabs in figure 1 was conceived to illustrate the weak dependence of the convergence on the number of size variables. The loading shown would have caused the truss-slab to both bend and twist. The displacement constraint of 60.00 in. prescribed at the loaded corner nodes prevented twist to develop resulting in a tailored displacement behavior. For uniformity 100 iterations were performed for all cases. An iteration consisted of a resizing based on previous analysis, a new single full analysis and then scaling to satisfy the constraints exactly. The first iteration was started with uniform sizes. As can be seen, about 25 iterations produced acceptable convergence in all cases independently of the number of size variables.

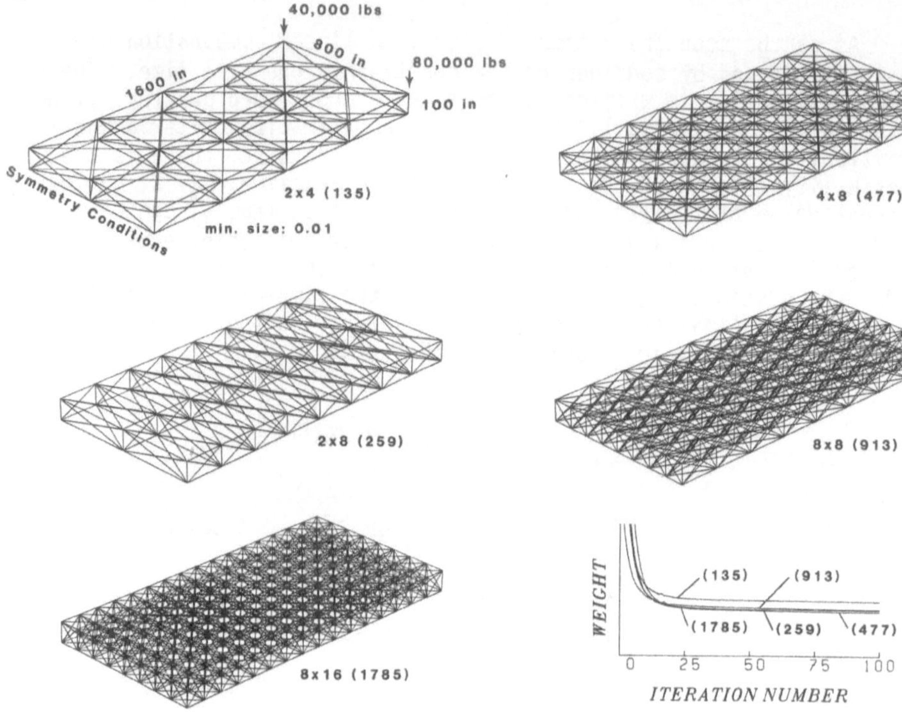

Figure 1. 3-D Truss Problems and Convergence Curves

REFERENCES

1. Berke, L: An Efficient Approach to the Minimum Weight Design of Deflection Limited Structures, AFFDL-TM-70-4, 1970.

2. Berke, L.: Convergence Behavior of Iterative Resizing Procedures Based on Optimality Criteria, AFFDL-TM-72-1-FBR, 1972.

3. Berke, L.; Khot, N. S.: 'Use of Optimality Criteria Methods for Large Scale Systems.' AGARD-LS-70, 1974, pp. 1-1 to 1-29.

4. Mota Soares, C. (ed): Computer Aided Optimal Design: Structural and Mechanical Systems. Proceedings of the NATO ASI held in Portugal, June 29-July 11, NATO ASI Series, Vol. F27, Springer-Verlag 1987,

5. Bertsekas, D. P.: Constrained Optimization and Lagrange Multiplier Methods, Academic Press, 1982, p. 104.

6. Venkayya, V. B.; Khot, N. S.; and Reddy, V. S.: Energy Distribution in an Optimum Structural Design, AFFDL-TR-68-156, Sept. 1968.

7. Lerner, E.: 'The Application of Practical Optimization Techniques in the Preliminary Structural Design of a Forward-Swept Wing.' Second International Symposium on Aeroelasticity and Structural Dynamics, Deutsche Gesellschaft fuer Luft und Raumfahrt, Bonn, Germany, 1985, pp. 381-392.

OPTIMAL SHAPE OF CABLE STRUCTURES

M. Cannarozzi
University of Bologna
Inst. of Struct. Mech.
Viale Risorgimento, 2
40136 Bologna - ITALY

C. Cinquini
University of Pavia
Dept. of Struct. Mech.
Via Abbiategrasso, 211
27100 Pavia - ITALY

R. Contro
University of Trento
Dept. of Engineering
Mesiano di Povo
38050 Trento - ITALY

ABSTRACT. The shape selection problem is formulated as the best approximation of a given not necessarily equilibrated configuration. This approximating configuration is achieved by minimizing the maximum value of suitably defined nodal distances under constraints which express equilibrium in terms of geometric and bounded pseudostress variables. The method seems to be very promising as a meaningful example shows.

1. INTRODUCTION

Design of cable structures takes a further degree of indeterminacy due to the intimate relation between shape and equilibrium. For this reason, at the beginning of the design process a trial and error procedure is inconceivable, even if it was based on analysis methods able to allow for both physical and geometric nonlinearities. In fact, such methods, even when theoretically well founded and computationally well developed, as those in Ref.s 1 and 2, must start from predetermined parameters that essentially concern layout, shape, pretension forces and cable member lengths. All these aspects, surveyed in Ref.3 and quite different from those pertaining e.g. beams, plates, shells, have been considered in papers each of which (e.g. see Ref.s 4,5,6) focuses only on one of the above features and assumes it as main objective, secondarily deriving the other ones. Attempts of meeting real needs and peculiarities of cable net design can be read in Ref.7, where numerical tools in Ref.s 1 and 6 are efficiently employed, and also in Ref.8, the by philosophy of which inspired the present paper. In order to design a cable networks, a typical approach consists on assuming a desired layout and shape that is to be in equilibrium with given loads and have each cable element in tension at a stress level suitably bounded. This goal is pursued in this paper: firstly the problem is formulated as the search for quantities which represent a computationally advantageous compromise between the approximation and the desired, not necessarily equilibrated, configuration; secondly an iterative algorithm, based on a sequence of linear programming problems, is discussed that implements the mathematical programming form which the previous general formulation is cast into. Finally, a numerical example is given.

G. I. N. Rozvany and B. L. Karihaloo (eds.), Structural Optimization, 47–56.

2. PRELIMINARY REMARKS

A cable net is considered in a Euclidean space $\mathbb{R}^3$ referred to a orthogonal coordinates $(0, x_i; i = 1,2,3)$.
The constrained nodes are supposed to be fixed and the other ones to be loaded by concentrated forces. The net consists of N cable members, mutually connected in L nodes the unknown coordinates of which are x_i^k, $k = 1,2...L$. The set which contains such nodes will be denoted as K. The set H comprises the remaining M nodes that are constrained and whose coordinates $\bar{x}_i^h$, $h = 1,2...M$, are prescribed. The nodes of K are subjected to the concentrated forces Q_i^k, $k = 1,2...L$.
The length of the cable element which links two nodes, p and q, will be denoted by as L^{pq} $(= L^{qp})$(*):

$$L^{pq} = \left[(x_i^p - x_i^q)(x_i^p - x_i^q)\right]^{1/2} = L^{qp} \tag{1}$$

while $T^{pq}(= T^{qp})$ denotes the relevant stress, positive when tensile. The set $F^k(C^k)$, which may be empty, contains the indeces p(q) of the free (constrained) nodes, directly connected to k, $\forall\ k \in K$. Equilibrium of the whole system requires that for any node $k \in K$ the following equations be fulfilled:

$$\sum_{1^p}^{L} r^{pk} x_i^p - \left[\sum_{1^p}^{L} r^{pk} + \sum_{1^q}^{M} r^{qk}\right] x_i^k + \sum_{1^q}^{M} r^{qk} \bar{x}_i^q + Q_i^k = 0, \quad p \neq k; \quad i = 1,2,3 \tag{2}$$

where

$$r^{pk} = \frac{T^{pk}}{L^{pk}}, \quad \text{if} \quad p \in F^k \ ; \quad r^{pk} = 0\ , \quad \text{if} \quad p \notin F^k \tag{3.1}$$

$$r^{qk} = \frac{T^{qk}}{L^{qk}}, \quad \text{if} \quad q \in C^k \ ; \quad r^{qk} = 0\ , \quad \text{if} \quad p \notin C^k \tag{3.2}$$

It follows from Rels. (3) that the values of r^{pk} and r^{qk} which may be non-zero in Eq.(2) concern the nodes $p \in k$ and $q \in H$ directly connected to the node k. Thus, in the equilibrium equations the numbers of the quantities r^{qk}, that may are non-zero and are actually different are equal to the numbers of the cable net members. Taking into consideration Eq.(2), the equilibrium equations in the i-direction for the L free nodes can be expressed in compact form as:

$$(\underline{A}(\underline{r}) - \underline{D}(\underline{r}))\ \underline{x}_i + \underline{B}(\underline{r})\ \bar{\underline{x}}_i + \underline{Q}_i = \underline{0} \tag{a}$$

where

$$\underline{x}_i = \left\{x_i^k\right\}, \quad \underline{Q}_i = \left\{Q_i^k\right\}, \quad k = 1,2...L; \quad \bar{\underline{x}}_i = \left\{\bar{x}_i^q\right\}, \quad q = 1,2...M \tag{b}$$

while the N-vector $\underline{r}$ collects the variables r^{pk} and r^{qk} which may be non-zero and actually different. The elements of the matrices $\underline{A}$, $\underline{D}$

(*) Index i, when repeated, implies summation

and $\underline{B}$ linearly depend on the vector $\underline{r}$. Matrix $\underline{A}$ is symmetric with zero diagonal elements. On the other hand $\underline{D}$ is a diagonal matrix with non-zero elements of each row given by the sum of the terms of the same row of the matrices $\underline{A}$ and $\underline{B}$. Therefore matrix $(\underline{A}-\underline{D})$ is a not strictly diagonally dominant matrix for any vector $\underline{r} > \underline{0}$ and, in addition, as it is also irriducible [8], it is non-singular [9, p. 48].
The matrix form of the 3L equilibrium equations is:

$$\underline{J}\ \underline{x} + \underline{C}\ \bar{\underline{x}} + \underline{Q} = \underline{0} \tag{4}$$

where

$$\underline{x}^T = \left[\underline{x}_1^T \,\vdots\, \underline{x}_2^T \,\vdots\, \underline{x}_3^T\right], \quad \bar{\underline{x}}^T = \left[\bar{\underline{x}}_1^T \,\vdots\, \bar{\underline{x}}_2^T \,\vdots\, \bar{\underline{x}}_3^T\right] \tag{c}$$

$$\underline{J} = \mathrm{Diag}\left[(\underline{A} - \underline{D}) \,\vdots\, (\underline{A} - \underline{D}) \,\vdots\, (\underline{A} - \underline{D})\right] \tag{d}$$

$$\underline{C}^T = \left[\underline{B}^T \,\vdots\, \underline{B}^T \,\vdots\, \underline{B}^T\right], \quad \underline{Q}^T = \left[\underline{Q}_1^T \,\vdots\, \underline{Q}_2^T \,\vdots\, \underline{Q}_3^T\right] \tag{e}$$

As indicated previously, the matrix $\underline{J}$ is non singular for any $\underline{r} > \underline{0}$. On the other hand Eqs. (4) can be conceived as a system of 3L bilinear equations in the $\underline{r}$ and $\underline{x}$ variables:

$$\underline{h}\ (\underline{r}, \underline{x}) = \underline{0} \tag{5}$$

where $\underline{h}(\underline{r}, \underline{x}) \in C^1$ and $\underline{J} \equiv [\nabla_x \underline{h}]$. Thus Eqs. (5) define one and only one vector of functions $\underline{x} = \underline{x}(\underline{r})$, $\underline{x}(\underline{r}) \in C^1$, $\forall\ \underline{r} > \underline{0}$, for the Implicit Function Theorem [9, p. 128].
It is worth noting that Eq.(4) yields for any vector $\underline{r} > \underline{0}$ the corresponding equilibrium configuration $\underline{x}$ of the net subjected to the loads $\underline{Q}$. Once the nodal coordinates $\underline{x}$ are obtained, the cable stresses can be evaluated by using definitions (1) and (3).

3. FORMULATION OF THE PROBLEM

Two different network configurations are characterized by the same coordinates of the constrained nodes and by different coordinates of the free nodes. The coordinates of the free nodes $k \in K$ in a given configuration (not necessarily in equilibrium in presence of $\underline{Q}$ loads) and the coordinates of the nodes in any configuration are contained in the L-vectors $\underline{x}^o$ and $\underline{x}$ respectively. The absolute value of the i-component of the vector pointing any node $k \in K$ in the configuration $\underline{x}$ from its own position in $\underline{x}^o$ is denoted by

$$\rho_i^k = \left|x_i^k - x_i^{ok}\right| \quad , \quad k = 1,2...L \tag{f}$$

Let the measure of the distance between a given and an approximating configuration be represented by the function

$$\varepsilon(\underline{x}) = \max_{k,i} \left\{ \rho_i^k \mid k = 1,2...L; \quad i = 1,2,3 \right\} \tag{g}$$

If we define the (non empty) bounded and closed set

$$D = \left\{ \underline{r} \mid \underline{1} \leq \underline{r} \leq \underline{u} , \quad \underline{1} > \underline{0} , \quad \underline{u} \geq \underline{1} \right\} , D \subset \mathbb{R}^N \tag{6.1}$$

where the vectors $\underline{1}$ and $\underline{u}$ contain some given lower and upper bounds on $\underline{r}$, the problem of minimizing the distance between the configuration $\underline{x}^o$ and $\underline{x}$ can be written as:

$$\min \left\{ \varepsilon \mid \varepsilon = \max_{i,k} \left[\rho_i^k \mid k = 1,2...L; \ i = 1,2,3 \right], \ \rho_i^k = |x_i^k - x_i^{ok}|, \right.$$

$$\left. \underline{h}(\underline{r}, \underline{x}) = \underline{0} , \quad \underline{r} \in D \right\} \tag{6.2}$$

It is to be noted that ε is a continuous function of the variables ρ_i^k that are in turn continuous in $\underline{x}$. On the other hand, the component of $\underline{x}$ are continuously dependent on $\underline{r}$, $\underline{r} \in D$ ($\underline{1} > \underline{0}$), via Eq.(5). Thus ε is a continuous function of $\underline{r}$ on the set D. As the set D is bounded and closed, the function ε is bounded from below and takes its minimum on D.

Problem (6) leads to a network shape which approximates the given one according to the previously assumed definition of distance, complying with the equilibrium conditions and the lower and upper bounds prescribed on the vector $\underline{r}$. Having in mind Eqs. (3), it is evident that such bounds do not affect directly the maximum and minimum values of the cable stresses in the solution, since the length of the cable members in the approximating configuration is not known a priori. However, the stress positiveness is ensured in such a configuration, which is an essential requirement because of incapability of resisting stresses in compression.

4. THE SOLVING ALGORITHM

By virtue of the fact that the equilibrium equations $(5)_1$ implicitly define one and only one system of functions $\underline{x} = \underline{x}(\underline{r})$ ($\in C^1$), for any $\underline{r} \in D$, the minimization problem (6) can be cast into the mathematical programming form(*):

$$\underline{g}(\underline{r}) - \underline{i}\,\varepsilon \leq \underline{0} \tag{7.1}$$

$$\underline{1} - \underline{r} \leq \underline{0} , \quad \underline{r} - \underline{u} \leq \underline{0} \tag{7.2,3}$$

(*) From now on $\underline{i}$ denotes a unit vector of such order that matrix operations be meaningful.

$$\min \varepsilon \tag{7.4}$$

where

$$\underline{g}^T(\underline{r}) = \left[\underline{x}^T(\underline{r}) - \underline{x}^{oT} \;\middle|\; \underline{x}^{oT} - \underline{x}^T(\underline{r})\right] \tag{8}$$

The feasible set Z ($Z \subset \mathbb{R}^{N+1}$), with boundary ∂Z, of the problem (7) consists of the intersection of the set $D^* = \{\underline{r}, \varepsilon | \underline{r} \in D\}$ (Ineqs.7.2,3) with the epigraphs of the functions $\underline{g}(\underline{r})$ (Ineq.7.1). Since the objective function is linear, it takes extremum or stationarity values on ∂Z. Moreover, since the functions $\underline{g}(\underline{r})$ are non-convex, the set Z is (possibly) non-convex and the characterization of the extremum or stationarity points cannot be a priori predicted.
Here a descent algorithm is proposed which, starting from a point on ∂Z and going on through a sequence of linear programming problems, reaches another point (also on ∂Z) where Kuhn-Tucker conditions are fulfilled. Being a descent algorithm, the solution point is not a relative maximum.
Before we discuss the above problem in detail, we introduce two Lemmas.

Lemma 1. Consider the vector $\underline{r}^* \in D \subset \mathbb{R}^N$ defining, via Eqs. (5), the configuration $\underline{x}(\underline{r}^*)$ characterized by the distance from $\underline{x}^o$:

$$\varepsilon^* = \varepsilon(\underline{r}^*) = \max_k \left\{g^k(\underline{r}^*); \quad k = 1,2...2L\right\} \tag{h}$$

Hence $\underline{r}^*$ and ε^* satisfy the relations:

$$g^i(\underline{r}^*) - \varepsilon^* = 0, \quad i \in I; \quad g^j(\underline{r}^*) - \varepsilon^* < 0, \quad j \in J \tag{9.1,2}$$

where I is a subset of the set containing the indeces of the $\underline{g}$ entries and J is its complementary set.
If a vector $\underline{\Delta r} \in \mathbb{R}^N$ exists such that (note that the functions g^i are continuously differentiable on D)

$$\left[\underline{\nabla}_{\underline{r}}\, g^i\right]_{\underline{r}^*} \underline{\Delta r} < 0 \,, \quad \forall\, i \in I \tag{10.1}$$

and once

$$\bar{\underline{r}}(\alpha) = \underline{r}^* + \alpha\, \underline{\Delta r} \tag{i}$$

has been defined, with α as real number, we will prove that it is possible to determine a real number $\beta > 0$ such that

$$\bar{\varepsilon}(\alpha) = \varepsilon(\bar{\underline{r}}(\alpha)) = \max_k \left\{g^k(\bar{r}(\alpha)), \quad k = 1,2...2L\right\} < \varepsilon^*, \quad \forall\, \alpha \in (0,\beta) \tag{10.2}$$

In this case every vector $\bar{\underline{r}}(\alpha)$, $\forall \alpha \in (0,\beta)$, defines configurations with distance $\bar{\varepsilon}(\alpha)$ from $\underline{x}^o$ less than ε^*. Hence, the function ε in

problem (6) is strictly decreasing along the direction $\underline{\Delta r}$. Proof of Rel.(10.2) is omitted for the sake of brevity.
With regard to problem (7), one can note that point $\underline{P}^* \equiv (\underline{r}^*, \varepsilon^*)$ belongs to ∂Z. On the other hand, setting $\zeta = \min(\beta, 1)$, if $\underline{r} + \underline{\Delta r} \in D$ also $\underline{r}(\alpha) \in D$, $\forall\ \alpha \in (0,\zeta)$, because of convexity of set D.
In this case every point $\underline{P}(\alpha) \equiv (\underline{r}(\alpha), \varepsilon(\alpha))$ belongs to ∂Z, $\forall\ \alpha \in (0,\zeta)$ and the value $\varepsilon(\alpha) < \varepsilon^*$ of the objective function is associated to each of these points $\underline{P}(\alpha)$.

Lemma 2. Consider the vector $\underline{r}^* \in D$ and the related point $\underline{P}^*(\underline{r}^*, \varepsilon^*) \in \partial Z$ fulfilling Rels. (9). At $\underline{r}^*$ the following linear programming problem is defined:

$$\left[\underline{\nabla}_{\underline{r}}\, g^i\right]_{\underline{r}^*} (\underline{r} - \underline{r}^*) - (\varepsilon - \varepsilon^*) \leq 0, \qquad i \in I \tag{11.1}$$

$$\left[\underline{\nabla}_{\underline{r}}\, g^j\right]_{\underline{r}^*} (\underline{r} - \underline{r}^*) - (\varepsilon - g^j(\underline{r}^*)) \leq 0, \qquad j \in J \tag{11.2}$$

$$(\underline{l} - \underline{r}) \leq \underline{0}, \qquad (\underline{r} - \underline{u}) \leq \underline{0} \tag{11.3}$$

$$\min\ \varepsilon \tag{11.4}$$

where rel.s (11.1,2) represent the linear approximations for the constraints (7.1) at $\underline{r} = \underline{r}^*$ (see rel.s 9.1). Let $\hat{\underline{P}} \equiv (\hat{\underline{r}}, \hat{\varepsilon})$ be the/a solution point of the problem (11). It must be $\hat{\varepsilon} \leq \varepsilon^*$ and it can be proved that: (a) if $\hat{\varepsilon} < \varepsilon^*$, then the function ε-probl.(6) strictly decreases along the direction $\underline{\Delta r} = (\hat{\underline{r}} - \underline{r}^*)$; (b) if $\hat{\varepsilon} = \varepsilon^*$, then the K.-T. conditions for problem (7) are met at $\underline{P}^*$. Proofs of the statements (a) and (b) are omitted for the sake of brevity.

The solution (a local solution) of the problem (7) is found for by means of the algorithm shown in the following and based on the previous considerations.

Step 1. Let a starting point vector $\underline{r}^* \in D$ be given. Matrices $\underline{J}(\underline{r}^*)$, $\underline{C}(\underline{r}^*)$ are formed and the system (4) is solved, determining in this way the configuration $\underline{x}^* : \underline{h}(\underline{r}^*, \underline{x}^*) = \underline{0}$. Then the distance between $\underline{x}^*$ and $\underline{x}^o$:

$$\varepsilon^* = \max_{k,i} \left\{ |x_i^{*k} - x_i^{ok}| \ ; \qquad k = 1,2 \ldots L; \qquad i = 1,2,3 \right\} \tag{1}$$

can be calculated. Point $\underline{P}^* \equiv (\underline{r}^*, \varepsilon^*)$ lies on ∂Z -problem (7).

Step 2. If $\varepsilon^* = 0$, the configuration $\underline{x}^*$ coincides with $\underline{x}^o$ and the procedure stops. If $\varepsilon^* > 0$, matrix

$$\underline{H} = - \left[\underline{\nabla}_{\underline{x}}\, \underline{h}\right]^{-1}_{\underline{r}^*} \left[\underline{\nabla}_{\underline{r}}\, \underline{h}\right]_{\underline{x}^*} \tag{m}$$

is formed and the linear program

$$\begin{aligned}
&\underline{H}(\underline{r} - \underline{r}^*) - \left[\underline{i}\,\varepsilon - (\underline{x}^* - \underline{x}^o)\right] \leq \underline{0} \\
&- \underline{H}(\underline{r} - \underline{r}^*) - \left[\underline{i}\,\varepsilon - (\underline{x}^o - \underline{x}^*)\right] \leq \underline{0} \qquad \text{(n)} \\
&\underline{l} \leq \underline{r} \leq \underline{u} \\
&\min \varepsilon
\end{aligned}$$

is solved. The above problem coincides with the problem (11), which can be shown by applying the differentiation chain rule to the components of $\underline{g}$ -definition (8)- and taking into consideration that the vector $\underline{x}$ implicitly depends on $\underline{r}$ -eq.s (5). The solution is indicated by $\hat{\underline{r}}$, $\hat{\varepsilon}$ and it must satisfy $\hat{\varepsilon} \leq \varepsilon^*$.

Step 3. If $(\hat{\varepsilon}_* - \varepsilon^*)_* = 0_*$ the K.-T. conditions for the problem (7) are fulfilled at $P^* \equiv (\underline{r}^*, \varepsilon_*)$ (Lemma 2) and the solution is achieved with $\underline{r} = \underline{r}^*$, $\underline{x} = \underline{x}^*$, $\varepsilon = \varepsilon^*$. If $(\hat{\varepsilon} - \varepsilon^*) < 0$, the smallest value of $\alpha \in (0,1]$, say $\bar{\alpha}$, is determined which (locally) minimizes the function:

$$\begin{aligned}
&\varepsilon(\alpha) = \max_{k,i} \left\{ |x_i^k(\alpha) - x_i^{ok}| \;:\quad k = 1,2...L; \quad i = 1,2,3 \right\} \qquad \text{(p)} \\
&\underline{x}(\alpha):\ \underline{h}(\underline{r}(\alpha), \underline{x}(\alpha)) = \underline{0}, \quad \underline{r}(\alpha) = \underline{r}^* + \alpha(\hat{\underline{r}} - \underline{r}^*)
\end{aligned}$$

As a consequence of Lemmas 1 and 2, it is $\bar{\alpha} > 0$ and $\varepsilon(\bar{\alpha}) < \varepsilon^*$. Hence the distance of the configuration $\underline{x}(\bar{\alpha})$ from $\underline{x}^o$ is less than the distance of $\underline{x}^*$. Point $P(\bar{\alpha}) \equiv (\underline{r}(\bar{\alpha}), \varepsilon(\bar{\alpha}))$ belongs to ∂Z -problem (7). Then set $\underline{r}^* = \underline{r}(\bar{\alpha})$, $\underline{x}^* = \underline{x}(\bar{\alpha})$, $\varepsilon^* = \varepsilon(\bar{\alpha})$ and go to Step 2.
The proposed algorithm yields points on ∂Z that cause a strictly decreasing sequence of values of the objective function (7).
Since the objective function is bounded from below (being non-negative) the sequence converges. In addition, since the algorithm terminates only if a point on ∂Z is reached where the K.-T. conditions for the problem (7) are fulfilled, the sequence converges to the value assumed by the objective function at this point.
It is worth emphasizing that the value $\bar{\alpha}$ of α (Step 3) can be numerically determined by calculating the function $\varepsilon(\alpha)$ at a finite number of points, starting from the left-hand bound of the interval (0,1]; high accuracy is not required for determining $\bar{\alpha}$. Moreover, solution of the system (4) (Steps 1 and 3) becomes less expensive if one considers that the matrix $\underline{J}$ is a three-block-diagonal matrix in which the blocks are equal and sparse matrices. Matrix $\underline{J} \equiv [\nabla_{\underline{x}}\, \underline{h}]_{\underline{r}^*}$ has not to be necessarily inverted; in fact matrix $\underline{H}$ can be directly obtained by solving the matrix equation $\underline{J}\,\underline{H} + [\nabla_{\underline{r}}\, \underline{h}]_{\underline{x}^*} = \underline{0}$. Finally, in practice the procedure is stopped when ε^* or the difference $(\hat{\varepsilon} - \varepsilon^*)$ is smaller than a suitably small given value.

5. A NUMERICAL EXAMPLE

In order to show the validity of the formulation and the effectiveness of the algorithm proposed above, the cable layout shown in Fig.1a,b is considered.
Such layout represents the shape of the cable net to be approximated when the loads in the Z direction and acting at each node consists on 250 forces and suitable bounds are imposed on the pseudo-stresses $\underline{r}$. These bounds can be determined by assuming as final length of the cable elements the lengths of the (expected) design configuration. In the present case, lower and upper bounds of 100 and 1000 force/length quantities are assumed respectively, which imply approximately bounds of 450 and 4500 on cable stresses in the design configuration.
The procedure starts by considering $\underline{r} = \underline{100}$ as initial point. Numerical procedure has been stopped when inequality $(\varepsilon_i^* - \hat{\varepsilon}_i)/\varepsilon_i^* \leq 1.x10^{-6}$ resulted to be satisfied. According to Section 4, ε_i^* and $\hat{\varepsilon}_i$ represent respectively the objective function of the Nonlinear Programming problem and of the Mathematical Programming problem which linearly approximates the previous one at the same iteration i. The termination criterion has been met at the 8th iteration.
In Fig.2 the values of the objective function are indicated. It is to be noted that in this case the current value of the objective function represents the distance of the coordinate z at the central node from the origin of axes where location of the same node has been fixed in its expected configuration. At 0 iteration, the value of such a distance is referred to as the starting point.
The value of the objective function at the final configuration shows that the corresponding shape is very close to the expected one. As a consequence, the lengths of the cable elements in the expected and final configuration are quite similar and hence the bounds on the cable element stresses, having in mind the expected configuration, are satisfactorily complied with in the final configuration.
The shapes of the network at the iterations 0, 1,2,3,8 are sketched in Fig.2.

6. CONCLUDING REMARKS

Intentionally, practical applications of the proposed method were not discussed in greater detail and the treatment was restricted to computational aspects.
The effectiveness of the numerical procedure is mainly due to algebraic properties of structural problems and to a strategy which matches theoretical and computational features. In particular, it is to be noted that using pseudostresses as variables permits the formulation of equilibrium equations in a bilinear form so that the current configuration is obtained through a system of linear equations. This fact together with the employement of a sequence of linear programming problems allow the optimization of relatively large cable networks.

ACKNOWLEDGEMENT

Financial support provided by Italian Ministry of Public Education is gratefully acknowledged. The Authors are also in debt with Mr. A. Cazzani for his collaboration in preparing the numerical example.

REFERENCES

1. Contro, R. and Maier, G., 'Inelastic Analysis of Pretensioned Cable Structures by Mathematical Programming', Steel Structures, Elsevier Publ., 85, 1986.
2. Cannarozzi, M. and Contro, R., 'A Minimum Principle on Displacements and a Related NLP Formulation for the Elastostatics of Cable Systems', Proc. of Int. Conf. on Comput. Mechanics, II, 79, Tokyo, 1986.
3. Jendo, S., 'Some Aspects on the Optimization of Cable Suspended Structures', SFB 64 Universität Stuttgart, 1984.
4. Rozvany, G.I.N., Wang, C.M. and Dow, M., 'Archgrids and Cable-networks of Optimal Layout', Comput. Methods Appl. Mech. Eng., 31, 91, 1982.
5. Argyris, J.H., Angelopoulos, T. and Bichat, B., 'A General Method for the Shape Finding of lightweight Tension Structures', Comput. Methods Appl. Mech. Eng., 3, 135, 1974.
6. Cinquini, C. and Contro, R., 'Prestressing Design Method for Cable Net Structures', Engng. Struct., 7, 183, 1985.
7. Contro, R., 'Analysis and Design of Tension Structures by Optimization Techniques', Proc. of Int.Symp. on Weitgespannte Flächentragwrke, 2, 95, Stuttgart 1985.
8. Cannarozzi, M., 'A Shape Finding Method for Cable-net Structures', Costruzioni Metalliche, 4, 181, 1981.
9. Ortega, J.M. and Rheinboldt, W.C., 'Iterative Solution of Nonlinear Equations in Several Variables, Academic Press, 1970.

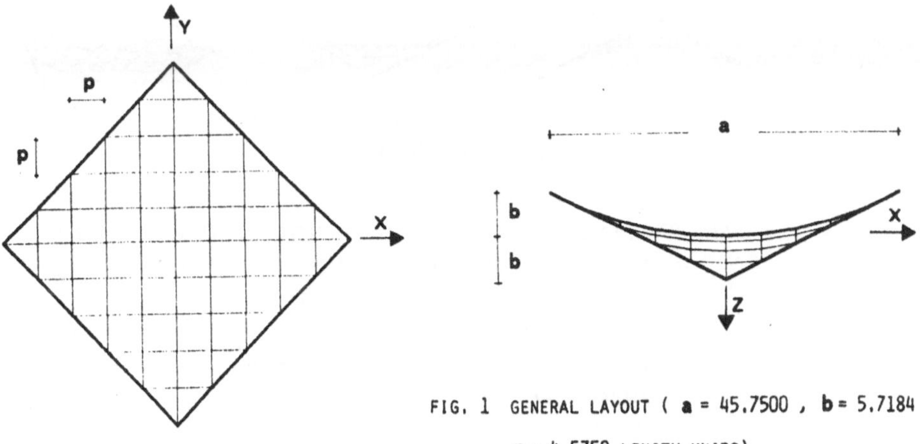

FIG. 1 GENERAL LAYOUT (a = 45.7500 , b = 5.7184 , p = 4.5750 LENGTH-UNITS)

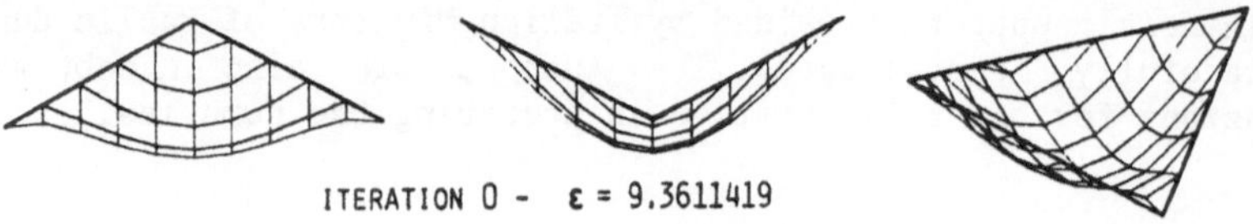

ITERATION 0 - ε = 9.3611419

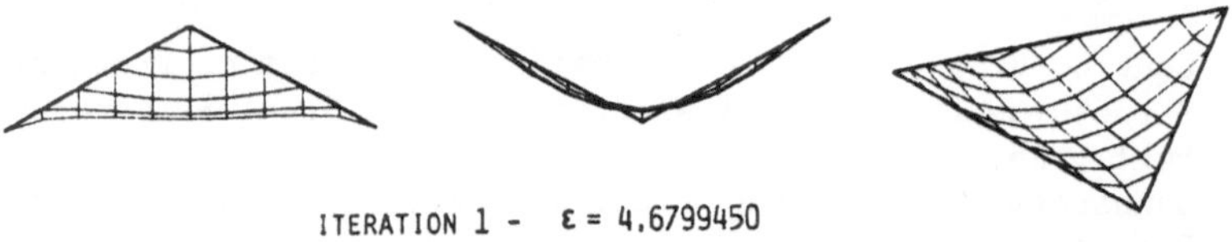

ITERATION 1 - ε = 4.6799450

ITERATION 2 - ε = 2.3399984

ITERATION 3 - ε = 1.1705903

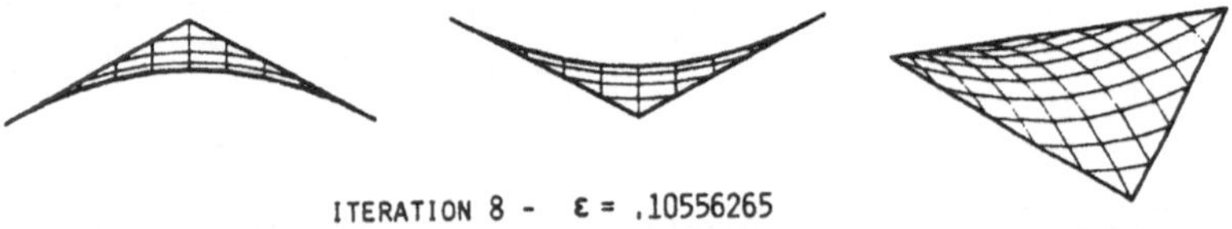

ITERATION 8 - ε = .10556265

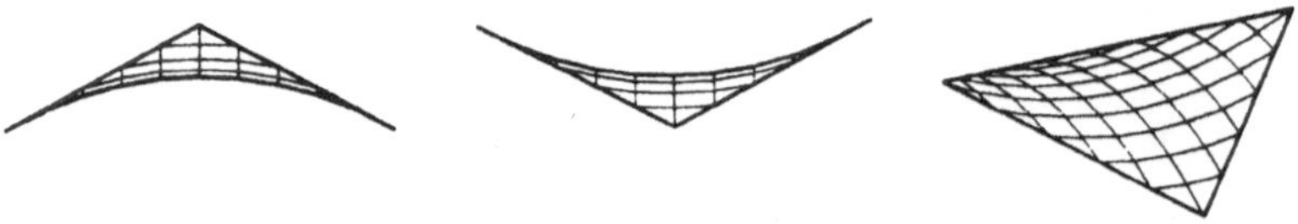

EXPECTED CONFIGURATION

FIG. 2 LATERAL AND PERSPECTIVE VIEWS OF CABLE NET CONFIGURATIONS AT THE INDICATED ITERATIONS. (VALUES OF THE OBJECTIVE FUNCTION ε AT ITERATIONS 4 TO 7 ARE RESPECTIVELY: .10557083, .10556743, .10556275, .10556267; THE CORRESPONDING DRAWINGS ARE MISSED BECAUSE OF UNAPPRECIABLE DIFFERENCES WITH RESPECT TO THE ITERATION 8)

SHAPE OPTIMIZATION OF CONTINUUM WITH CRACK

GENGDONG CHENG
Research Institute of Engineering Mechanics
Dalian Institute of Technology, Dalian, 116024, China

BIAO FU
Beijing Institute of Technology, Beijing, China

ABSTRACT. The present paper studies the problem of shape optimization from the viewpoint of structural design philosophy based on durability and damage tolerance. Initial cracks are assumed to exist or to occur at an early stage of fatigue life. The objective is to minimize the crack propagation rate, or the stress intensity factor range. Quadratic boundary elements are applied to discretize the continuum to be optimized. To obtain the stress intensity factor range, quarter-point singular elements are placed at the tip of the crack. The sensitivity of the stress intensity factor with respect to the structural shape is derived. A numerical example is presented and dicussed.

1. INTRODUCTION

In recent years, a number of new design concepts such as durability, damage tolerance, fuzzy design and reliability based design have emerged. All these have had an impact on structural optimization. Most of the research on structural shape optimization in the literature [1, 2] aims at minimizing the maximum stress in an elastic continuum which is subject to external loads and relevant constraints in order to prevent structures from crack initiation or plastic deformation. However, for structural components made of ductile material, the fatigue life of structures consists of the life of crack initiation as well as crack propagation, the latter often being more important than the former. Based on the above observation, the concept of durability and damage tolerance is adopted in aircraft design. Within the context of this design concept, one of the interesting problems is to consider shape optimization from the viewpoint of the resistance of the material against crack propagation. Ideally, the objective of optimization is to maximize the crack propagation life to the structural component which has an existing crack and is subject to the given service loading conditions.

Based on research in the field of fatigue cracks, the crack growth rate can be expressed in terms of a continuous functional relationship with the stress cycle parameters, i.e.,

$$da/dn = G(dk, r) \tag{1}$$

where r is the stress ratio, dk the stress intensity factor range, a the length of the existing crack and n the number of the loading cycles. When the stress intensity factor range occurences occupy only a small region on the da/dn diagram, the crack growth rate is governed by a simple power law:

$$da/dn = c(dk)^m \tag{2}$$

G. I. N. Rozvany and B. L. Karihaloo (eds.), Structural Optimization, 57–62.

in which c and m are material dependent constants, whereas the stress intensity factor range dk depends on the current crack length "a" as well. To obtain the crack propagation life from Eq. (2) numerically, one should apply the following procedure. At any given crack length "a", dk is calculated numerically by FEM or BEM. At this same crack length, the value of (da/dn) is then determined from Eq. (2). If the structural component experiences Δn loading cycles, the crack length increases by $\Delta a = (da/dn)\Delta n$. Now we have a new crack of length $a + \Delta a$, and can follow the above steps again until the useful life of the structural component is reached.

It can be seen from the above discussion that the shape optimization problem is extremely difficult if the objective function in the formulation is the crack propagation life. As a first step towards solving the problem, the present paper studies shape optimization of a planar continuum which is subject to external surface loading, thermal loading, body and centrifugal force. The objective of optimization is to minimize the crack propagation rate, or the stress intensity factor range under the given service loading conditions. An initial crack of a given length and location in the structural component is assumed to exit or to occur at an early stage of fatigue life. Now, the problem is formulated as follows:

$$\begin{gathered}\text{To find design variables } X \\ \min\, dk(X) \\ \text{subject to } \underline{X} \le X \le \overline{X}\end{gathered} \tag{3}$$

where X stands for the properly chosen design variables describing the shape of the structural component.

2. BOUNDARY ELEMENT METHOD

The boundary element method for elasticity is based on the following boundary integral equation:

$$C_{ki}(P)u_i(P) + \int_{S_Q} T_{ki}(P,Q)u_i(Q)ds_Q = \int_{S_Q} U_{ki}(P,Q)d_i(Q)ds_Q +$$

$$+\int_{S_Q} D_k(P,Q)ds_Q + \int_{S_Q} E_k(P,Q)h(Q)ds_Q + \int_{S_Q} F_k(P,Q)[\partial h(Q)/\partial n]ds_Q \tag{4}$$

in which P and Q are the two points on the boundary s_Q (see Fig. 1) and $u_i(Q)$, $d_i(Q)$, $h(Q)$ and $\partial h(Q)/\partial n$ are the displacement and traction along the i-th direction, the temperature and its gradient at point Q, respectively. $T_{ki}(P,Q)$ and $U_{ki}(P,Q)$ are the displacement and tractions due to a unit concentrated force in an elastic infinite space. $D_k(P,Q)$ $E_k(P,Q)$ and $F_k(P,Q)$ are the tractions at point P contributed by the centrifugal force, temperature and its gradient at the point Q. Their algebraic expressions in two-dimensional elasticity can be found in the literature [3]. The coefficients $C_{ki}(P)$ depend on the local shape of the boundary s_Q near the point P.

In the boundary element method one divides the boundary s_Q into n elements. For a quadratic element each element has two end-point nodes and one mid-point node, and

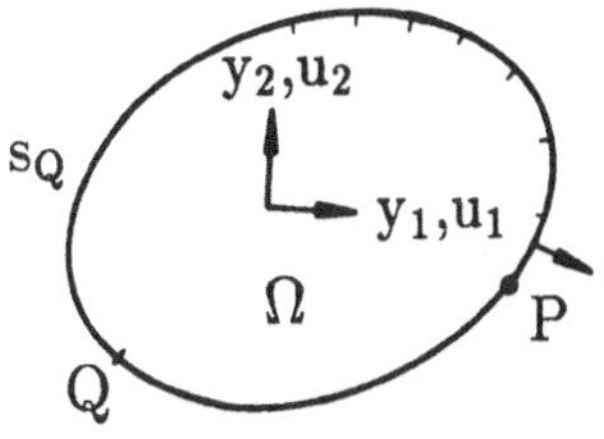

Fig. 1
Notation for Eq. (4).

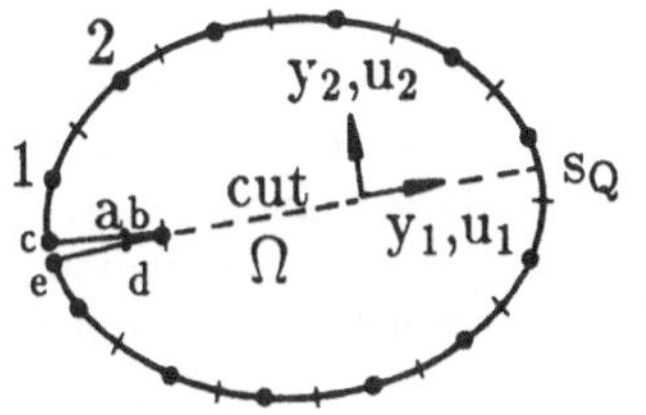

Fig. 2
Singular elements at the crack tip.

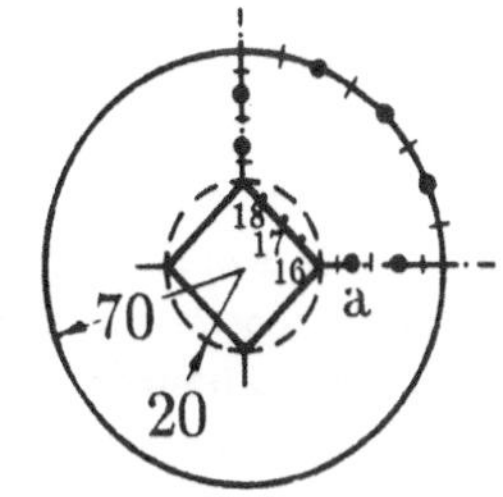

Fig. 3
Example and its BEM mesh.

within each element the boundary curve and the functions u_i, d_i, h and $\partial h/\partial n$ are interpolated by the quadratic functions

$$f = f^1 N_1(t) + f^2 N_2(t) + f^3 N_3(t) \tag{5}$$

where f_1, f_2, f_3 are values for the function f at nodes 1, 2 and 3. The shape functions $N_1(t)$, $N_2(t)$ and $N_3(t)$ are given as follows,

$$N_1(t) = t(t-1)/2, \quad N_2(t) = 1 - t^2, \quad N_3(t) = t(t+1)/2 \tag{6}$$

The function f represents any of the parameters y_1, y_2, u_1, u_2 etc. and t is the local nondimensional coordinates $(-1 \le t \le 1)$. Substituting Eqs (5) and (6) into (4), we obtain the following equation for point P:

$$Cu^p + \sum\left(\int_{-1}^{1} TN|J|dt\right)u^e = \sum\left(\int_{-1}^{1} UN|J|dt\right)d^e + \sum\left(\int_{-1}^{1} D|J|dt\right) +$$
$$+ \sum\left(\int_{-1}^{1} EN|J|dt\right)h^e - \sum\left(\int_{-1}^{1} FN|J|dt\right)\partial h^e/\partial n \tag{7}$$

where J is the value of the Jacobian, C, T and U are 2*2 matrices, D, E and F are 2*1 vectors, the superscripts p and e refer to quantities at the point P and element, respectively. Taking all the nodal points on the boundary s_Q as the source point in turn, we obtain the following equation from Eq. (4),

$$Hu = Gd + D + Eh - F(\partial h/\partial n) \tag{8}$$

Alternatively, gathering all the unknowns in z, we obtain the following equation:

$$Rz = b \tag{9}$$

To cope with a given crack in the continuum and to calculate the stress intensity factor at the crack tip, the above procedure should be modified [2]. Two quarter-point singular boundary elements, whose internal node is at the quarter point of the element length instead of the mid-point (see Fig. 2), should be placed at the tip of the crack. Once the displacements at the faces of the crack are obtained, the stress intensity factors for mode I and mode II fracture are given as follows:

$$K_I = 8A\sqrt{\frac{2\pi}{a}}[4(u_{2b} - u_{2d}) + (u_{2e} - u_{2c})], \quad K_{II} = 8A\sqrt{\frac{2\pi}{a}}[4(u_{1b} - u_{1d}) + (u_{1e} - u_{1c})] \tag{10}$$

where A is E (Young's modulus) for the plane stress problem, $E/(1-\nu^2)$ for the plane strain problem.

When the continuum is unloaded, the nodes on the upper and lower face of the crack merge and the corrresponding nodes share the same coordinates. If one treated the upper and lower face of the crack as part of the boundary only, the resulting equation (9) would be deficient. To overcome this difficulty, we cut the continuum along the crack into two parts, and apply the boundary element method to the two parts separately. At the nodes along the artificial cut which have the same coordinates, the displacements should be continuous and the tractions should be in equilibrium. With these conditions, the two sets of equations are solved simultaneously. After solving for the displacements at the crack tip and the two faces, we can calculate the stress intensity factor from Eqs (10).

3. SENSITIVITY ANALYSIS AND OPTIMIZATION

For the sake of simplicity, it is assumed that the location and length of the crack, the temperature and its gradient, angular speed and the material properties of the continuum are not modified during the process of shape optimization. Only the shape of the part of the boundary which is not the face of the crack is designed in this paper. Various choices of shape design variables have been proposed [1, 4] but they relate to the boundary nodal coordinates Y_i $(i = 1, 2)$ implicitly or explicitly. From Eqs (10), the sensitivity of the stress intensity factors with respect to the nodal coordinates Y_i

$$\partial K_i/\partial Y_i = 8A\sqrt{2\pi/a}[4(\partial u_{2b}/\partial y_i - \partial u_{2d}/\partial Y_i) + (\partial u_{2e}/\partial y_i - \partial Y_i - \partial u_{2c}/\partial Y_i)]$$

$$\partial K_{ii}/\partial Y_i = 8A\sqrt{2\pi/a}[4(\partial u_{1b}/\partial y_i - \partial u_{1d}/\partial Y_i) + (\partial u_{1e}/\partial y_i - \partial Y_i - \partial u_{1c}/\partial Y_i)] \quad (11)$$

are expressed in terms of the sensitivities of the nodal displacements at the crack faces with respect to the nodal coordinates $\partial u_j/\partial Y_i$. To obtain $\partial u_j/\partial Y_i$ Eq. (5) is differentiated with respect to Y_i:

$$\partial C/\partial Y_i u^p + C\partial u^p/\partial y_i + \sum\left[\int_{-1}^{1}(\partial T/\partial Y_i N|J| + TN\partial|J|/\partial Y_i)dt\right]u^e +$$

$$+\sum\left(\int_{-1}^{1} TN|J|dt\right)\partial u^e/\partial Y_i = \sum\left[\int_{-1}^{1}(\partial U/\partial Y_i N|J| + UN\partial|J|/\partial Y_i)dt\right]d^e +$$

$$+\sum\left(\int_{-1}^{1} UN|J|dt\right)\partial u^e/\partial Y_i + \sum\left[\int_{-1}^{1}(\partial D/\partial Y_i|J| + D\partial|J|/\partial Y_i)dt\right] +$$

$$+\sum\left[\int_{-1}^{1}(\partial E/\partial Y_i N|J| + EN\partial|J|/\partial Y_i)dt\right]h^e +$$

$$+\sum\left[\int_{-1}^{1}(\partial F/\partial Y_i N|J| + FN\partial|J|/\partial Y_i)dt\right]\partial h^e/\partial n \quad (12)$$

To simplify the above formulae, we impose an arbitrarily rigid motion to Eq. (7) and obtain an alternative expression for the matrix C as follows:

$$C = -\sum_e \left(\int_{-1}^{1} TN|J|dt \right) \cdot \begin{bmatrix} 1 & 0 \\ 0 & 1 \\ 1 & 0 \\ 0 & 1 \\ 1 & 0 \\ 0 & 1 \end{bmatrix} \tag{13}$$

Differentiating it with respect to Y_i, we obtain the sensitivity of the matrix C

$$\partial C/\partial Y_i = -\sum_e \left[\int_{-1}^{1} (\partial T/\partial Y_i N|J| + TN\partial|J|/\partial Y_i)dt \right] \cdot \begin{bmatrix} 1 & 0 \\ 0 & 1 \\ 1 & 0 \\ 0 & 1 \\ 1 & 0 \\ 0 & 1 \end{bmatrix} \tag{14}$$

Substituting Eq. (14) into Eq. (12) and comparing the result with Eq. (7), we have

$$R\partial z/\partial Y_i = -\sum_e \left[\int_{-1}^{1} (\partial T/\partial Y_i N|J| + TN\partial|J|/\partial Y_i)dt \right] \cdot \begin{bmatrix} u^1 & -u^p \\ u^2 & -u^p \\ u^3 & -u^p \end{bmatrix} +$$

$$+\sum_e \left[\int_{-1}^{1} (\partial U/\partial Y_i N|J| + UN\partial|J|/\partial Y_i)dt \right] d^e + \sum_e \left[\int_{-1}^{1} \partial D/\partial Y_i |J| + D\partial|J|/\partial Y_i)dt \right] +$$

$$+\sum_e \left[\int_{-1}^{1} (\partial E/\partial Y_i N|J| + EN\partial|J|/\partial Y_i)dt \right] h^e -$$

$$-\sum_e \left[\int_{-1}^{1} (\partial F/\partial Y_i N|J| + FN\partial|J|/\partial Y_i)dt \right] \partial h^e/\partial n \tag{15}$$

Since some of the integrals which appear in Eq. (15) contain singular integrands, a special treatment is needed. Because of the limited space, we omit the detailed derivation.

With the sensitivity information available, any first-order algorithm of mathematical programming is applicable to solve the shape optimization problem (3). Here, we have constructed a special version of the Gill-Murry Quasi-Newton method for solving (3).

4. NUMERICAL EXAMPLES

A computer program has been developed for minimizing the stress intensity factor based on the above study. The program is written in Fortran and runs on IBM/PC compatible computers. A classical example from [3] is tested to ensure the correctness of the stress intensity factor calculation. Since the formulae and program for sensitivity analysis are very complicated, we have applied the global finite difference method (GFD) to obtain sensitivities for a number of examples. In the GDF method, the stress intensity factors at the original design and a number of its neighbouring designs are calculated and then used to approximate the sensitivities. The sensitivities obtained from these two methods for the examples are in good agreement.

Fig. 3 shows one interesting example and BEM mesh. It is an artificial structure which has four symmetric cracks and whose material properties are as follows:

$$E = 200000.0N/\text{mm}^2, \quad \nu = 0.3, \quad \rho = 0.0078\,\text{kg/mm}^3$$

The structure rotates around its center at an angular speed of $\omega = 10$/sec. The positions of nodes 16, 17 and 18 along the inward normal of the boundary are the design variables. Two cases are considered: $a = 2$mm, $a = 6$mm. The initial and optimal designs associated with the stress intensity factors for the two cases are listed in Table 1. The optimal design is shown in Fig. 3 by a dashed line. It is interesting to see that the optimal designs for the two cases are the same and have a nearly circular hole.

5. DISCUSSION

Obviously, different design concepts will lead to different optimal designs. Since many new design concepts have emerged, more research is needed to understand the impact of these on formulation of the optimization problem and the resulting designs. The present paper has considered shape optimization from the design philosophy based on durability and damage tolerance. The numerical investigation given here is only a first step towards understanding the problem.

Table I

a	design	x_{16}	x_{17}	x_{18}	K
2mm	initial	0	0	0	30666
	optimal	3.535	7.07	3.535	27904
6mm	initial	0	0	0	37474
	optimal	3.535	7.07	3.535	36274

REFERENCES

1. Pedersen, P. and Laursen, C. (1982) "Design for minimum stress concentration by finite elements and linear programming", *J. Struc. Mech.*, 10, 375-391.

2. Cheng, G.D. and Wang, L. (1984) "A numerical method of shape optimization and the course line search strategy", *Eng. Opt.*, 8, 69-82.

3. Brebbia, C.A. (Ed.) (1981) "Progress in boundary element methods", Vol. 2. Pentech Press.

4. Cheng, G.D. "Some improvement of numerical method for shape optimization", Proc. China-US Workshop on Advances of Comp. Eng. Mech., Dalian, China, 1983.

LINEAR COMPLEMENTARITY PROBLEMS: A CUTTING PLANE METHOD BASED ON THE CONVEX HULL OF POLYHEDRA

V. De Angelis
Dept of Statistics, Probability and Applied Statistics
University "La Sapienza", Rome, Italy

ABSTRACT. Some structural engineering issues leading to linear complementarity problems (l.c.p.) are presented. Then the set of solutions of a linear complementarity system (l.c.s.) is represented as the union of a finite - but, unfortunately, large - number of polyhedra and the l.c.p. is reformulated as the problem of determining a supporting hyperplane of the closure of the convex hull of such union. The supporting hyperplane of any single polyhedron - which is obtained by means of parametric linear programming - generates a cut in the space of parameters, which can be so powerful as to eliminate half of the polyhedra that are to be considered in the resolution of the l.c.p..Besides, the parametric linear programming problem which has to be solved after the introduction of the cut has a smaller dimension.

1.INTRODUCTION

Linear and non-linear complementarity systems have been known up to the years '70 in connection with the study of stationary points. Since then, problems of structural engineering have been formulated, especially by initiative of the group lead by Prof. Maier in Milan, which included complementarity constraints obtained not in the context of stationarity but as a mathematical model of some real situations, such as the elasto-plastic behaviour of a structure.

The opening of this new field of applications has provoked a great increase of interest towards the complementarity systems.

2.SOME PROBLEMS LEADING TO A COMPLEMENTARITY SYSTEM

The mathematical description of unilateral contact generally involves a complementarity relation: corresponding variables (support reaction, distance between structure and support) are sign-constrained and their

G. I. N. Rozvany and B. L. Karihaloo (eds.), Structural Optimization, 63–67.

product is zero. When such relations are among the constraints of a structural optimization, this becomes a rather unusual problem in non-linear, non-convex programming.

(a) A rough seabottom supporting a pipeline can be modified along its route by means of excavations (trenching) and/or artificial supports (trestles) in order to keep the bending stresses in the pipe below admissible thresholds [4], [5]. Assuming that the total cost is a quadratic function of the design variables (trenching depths and/or trestle heights), the problem is to find the cheapest design of the supporting profile modifications. A practically important generalization of this problem concerns the presence of "fixed charges" for equipment mobilization in the cost function.

(b) An elastic - perfectly plastic discretized structure subjected to given proportional loads undergoes displacements, some of which are measured. The identification of local resistances (yield limits, buckling loads) on the basis of measured displacements can be formulated as the minimization, with respect to plastic multipliers and unknown parameters, of a convex quadratic error function subject to linear constraints and a (non-linear, non-convex) complementarity condition [6].

(c) More generally, the elasto - plastic behaviour of a complex engineering structure can be formulated by means of a complementarity relationship [1].

3. A CUTTING PLANE METHOD BASED ON THE CONVEX HULL OF POLYHEDRA

Consider a linear complementarity problem of the form:

$$\begin{cases} \text{minimize: } \langle c,v\rangle + \langle d,z\rangle \\ \text{subject to the l.c.s.:} \\ Lv + Mz - q \geqslant 0;\ v \geqslant 0;\ z \geqslant 0;\ \langle z\ ,\ Lv + Mz - q\rangle = 0 \end{cases} \tag{1}$$

where L, M, q are of order s x t, s x s, s x 1, respectively, and let R be the set of solutions of (1).

Let ϱ be a positive large enough real; set $e^T = (1,\dots,1)$ and denote by $\alpha = (\alpha_1,\dots,\alpha_s)$ a vector of binary parameters, i.e. $\alpha_i \in \mathbb{B} = \{0\ ,\ 1\}$ [3]. Introduce the polyhedron $P(\alpha)$ defined by:

$$\{Lv + Mz \geqslant q;\ -Lv - Mz \geqslant -q - \varrho(e - \alpha);\ \ -z \geqslant -\varrho\alpha;\ v \geqslant 0;\ z \geqslant 0\}$$

It is easy to note that $R = \bigcup_{\alpha\in\mathbb{B}^s} P(\alpha)$.

By setting:

$$A = \begin{bmatrix} L & M \\ -L & -M \\ 0 & -I \\ I & 0 \\ 0 & I \end{bmatrix}; \; b(\alpha) = \begin{bmatrix} q \\ -q-\varrho(e-\alpha) \\ -\varrho\alpha \\ 0 \\ 0 \end{bmatrix}, \alpha \in B^s; \; x = \begin{bmatrix} v \\ z \end{bmatrix}; \; w = \begin{bmatrix} c \\ d \end{bmatrix}$$

where A, $b(\alpha)$ are of order m x n, m x 1, respectively, with $m = 4s + t$, $n = s + t$, we have:

$$P(\alpha) = \{ x : Ax \geqslant b(\alpha) \}$$

and (1) can be stated as:

$$\begin{aligned} &\text{minimize: } \langle w,x \rangle \\ &\text{subject to: } x \in \bigcup_{\alpha \in \mathbb{B}^s} P(\alpha) \end{aligned} \tag{2}$$

Let $H^+(w, \omega)$ be the halfspace defined by $\langle w,x \rangle \geqslant \omega$, with $w \in R^n \setminus \{0\}$ and $\omega \in R$; and $H^\circ(w, \omega)$ be the hyperplane defined by $\langle w,x \rangle = \omega$. Then solving (2) is equivalent to finding the value $\omega_\circ$ such that $H^+(w, \omega_\circ) \supseteq R$ and $H^\circ(w, \omega_\circ)$ is a supporting hyperplane of clconv R [2].

By setting:

$$\omega(\alpha) = \min_{x \in P(\alpha)} \langle w,x \rangle \tag{3}$$

we have:

$$\omega_\circ = \min_{\alpha \in \mathbb{B}^s} \omega(\alpha)$$

Alternatively, making use of the duality in linear programming, $\omega(\alpha)$ can be defined as:

$$\omega(\alpha) = \max_{\substack{A^T\lambda = w \\ \lambda \geqslant 0}} \langle b(\alpha), \lambda \rangle \tag{4}$$

so $\omega_\circ$ becomes:

$$\omega_\circ = \min_{\alpha \in \mathbb{B}^s} \left\{ \max_{\substack{A^T\lambda = w \\ \lambda \geqslant 0}} \langle b(\alpha), \lambda \rangle \right\} \tag{5}$$

The difficulty in solving (5) stands in the cardinality of $\mathbb{B}^s$, which is 2^s. Fortunately, we are able to introduce some cutting hyperplanes, on the basis of the following theorem.

Theorem. Let α^i, b^i, ω_i, for $i = 1,\dots,2^s$, be, respectively, the elements of $\mathbb{B}^s$, the vectors $b(\alpha^i)$ and the scalars $\omega(\alpha^i)$. Let λ^i be an optimal solution of (4) for $\alpha = \alpha^i$ and T_i be the subset of $\mathbb{B}^s$ such that $\langle b^j, \lambda^i \rangle \geqslant \omega_i$ for $\alpha^j \in T_i$. Then we have: $\omega_\circ = \min\{\omega_i; \min_{\alpha^j \in \mathbb{B}^s \setminus T_i} \omega_j\}$.

Proof: Obvious. Since λ^i is a feasible solution of (4) for any $\alpha^j \in \mathbb{B}^s$, $\omega_j \geqslant \omega_i$ follows for any $\alpha^j \in T_i$.

As a consequence of such theorem, we can introduce the cutting hyperplane $\langle b(\alpha), \lambda^i \rangle = \omega_i$ in order to eliminate some values of $\alpha \in \mathbb{B}^s$.

The left-hand side of the equation of the cutting hyperplane is a linear function of α, which is represented from here on by $\langle c, \alpha \rangle + k$.

The power of the cut $\langle c, \alpha \rangle \geqslant \omega_i$ depends on the coefficients c_r and on the values α^i_r, for $r = 1,\dots,s$.

Define the following subsets of $S = \{1,\dots,s\}$: $C_- = \{r \in S: c_r < 0\}$, $C_\circ = \{r \in S: c_r = 0\}$, $C_+ = \{r \in S: c_r > 0\}$; $A_0 = \{r \in S: \alpha^i_r = 0\}$, $A_1 = \{r \in S: \alpha^i_r = 1\}$. Set $I = (A_0 \cap C_-) \cup (A_1 \cap C_+)$ and let $c(I)$ denote the cardinality of I.

The power of the cut depends on $c(I)$.

If $c(I) = 0$, i.e. if $I = \emptyset$, we have $T_i = \mathbb{B}^s$; if $c(I) = 1$, T_i contains half of the elements of $\mathbb{B}^s$, i.e. $T_i = \{\alpha: \alpha_r = \alpha^i_r, r \in I\}$; if $c(I) = s$, then $T_i = \{\alpha^i\}$. In general, $T_i = \{\alpha: \alpha_r = \alpha^i_r, r \in I\}$ has got a cardinality equal to $2^{s-c(I)}$.

A rudimentary algorithm for the solution of (5) is easily derived:

Step 0. Set $T_\circ = \emptyset$; $i = 1$; $\omega_\circ = +\infty$.

Step 1. Solve (4) by means of parametric linear programming for any value $\alpha^i \in \mathbb{B}^s \setminus T_\circ$ and determine T_i (if $\omega_i = +\infty$, set $T_i = \{\alpha^i\}$). Calculate $T_\circ = T_\circ \cup T_i$ and set $\omega_\circ = \min\{\omega_\circ, \omega_i\}$.

Step 2. If $T_\circ = \mathbb{B}^s$: stop, we have found the optimal solution. Otherwise set $i = i + 1$ and repeat step 1.

The construction of an algorithm taking the greatest possible advantage of the properties of the cutting plane introduced here and numerical experiments are in progress.

4. FINAL REMARKS

It is worth noticing that the above algorithm leads naturally to an interactive solution process, especially when applied to problems of structural engineering. In fact, every $\alpha \in \mathbb{B}^s$ represents a possible design, so, finding out that all solutions with, for example, $\alpha_1 = 0$, can be eliminated and only solutions with $\alpha_1 = 1$ must be considered, means to be able to establish at an early stage that a structure must have a determinate feature.

Starting from this knowledge, consistent specifications on other features of the design, i.e. reasonable suggestions on the value of other components of α, may be supplied by the planner, which are likely to shorten the computational path.

Obviously, as the solution process progresses, more and more features of the structure considered are likely to have been determined and more easily the planner may intervene with his experience.

REFERENCES

[1] Cohn M. Z., Maier G (eds) Engineering Plasticity by Mathematical Programming. Proceedings of the NATO Advanced Study Institute. Pergamon Press, N. Y., 1979

[2] De Angelis V. Characterization of the convex hull of the union of a finite set of polyhedra by means of linear programming. Dept of Statistics, Probability and Applied Statistics, Univ. La Sapienza, Rome, 1987

[3] Giannessi F.'A characterization of the convex hull of the union of polyhedra. Applications to linear complementarity problems'. Nordic Symposium on Linear Complementarity Problems and Related Areas. Dept of Mathematics, Institute of Technology, Linköping Univ., 1982

[4] Giannessi F., Jurina L., Maier G. 'Optimal excavation profile for a pipeline freely resting on the sea floor'. Engineering Structures 1 (1979) 81-91

[5] Maier G., Andreuzzi F., Giannessi F., Jurina L., Taddei F. 'Unilateral contact, elastoplasticity and complementarity with reference to offshore pipeline design'. Computer Methods in Applied Mechanics and Engineering 17/18 (1979) 469-495

[6] Maier, G., Giannessi, F., Nappi, A., 'Indirect identification of yield limits by mathematical programming'. Engineering Structures 4 (1982) 86-97

OPTIMIZATION OF VIBRATING THIN-WALLED STRUCTURES

R. de Boer
Universität Essen
Institut für Mechanik
FB Bauwesen
4300 Essen 1, FRG

ABSTRACT. Thin-walled structures play an important role in structural engineering. These structures - due to the thinness of the walls - are very sensitive to external excitation which can be caused either by harmonically fluctuating loading or by impact. The aim of the paper is to determine optimality criteria for vibrating thin-walled structures. In deriving these conditions, one must be very careful since thin-walled structures exhibit a different mechanical behaviour to solid sections. These special effects are the influence of shear deformations and of the deformation of the cross-section. Moreover, for special cross-sections there is a coupling of bending and torsion. In order to consider all these effects, a system of differential equations is introduced. Then, optimization problems are discussed and optimality conditions are derived, including some known and several new results.

1. Introduction

The field of optimization of vibrating thin-walled structures was pioneered by Ashley and McIntosh [1] in 1968. They treated the free axial and torsional vibration of the thin-walled bars with closed cross-section. The author dealt with the optimization of vibrating thin-walled structures with open and closed cross-sections in 1972, with particular emphasis on flexural and torsional vibration including warping effects [2]. In several papers [3 - 5] from 1978 onwards, Szymczak presented the optimal design of thin-walled bars with bisymmetric open cross-sections that undergo torsional vibration; he also considered warping stresses. Also in 1978 Vavrick and Warner [6] studied the duality between optimal design problems for torsional vibration of thin-walled cylinders. Further research in this field was done by Bratus and Kartvelishvili [7], Banichuk et al. [8] between 1981 and 1985. The papers of Hanagad and Smith [9, 10] from 1985 and 1986 about the optimal design of a vibrating bar with coupled bending and torsion contain interesting aspects of optimization for natural frequency.
A general theory of optimization of vibrating thin-walled structures has not yet been developed. This theory should include all the effects men-

G. I. N. Rozvany and B. L. Karihaloo (eds.), Structural Optimization, 69–76.

tioned in the ABSTRACT. The aim of this paper is to suggest an approach to this complicated field. We will start with a system of differential equations for the vibration problem which is based on an extended form of Wlassow's method [11] for the treatment of thin-walled structures. Then the optimization problem will be discussed and explicit optimality conditions will be given for fundamental frequency. The system of equations of motion and the optimality condition involve several special problems. The theory will be illustrated with examples. In this paper, we consider only the geometrically linear theory and elastic behaviour of structures.

2. Equations of motion and solution

The equations of motion of thin-walled structures and the research into special problems associated with these structures have been developed in several papers [12 - 15] by the author via an extended version of Wlassow's method [11]. The latter is a well-known technique for the

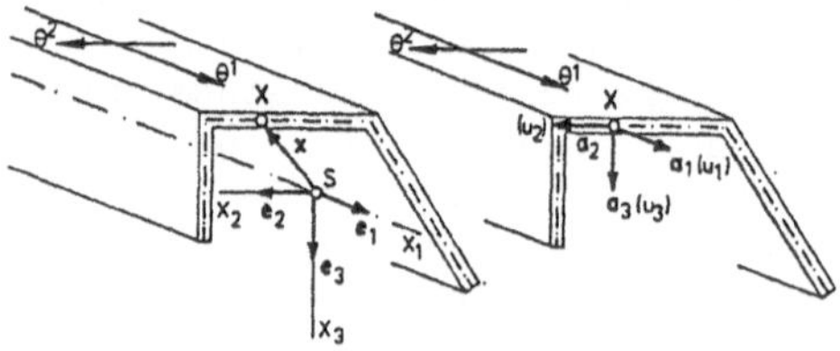

Fig. 1: Thin-walled rod

treatment of thin-walled structures. The main idea of his approach is to develop the displacements of the cross-section in series in such a way that the displacement components u_1, u_2 and u_3 in the $\underset{\sim}{a}_1$-, $\underset{\sim}{a}_2$- and $\underset{\sim}{a}_3$-directions (see Fig. 1) which depend on the parameters θ^1 and θ^2 (see Fig. 1) are decomposed into θ^1- and θ^2-dependent functions:

$$u_i(\theta^1, \theta^2) = \overset{i}{\underset{\sim}{v}}(\theta^1) \cdot \overset{i}{\underset{\sim}{\varphi}}(\theta^2) \quad , \tag{2.1}$$

where i stands for 1, 2 or 3. The vectors $\overset{i}{\underset{\sim}{v}}$ describe, in general, the unknown rotations and displacements of the center line of the bar or of special points of the cross-section, whereas the vectors $\overset{i}{\underset{\sim}{\varphi}}$ represent the course of displacements - shape functions - over the cross-section. Then by means of strain displacement relations, Hooke's law, equations of motion, the principle of virtual work, one arrives at a system of differential equations [13, 16]. The complete system of differential equations to describe the dynamical behaviour of thin-walled bars is given by

$$\underset{\sim}{A}\,\underset{\sim}{z}'' + \underset{\sim}{B}\,\underset{\sim}{z}' + \underset{\sim}{C}\,\underset{\sim}{z} + \underset{\sim}{D}\,\ddot{\underset{\sim}{z}} = \underset{\sim}{q} \quad , \tag{2.2}$$

where primes denote derivatives with respect to θ^1, a parameter which is parallel to the centerline of the beam (see Fig. 1), and dots denote the time derivation. The vector $\underset{\sim}{z}$ contains the displacements and rotations of some characteristic points of the cross-section. The matrices $\underset{\sim}{A}$, $\underset{\sim}{C}$ and $\underset{\sim}{B}$ contain cross-section quantities and material dependent numbers; in addition the matrix $\underset{\sim}{D}$ contains the density ρ and $\underset{\sim}{q}$ denotes the load of the bar. For the solution of the system of differential equations we use the mode-superposition analysis. The essential operation of this analysis is the transformation from the general geometric deformations $\underset{\sim}{z}$ to the modal-amplitudes $\bar{\underset{\sim}{z}}_i$:

$$\underset{\sim}{z}(\theta^1, t) = \sum_{i=1}^{\infty} \phi_i(t)\, \bar{\underset{\sim}{z}}_i(\theta^1) \quad . \tag{2.3}$$

The displacements and rotations of the bar $\underset{\sim}{z}(\theta^1, t)$ are developed as the sum of infinite unknown time functions $\phi_i(t)$ and a set of modal components. The application of the mode-superposition analysis is only possible if the problem is self-adjoint and the matrices involved positive--definite. This means, that for the two modal components $\bar{\underset{\sim}{z}}_1$ and $\bar{\underset{\sim}{z}}_2$ and e.g. for the matrix $\underset{\sim}{A}$ the following relations must be hold:

$$\int_0^1 (\bar{\underset{\sim}{z}}_1 \cdot \underset{\sim}{A}\,\bar{\underset{\sim}{z}}_2'' - \bar{\underset{\sim}{z}}_2 \cdot \underset{\sim}{A}\,\bar{\underset{\sim}{z}}_1'')\, d\theta^1 = 0 \;, \quad \underset{\sim}{v} \cdot \underset{\sim}{A}\,\underset{\sim}{v} > 0 \quad , \tag{2.4}$$

where the integration has to be taken over the length l of the bar and $\underset{\sim}{v}$ is an arbitrary non-zero vector.
These suppositions are fulfilled for all bending and torsion problems which we have treated in the past.
In order to determine the unknown time functions ϕ_i, the set-up for $\underset{\sim}{z}$ is introduced in the system of differential equations (2.2). Then, this system is multiplied by the modal component $\bar{\underset{\sim}{z}}_k$ and integrated with the result:

$$\ddot{\phi}_k \int_0^1 D\,\bar{\underset{\sim}{z}}_k \cdot \bar{\underset{\sim}{z}}_k\, d\theta^1 + \phi_k \Big[\int_0^1 (\underset{\sim}{A}\,\bar{\underset{\sim}{z}}_k'' \cdot \bar{\underset{\sim}{z}}_k + \underset{\sim}{B}\,\bar{\underset{\sim}{z}}_k' \cdot \bar{\underset{\sim}{z}}_k + \underset{\sim}{C}\,\bar{\underset{\sim}{z}}_k \cdot \bar{\underset{\sim}{z}}_k)\, d\theta^1 \Big]$$

$$= \int_0^1 \underset{\sim}{q} \cdot \bar{\underset{\sim}{z}}_k\, d\theta^1 \quad , \tag{2.5}$$

where only one term remains of the infinite series by virtue of the orthogonality condition. Equation (2.5) can be written in simpler form

$$\ddot{\phi}_k + \omega_k^2 \phi_k = \frac{q_k}{m_k} \tag{2.6}$$

with the circular frequency

$$\omega_k^2 = \frac{\int_0^1 \bar{\underset{\sim}{z}}_k \cdot (\underset{\sim}{A}\,\bar{\underset{\sim}{z}}_k'' + \underset{\sim}{B}\,\bar{\underset{\sim}{z}}_k' + \underset{\sim}{C}\,\bar{\underset{\sim}{z}}_k)\, d\theta^1}{m_k} \tag{2.7}$$

the generalized mass

$$m_k = \int_0^1 \underset{\sim}{D}\, \bar{\underset{\sim}{z}}_k \cdot \bar{\underset{\sim}{z}}_k \, d\theta^1 \tag{2.8}$$

and the generalized load associated with the modal component $\bar{\underset{\sim}{z}}_k$

$$q_k = \int_0^1 \underset{\sim}{q} \cdot \bar{\underset{\sim}{z}}_k \, d\theta^1 \quad . \tag{2.9}$$

The Duhamel integral expression - solution of (2.6) - gives

$$\phi_k(t) = \frac{1}{\omega_k m_k} \int_0^t q_k \sin \omega_k (t - \tau) d\tau \ , \tag{2.10}$$

where it is assumed that the bar does not have any motion when time t equals to zero. Now, we return to the mode-superposition analysis. With the known time functions ϕ_k (2.10) we obtain instead of (2.3):

$$\underset{\sim}{z}(\theta^1, t) = \sum_{i=1}^{\infty} \frac{1}{\omega_i m_i} \bar{\underset{\sim}{z}}_i \int_0^t q_i \sin \omega_i (t - \tau) d\tau \quad . \tag{2.11}$$

If we are interested in displacements or rotations in a special point θ_0^1 of the bar and at a special time t_0, then we arrive from (2.11) at

$$\bar{\underset{\sim}{z}}(\theta_0^1, t_0) = \sum_{i=1}^{\infty} \frac{1}{\omega_i m_i} \bar{\underset{\sim}{z}}_i(\theta_0^1) \int_0^{t_0} q_i \sin \omega_i (t_0 - \tau) d\tau \quad . \tag{2.12}$$

3. Optimization

3.1 General treatment

From (2.12), we can derive e.g. the displacement at the end of a cantilever subject to an impact

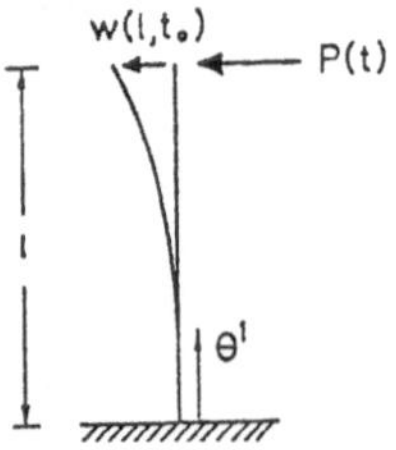

$$w(1, t_0) = \sum_{i=1}^{\infty} \frac{1}{\omega_i m_i} \bar{w}_i(1) \int_0^{t_0} P(t)\, \bar{w}_i(1) \sin \omega_i (t_0 - \tau) d\tau \ . \tag{3.1}$$

Fig. 2: Impact on a beam

If we wish to minimize the deflection at the end of the cantilever, we have to start with (3.1). However, since (2.12) contains the whole spectrum of displacements and rotations due to bending and rotation

with allowance for shear- and cross-section deformations, the relation (2.12) opens up a wide field of optimization problems including the optimal arrangement of transverse stiffeners to reduce cross-sectional deformations.
If the bar is loaded by harmonically fluctuating loading

$$\underset{\sim}{q}(\theta^1, t) = \hat{\underset{\sim}{q}}(\theta^1) \sin \Omega t \tag{3.2}$$

where $\hat{\underset{\sim}{q}}$ is the amplitude and Ω the circular frequency, then the solution of (2.2) becomes much simpler. The response to this loading can be assumed to be harmonic and in phase with the loading:

$$\underset{\sim}{z}(\theta^1, t) = \hat{\underset{\sim}{z}}(\theta^1) \sin \Omega t \quad . \tag{3.3}$$

Then, $\hat{\underset{\sim}{z}}$ can be calculated from the differential equation

$$\underset{\sim}{A}\hat{\underset{\sim}{z}}'' + \underset{\sim}{B}\hat{\underset{\sim}{z}}' + \underset{\sim}{C}\hat{\underset{\sim}{z}} - \Omega^2\hat{\underset{\sim}{z}} = \hat{\underset{\sim}{q}} \quad . \tag{3.4}$$

The response $\underset{\sim}{z}$ on the harmonically fluctuating loading is determined by (3.3) and (3.4).
For practical reasons, it is often necessary to maximize the fundamental frequency. However, as mentioned earlier, only some special problems have been treated in the literature. Using the above theory as a basis, we can develop a unified approach including shear- and cross--section deformations.
In order to maximize the fundamental frequency for thin-walled structures we proceed from (2.2) with the external load $\underset{\sim}{q}$ equal to zero.
It is assumed that the free vibration motions consist of a time independent mode $\overset{*}{\underset{\sim}{z}}(\theta^1)$ the amplitude of which is varying with the time according to $\phi(t)$:

$$\underset{\sim}{z} = \overset{*}{\underset{\sim}{z}}(\theta^1)\,\phi(t) \quad . \tag{3.5}$$

Then, if we follow the procedure described before, we obtain an expression for the fundamental circular frequency:

$$\omega^2 = \frac{\int_0^1 [(\underset{\sim}{A}\overset{*}{\underset{\sim}{z}}'' + \underset{\sim}{B}\overset{*}{\underset{\sim}{z}}' + \underset{\sim}{C}\overset{*}{\underset{\sim}{z}}) \cdot \overset{*}{\underset{\sim}{z}}]\, d\theta^1 -}{\int_0^1 (\underset{\sim}{D}\overset{*}{\underset{\sim}{z}} \cdot \overset{*}{\underset{\sim}{z}})\, d\theta^1} \tag{3.6}$$

The mass M of the rod is given by

$$M = \int_0^1 \rho\, A(\theta^1)\, d\theta^1 \quad , \tag{3.7}$$

where ρ denotes the density and $A(\theta^1)$ the area of the cross-section. If the mass is constant, the variation of the mass vanishes and we arrive at the relation

$$\delta M = \rho \int_0^1 \delta A \, d\theta^1 = 0 \quad , \tag{3.8}$$

provided that the density ρ is constant.
We assume that all matrices $\underset{\sim}{A}$ to $\underset{\sim}{D}$ in (3.5) are functions of the area A. Then, by well-known procedures (see [17]) we obtain the optimality condition

$$(\underset{\sim}{A}\,\overset{*}{\underset{\sim}{z}}{}'' - \underset{\sim}{B}\,\overset{*}{\underset{\sim}{z}}{}' - \underset{\sim}{C}\,\overset{*}{\underset{\sim}{z}}) \cdot \overset{*}{\underset{\sim}{z}} - \omega^2 (\underset{\sim}{D}\,\overset{*}{\underset{\sim}{z}} \cdot \overset{*}{\underset{\sim}{z}}) = \underset{\sim}{Q}^2 \qquad (3.9)$$

where Q^2 is a constant. The optimality condition (3.9) serves as an equation to determine $\overset{*}{\underset{\sim}{z}}$. Then, the equation of motion (2.2) with $\underset{\sim}{q} = \underset{\sim}{o}$ yields the unknown area, taking into account the boundary conditions.

3.2 Torsional vibration of a bar with bisymmetric cross-section

Next, a special problem, namely the torsional vibration of a thin-walled cantilever with I cross-section and with a single mass at the end, will be discussed. The thickness of the wall $h(\theta^1)$ is considered as a design parameter. We choose the following shape functions

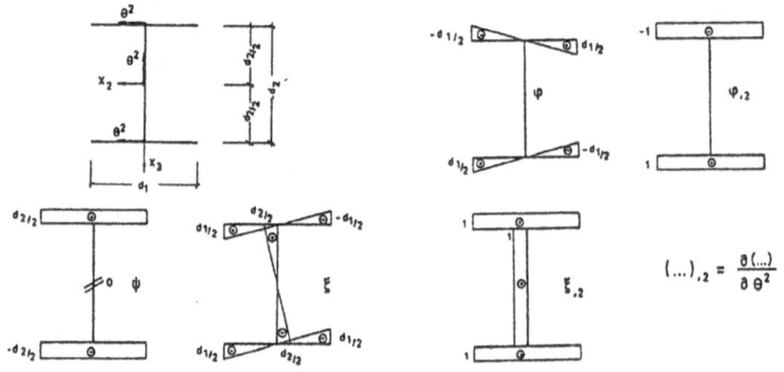

Fig. 3: Shape functions for the torsional problem

With these shape functions we derive from (2.2) the system of equations of motion, where u and χ denote warpening and rotation respectively.

$$E\,h\,u''\int \varphi\,\varphi\, d\theta^2 + E\,h'\,u'\int \varphi\,\varphi\, d\theta^2 - G\,h\,\chi'\int \psi\,\varphi_{,2}\, d\theta^2 -$$

$$- G\,h\,u\int \varphi_{,2}\,\varphi_{,2}\, d\theta^2 - \rho\,h\,\ddot{u}\int \varphi\,\varphi\, d\theta^2 = 0 \;,$$

$$G\,h\,\chi''\Big(\int \psi\,\psi\, d\theta^2 + \frac{h^2}{3}\int \xi_{,2}\,\xi_{,2}\, d\theta^2\Big) + G\,h'\chi'\Big(\int \psi\,\psi\, d\theta^2 + h^2\int \xi_{,2}\,\xi_{,2}\, d\theta^2\Big) + \qquad (3.10)$$

$$+ G\,h\,u'\int \varphi_{,2}\,\psi\, d\theta^2 + G\,h'u\int \varphi_{,2}\,\psi\, d\theta^2 - \rho\,h\,\ddot{\chi}\Big(\int \psi\,\psi\, d\theta^2 + \int \xi\,\xi\, d\theta^2\Big) = 0 \;.$$

By some rearrangements we arrive at the following differential equation for the torsional vibration where warpening effects are included:

$$B^2[h\,\chi'']'' - [h^3\,\chi']' - 3\gamma^2 h\,\chi = 0 \qquad (3.11)$$

with

$$B^2 = \frac{E\,C_{TO}}{G\,I_{TO}} \quad , \quad \gamma^2 = \frac{\omega^2\,\rho\,I_{OO}}{3\,G\,I_{TO}} \quad ,$$

$$C_{TO} = \frac{d_1^{\,3}}{6}\left(\frac{d_2}{2}\right)^2 \quad , \quad I_{TO} = \frac{1}{3}(2d_1 + d_2) \quad , \quad I_{OO} = \frac{d_1^{\,3}}{6} + \frac{d_2^{\,3}}{12} + 2d_1\left(\frac{d}{2}\right) \tag{3.12}$$

The optimality condition is given by

$$B^2\,\chi''^2 + h^2\,\chi'^2 - \gamma^2\,\chi^2 = Q^2 \;. \tag{3.13}$$

The optimality criteria (3.13) are coupled in the rotation and the thickness h. With the equation of motion (3.11) the thickness h can be eliminated. This procedure leads to a non-linear ordinary differential equation for the mode χ, which can be solved approximately with the help of the perturbation method. If namely B^2 is assumed as a small parameter, the equation of motion and the optimality criteria are reduced to

$$[h^3\,\chi']' - 3\,\gamma^2\,h\,\chi = 0 \quad , \quad h^2\,\chi'^2 - \gamma^2\,\chi = 0 \;. \tag{3.14}$$

In this case an exact solution is possible (see [2]). We treat the thin-walled structure represented in Fig. 4 with an open cross-section and assume

$$l = 5\ \text{m},\ R = 1\ \text{m},\ b = 20\ \text{cm},\ H = 50\ \text{cm},\quad k = \frac{M_1}{M}$$

$$G = 8{,}1 \cdot 10^6\ \text{kN/m}^2;\ \rho = 800{,}2\ \text{kp/m}^3.$$

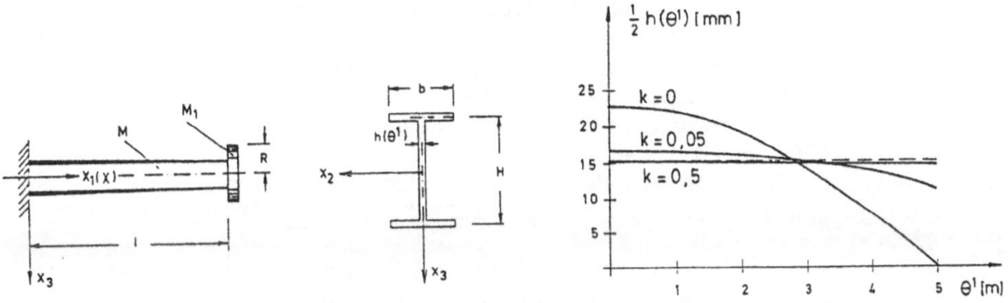

Fig. 4: System and the thickness of the wall

We can see from Fig. 4 the well-known fact that, for a large single mass at the end of the beam, the optimization is not really effective, because the behaviour of the vibrating bar is less influenced by the mass of the bar. However, for a vibrating bar without a single mass at the end of the beam (k = 0) the fundamental frequency can be increased, so e.g. in comparison with a vibrating beam with the thickness h = 3 cm by about 65 %.

References:

1. Ashley, H. and McIntosh, jr. S.C.:"Application of Aeroplastic constraints in Structural Optimization". *Proc. 12 th Int. Cong. Appl. Mech. Stanford, 1968.* Eds.: M. Hetényi and W.G. Vincenti. Springer--Verlag, Berlin 1969.
2. de Boer, R.:"Optimierung von Stabschwingern mit dünnwandigem Querschnitt". *Der Stahlbau* 41 (1972), pp. 245-249.
3. Szymczak, C.:"Torsional Vibrations of Thin-Walled Bars with Bisymmetric Open Cross-Sections". *Rozprawy Inzynierskie* 26, 2, (1978), pp. 267-274.
4. Szymczak, C.:"Optimal Design of Thin-Walled I Beams for Extreme Natural Frequency of Torsional Vibrations". *J. of Sound and Vibration* 86(2) (1983), pp. 235-241.
5. Szymczak, C.:"Optimal Design of Thin-Walled I Beams for a given Natural Frequency of Torsional Vibrations". *J. of Sound and Vibration* 97(1) (1984), pp. 137-144.
6. Vavrick, D.J. and Warner, W.H.:"Duality among Optimal Design Problems for Torsional Vibration". *J. Struct. Mech.,* 6(2) (1978), pp. 233-246
7. Bratus, A.S. and Kartvelishvili, V.M.:"Approximate Analytic Solutions in Problems of Optimizing the Stability and Vibrational Frequencies of Thin-Walled Elastic Structures". *Izv. AN SSSR. Mekhanika Tverdogo Tela,* 16, No. 6 (1981) pp. 119-139
8. Banichuk, N.V., Ivanova, S.Y. and Sharanyuk, A.V.:"Sensitive Analysis and Optimal Design of Structures to be Subjected to Dynamic Stimuli". *Izv. AN SSSR. Mekhanika Tverdogo Tela,* 20, No. 4 (1985), pp. 166-172.
9. Hanagud, S. and Smith, C.V.:"Optimal Design of a Vibrating Beam with Coupled Bending and Torsion". Publ. by AIAA (CP851) 1985
10. Chattopadhyay, A., Hanagud, S.V., Smith, C.V.:"Minimum Weight Design of a Structure with Dynamic Constraints and a Coupling of Bending and Torsion". Publ. by AIAA (CP851) 1986
11. Wlassow, W.S.: *Dünnwandige elastische Stäbe.* Berlin: VEB Verlag für Bauwesen 1964. Band 1 und 2.
12. de Boer, R.:"Der Einfluß der Querschnittsverformungen auf die Eigenschwingungen gerader dünnwandiger Stäbe". *ZAMM* 50 (1970), pp. T222-T225.
13. de Boer, R.:"Der gerade Stab mit geschlossenem dünnwandigem Profil unter näherungsweiser Berücksichtigung der Schub- und Querschnittsdeformationen". *Ing.-Archiv* 39 (1970), pp. 53-62.
14. de Boer, R.:"Die näherungsweise Ermittlung der mittragenden Breite bei geraden prismatischen Stäben mit geschlossenen dünnwandigen Profilen". *Der Stahlbau* 39 (1970), pp. 16-20.
15. de Boer, R. und Sass, H.H.:"Der Stoß auf gerade dünnwandige Träger". *Ing.-Archiv* 44 (1975), pp. 177-188.
16. de Boer, R. und Walther, W.: *Dünnwandige elastische Stäbe - Theorie und Anwendung.* Forschungsberichte aus dem Fachbereich Bauwesen der Universität-GH-Essen 1988/89, to appear.
17. Prager, W. and Taylor, J.E.:"Problems of Optimal Structural Design". *J. Appl. Mech.* (1968), pp. 102-106.

SHAPE OPTIMIZATION OF INTERSECTING PRESSURE VESSELS

Fernand Ellyin
Department of Mechanical Engineering
University of Alberta
Edmonton, Alberta T6G 2G8
Canada

ABSTRACT. A profile of variable thickness is sought which connects a spherical shell to a cylindrical one. The geometry of the mid-surface of the connecting shell of revolution is not known *a priori*, neither is the thickness variation. The profile and its thickness is obtained based on minimum material volume and a strength criterion. The 'optimum' shape is obtained through a direct variation procedure - Rayleigh-Ritz method.

1. INTRODUCTION

The optimum geometry with least material (weight) for pressure loading is a sphere. However, it must have openings for functional requirements. These openings are regions of high stress/strain concentration. Cracks are initiated at these locations which eventually lead to failure of the pressure vessel. The design requirement for these vessels is to reinforce the junction so as to obtain a strength equal to that of the unpierced vessel. One major problem is the amount of the reinforcing material and its distribution.

In an earlier paper [1], the optimum shape of a shell of revolution connecting a cylinder (nozzle) to a spherical pressure vessel was obtained based on a direct variational method. The optimized configuration of variable thickness was determined on the basis of the membrane theory and von Mises yield condition. However, due to considerable thickness around the opening (two to ten times that of the vessel) a purely membrane state of stress is an idealized, though a desirable, situation and may not be achieved.

It is the objective of this paper to present an optimized configuration based on the general shell theory. Neither the profile nor the thickness of the connecting shell is known *a priori*.

2. STATEMENT OF PROBLEM

Find a transitional shell of revolution which geometrically matches the spherical and cylindrical shell parts, Fig. 1. The profile and

G. I. N. Rozvany and B. L. Karihaloo (eds.), Structural Optimization, 77–84.

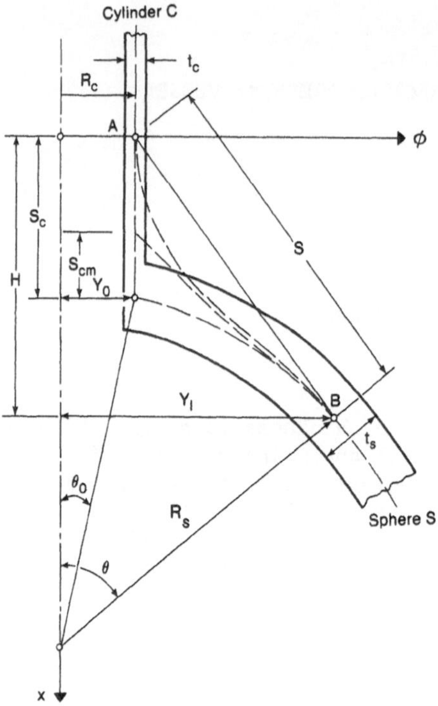

Figure 1 Geometry of an intersecting pressure vessel and nomenclature.

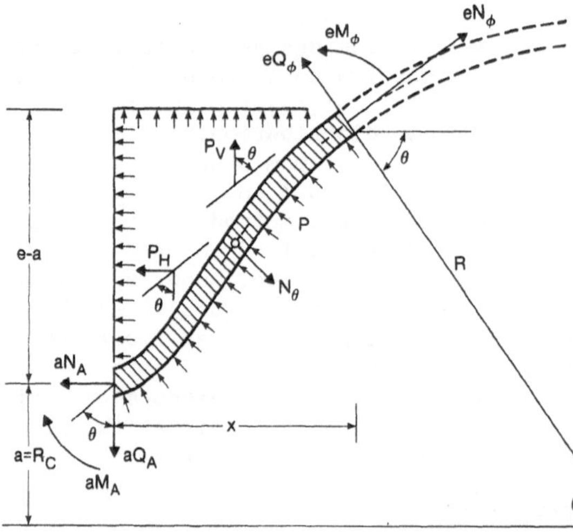

Figure 2 Free-body diagram of connection shell of revolution and prescribed boundary forces at the cylindrical section, point A.

thickness of such a connecting shell must satisfy a minimum volume and a maximum strength criterion.

2.1 Objective Function

Referring to Fig. 1, the problems stated above is one of shape and material volume optimization. In addition, practical consideration would require that the minimum volume of the initial sphere/cylinder intersecting shell be maintained.

Denoting the mid-surface position by $\phi_n = \phi_n(x)$, Fig. 1, the material volume optimization is described by minimizing

$$J = 2\pi \int_0^H t\, \phi_n\, (\phi_n'^2 + 1)^{1/2} dx \tag{1}$$

subject to constraints,

$$V = \pi \int_0^H \phi_n^2 dx = V_{min} \tag{2}$$

$$A_m = \int_0^H t\, (\phi_n' + 1)^{1/2} dx = A_{min} + \alpha A_r \tag{3}$$

where V_{min} is the inside volume of the initial cylinder/sphere intersection, A_{min} is the meridional material area of revolution of the direct cylinder-sphere connection, A_r is the material removed from the sphere by the opening, and α is a chosen positive limiting factor.

To solve the optimization problem defined above, we chose a combined objective function, i.e.

$$I = \int_0^H [2\pi t\, \phi_n(\phi_n'^2 + 1)^{1/2} + \psi_1 \pi \phi_n^2 + \psi_2 t(\phi_n'^2 + 1)^{1/2}] dx \tag{4}$$

where ψ_1, ψ_2 are Lagrange multipliers. Note that in the above equation t is neither constant nor known. Therefore, our first task is to express t as a function of the applied load and geometry.

3. FORMULATION

Let us consider the shell of revolution shown in Fig. 1, subject to the internal pressure P. For sufficiently small values of P, stresses everywhere will be below the yield stress σ_0, so that the stresses and strains are determined from an elastic solution. When the pressure reaches a certain value, there will be plastic deformation at the intersection (localized region of high stress concentration). The circumferential stress, σ_θ, at this region will reach σ_0. Most pressure vessels are constructed from carbon steels whose stress-strain curve can be reasonably approximated by an elastic-perfectly plastic curve. An exact limit analysis solution was given by Ellyin [2] for the cylinder/sphere intersecting shell. The stress distribution at the limit load is $\sigma_\theta = \sigma_0$ throughout the thickness, so that

$$N_\theta = \sigma_0 t, \quad \text{and} \quad M_\theta = 0 \tag{5}$$

Using the Tresca yield condition, it can easily be shown that yielding in the meridional direction is given by:

$$M_\phi/M_0 = 2(1 - N_\phi/N_0)(N_\phi/N_0), \tag{6}$$

where $M_0 = 1/4\ \sigma_0 t^2$ and $N_0 = \sigma_0 t$, are the plastic limits of the cross-section. From (6) we can get the thickness in terms of the stress resultants in the meridional directional, i.e.

$$t = (2N_\phi/P)(t_s/R_s) + 2M_\phi/N_\phi \tag{7}$$

where $\sigma_0 = P\ R_s/2t_s$ has been used to indicate yielding in the circumferential direction for the spherical part. A subscript 's' refers to the spherical, and 'c' to the of cylindrical shell. For a given spherical shell geometry, (7) requires specification of N_ϕ and M_ϕ.

3.1 Equilibrium Conditions

For a shell of revolution, equations of equilibrium obtained by considering a small element [3] are not convenient for our purpose, since the shell thickness is not constant. An alternative approach is to follow Lind's use of a free-body diagram [4].

Figure 2 shows a shell profile starting from the intersection of the cylindrical part (point A) and ending on the spherical shell. Forces acting on any section can be determined in terms of those acting on A, the circumferential force and pressure. Projecting forces on a plane tangent to the meridional curve at point B, we get

$$N_\phi = \frac{1}{R}\,[aN_A + P(e^2 - a^2)/2] + [aQ_A + N_\theta - P(a + e)x/2](\cos\phi/e)$$

$$M_\phi = [M_A - Q_A x + N_A(e-a)]\,\frac{a}{e} + \frac{P}{4}\,(1 + \frac{a}{e})[x^2 + (e-a)^2] - N_\theta x/2e \tag{8}$$

At the spherical boundary, the boundary conditions are:

$$N_\phi = PR_s/2 \quad \text{and} \quad M_\phi = 0 \quad \text{at} \quad x = H \tag{9}$$

Substitution from (9) into (8) and noting that $N_A = PR_c/2$, we can determine Q_A and M_A in terms of geometry and pressure. Finally, the resultant force and moments along the meridian are given by:

$$N = \phi_n^2/R_1 + [R_c Q_A' + (R_s/t_s)\int_0^x t\,dl - (\phi_n + R_c)x](Q_n^{-2} - R_1^{-2})^{1/2}$$

$$M = (R_c/\phi_n)(M_A' - Q_A' x) + R_c^2(1 - R_c/\phi_n) + \frac{1}{2}\,(1 - R_c/\phi_n)$$
$$[x^2 + (\phi_n - R_c)^2] - (R_c x/t_c\phi_n)A \tag{10}$$

where $N = 2N_\phi/P$, $M = 2M_\phi/P$, $Q'_A = 2Q_A/P$, $M'_A = 2M_A/P$, $A = \int_0^H t(\phi'_n+1)^{1/2}dx$, and R_1 = meridional radius of curvature. Thus t can be expressed in terms of geometric variables by substituting from (10) into (7).

4. OPTIMIZATION PROCEDURE

We use Rayleigh-Ritz method to minimize (4) by expressing ϕ_n as:

$$\phi_n = U(x) + \sum_{i=1}^{n} \lambda_i W_i(x) \tag{11}$$

where U(x) and its first two derivatives satisfy all non-homogeneous boundary conditions, and $W_i(x)$ and its first and second derivatives satisfy the homogeneous boundary conditions for any λ_i. Details of specific forms of U(x) and $W_i(x)$ are given in Ref. [1]. The extremum solution of the problem posed by (4) with the equality constraints (2) and (3) may be obtained from the solution of the non-linear system:

$$\frac{\partial I}{\partial \lambda_i} + \sum_{j=1}^{2} \psi_j \frac{\partial G_j}{\partial \lambda_i} = 0 \qquad i = 1,2, \dots n, \tag{12}$$

$$\text{and} \qquad G_j = 0 \qquad j = 1,2$$

where G_j represent the equality constraints (2) and (3). The above system can be transformed into an equivalent single function of the form (e.g. see [5])

$$I_1(\lambda, \psi) = \sum_{i=1}^{n} \left\{ \frac{\partial I}{\partial \lambda_i} + \sum_{j=1}^{2} \psi_j \frac{\partial G_j}{\partial \lambda_i} \right\}^2 + \sum_{j=1}^{2} G_j^2. \tag{13}$$

The minimization technique used requires derivative of (13) with respect to all λ_i's and ψ_j's [1]. The solution of the resulting set of nonlinear equations has to be obtained through an iterative method and numerical integration procedure.

4.1 Example

Referring to Fig. 1, there are five geometric variables: R_s, t_s, R_c, S_c and θ. The cylindrical shell thickness, t_c, is found from assuming that the nozzle and pressure vessel have the same membrane strength, i.e.

$$t_c = 2(R_c/R_s)t_s, \tag{14}$$

and once R_c and R_s are specified, we obtain the opening angle θ_o (Fig. 1) from:

$$\theta_o = \sin^{-1}(R_c/R_s) \tag{15}$$

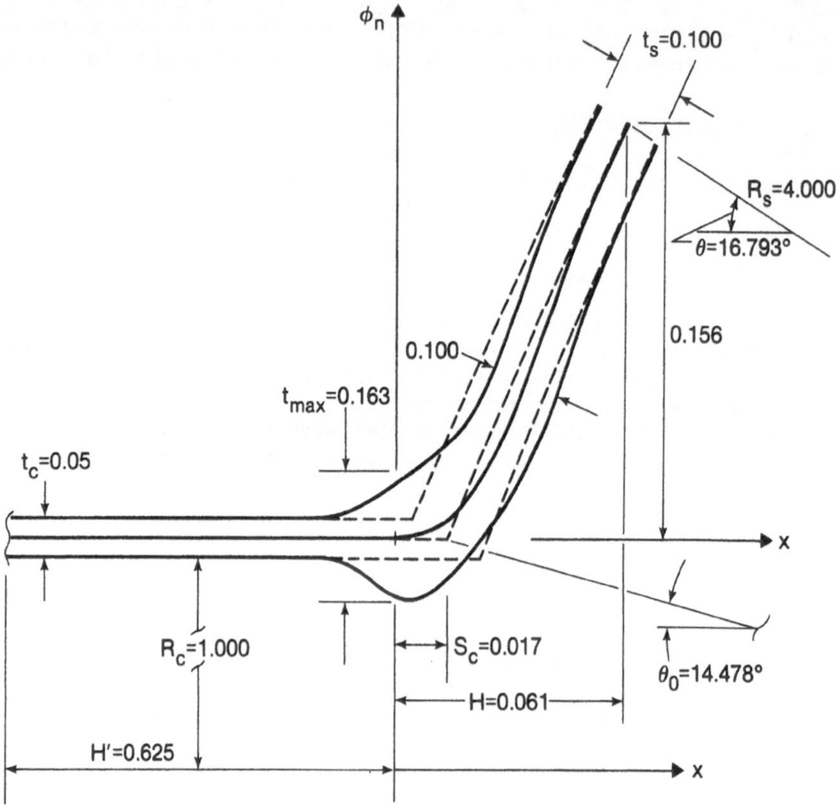

Figure 3 An example of the optimized shape for the specified parameters. $t_s/R_c = 0.1$, $R_s/R_c = 4.0$, $S_c/R_c = 0.017$

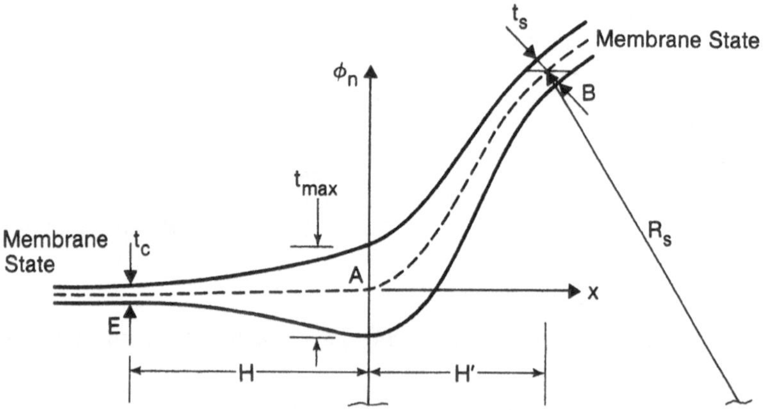

Figure 4 Illustration of the complementary solution to reach a membrane state at the cylindrical part.

The angle θ indicating the extent of the optimized profile, is an implicit function of t_s and R_s. Thus, there are three dimensionless variables t_s/R_c, R_s/R_c, and S_c/R_c to be investigated in a poramteric study. Once these parameters are specified, the shape function ϕ_n is evaluated by assuming trial values for λ_i and ψ_j. Subsequently, the meridional radius of curvature,

$$R_1 = \phi_n(\phi_n'^2 + 1)^{1/2}, \tag{16}$$

and N and M are calculated from (10). Substitution into (7) gives the thickness in terms of the ϕ_n and other geometric parameters. Through an iterative process [6] and using Newton-Cotes integration procedure, the solution of (13) finally converges towards the optimum shape. Detailed procedure and accompanying computer program is given in [7].

Figure 3 shows an optimized profile for the shell geometry, $t_s/R_c = 0.1$, $R_s/R_c = 4.0$, $S_c/R_c = 0.017$, and $\theta = 16.793°$. Note that the maximum thickness, $t_{max} = 1.63\ t_s$. A solution based on the membrane theory will produce a maximum thickness more than three times that of Fig. 3. However, a direct comparison is not possible since the optimized profiles are not the same.

It is to be noted that the procedure described here gives the profile between AB, see Fig. 1. However, the imposed boundary condition at A (Fig. 2) involves moments and forces: M_A, Q_A and N_A. If the cylindrical shell (nozzle) is long, then the thickness obtained from (14) will be much less than t_{max}. Therefore, a complementary solution is required to provide a transitional thickness profile, Fig. 4. Note that in this case the mid-surface profile is known (cylindrical shell) and we seek the optimum variation of the thickness between t_{max} at $x = 0$ and t_c at $x = -H'$. Such a solution is provided by Freiberger [8], and we will not repeat the development herein. The solution leads to the thickness variation of the form,

$$t = t_c + (t_{max} - t_c)(\cos \omega x - \cot \omega H' \sin\omega x) \tag{17}$$

where $\omega^2 = 2\sigma_0/N_x R_c$. The interval H' is found from the equilibrium of a free-body diagram similar to Fig. 2, for the cylindrical part.

5. CONCLUSIONS

The principal objective of this study was to find a profile which eliminates the discontinuity at the outlets of spherical pressure vessels, and provides the same strength as the unpierced sphere. The optimum shape of variable thickness is obtained through a direct variational formulation - Rayleigh-Ritz method. The objective function is set up in terms of the predefined boundary condition, and the method of solution results in a system of nonlinear equations.

The optimum profile obtained by the general shell theory is superior to that of the membrane theory, and results in a well-balanced reinforcement. The thickness around the junction is reduced considerably which is desirable from the thermal loading point of view. In addition, such a profile is easier to construct, for example through forging or machining.

For long nozzles (cylindrical part) where a membrane state of stress may exist, a complementary solution is required to obtain a thickness transition from t_{max} to t_c (see Fig. 4). The necessity for this additional part stems from the nature of the formulation, i.e. the prescribed force and moment boundary conditions at the cylinder and connecting shell of revolution junction. Further investigation would be necessary to confirm the optimality of such a combined procedure.

6. ACKNOWLEDGEMENT

The results presented here form part of a general investigation into safety and reliability of pressure retaining components. The research is supported, in part, by the Natural Sciences and Engineering Research Council of Canada (NSERC Grant No. A-3808). We wish to acknowledge the contribution of Dr. B. Diallo to this investigation and hereby express our appreciation to him.

7. REFERENCES

[1] Diallo, B. and Ellyin, F., 'Optimization of Connecting Shell', *J. Engrg. Mech.* ASCE, **109**, 1983, 111-126.

[2] Ellyin, F., 'The Effect of Yield Surfaces on the Limit Pressure of Intersecting Shells', *Int. J. Solids Structures*, **5**, 1969, 713-725.

[3] Flugge, W., *Stresses in Shells*, 2nd Ed., Springer-Verlag, New York, NY, 1973.

[4] Lind, N.C., 'Plastic Analysis of Radial Outlets from Spherical Pressure Vessels', *J. Engrg. for Ind.*, ASME, **86**, 1969, 713-725.

[5] Ortega, J.M. and Rheinholt, W.C., *Iterative Solution of Nonlinear Equations in Several Variables*, Academic Press, New York, NY, 1970.

[6] Fletcher, R. and Reeves, C.M., 'Function Minimization by Conjugate Gradients', *Computer J.*, **7**, 1964, 149-154.

[7] Diallo, B., 'Optimisation Par le Calculs de Variation Directe d'un Vaisseau Sous Pression Axisymetric Compose', *Ph.D. Thesis*, Universite de Sherbrooke, Sherbrooke, PQ, IV - 218, 1978.

[8] Freiberger, W., 'Minimum Weight Design of Cylindrical Shells', *J. Appl. Mech.* ASME, **23**,1956, 576-580.

OPTIMIZATION PROCEDURE S A P O P APPLIED TO OPTIMAL LAYOUTS OF COMPLEX STRUCTURES

H.A. Eschenauer and P.U. Post
University of Siegen
Research Laboratory for Applied
Structural Optimization
5900 Siegen/FR Germany

ABSTRACT. For the treatment of problems of structural optimization, the "Three-Columns-Concept" is introduced along with the software system SAPOP (Structural Analysis Program and Optimization Procedure). SAPOP consists of an input module, a graphic module and a main module with subordinate sections containing several interchangeable software packages. Thus, it is possible to link various optimization algorithms via generally formulated optimization models with several structural analysis programs. The solution concept and the introduced definitions are illustrated by the example of a composite cantilever.

1. INTRODUCTION

The verbal formulation of the design problem reads as follows [1, 2, 3]:

> The values of the design variables must be selected with regard to constraints in such a way that an objective function $f=f(\underline{x})$ attains an optimum value.

This is expressed mathematically:

$$\min_{\underline{x} \in \mathbb{R}^n} \{ f(\underline{x}) \mid \underline{h}(\underline{x}) = \underline{0};\ \underline{g}(\underline{x}) \le \underline{0} \} \tag{1}$$

with $\mathbb{R}^n$ n-dimensional set of real numbers, f objective function, $\underline{x}$ vector of n design variables, $\underline{g}$ vector of p inequality constraints, $\underline{h}$ vector of q equality constraints. The feasible domain or design space is defined as follows:

$$X: = \{ \underline{x} \in \mathbb{R}^n \mid \underline{h}(\underline{x}) = \underline{0};\ \underline{g}(\underline{x}) = \underline{0} \} \ . \tag{2}$$

Due to intensive world-wide research activities in the field of structural optimization during the last decade, various software systems have been developed usually for special applications [4, 5, 6]. With few exceptions, just sizing optimization with regard to minimum weight can be carried out. One or two optimization algorithms are linked to special structural analysis modules. Therefore, they are only of limited use for general optimization problems.

G. I. N. Rozvany and B. L. Karihaloo (eds.), Structural Optimization, 85–92.

2. DEFINITIONS AND BASIC IDEAS

We have two main demands on a software system for solving problems in the field of engineering resp. structural optimization [7, 12]:

a. Goals for optimization techniques
- modular architecture, standard interfaces between the modules,
- application of several optimization algorithms to solve an optimization problem,
- application of several algorithms for structural analysis,
- automatic design- and evaluation-models for standard problems of structural optimization,
- effortless extension to special problems.

b. Goals for software efficiency
- efficient solution techniques (linear and nonlinear equation solver, sensitivity analysis etc.),
- efficient management of a large amount of data for large-scale-systems, use of modern programming techniques,
- supporting facilities for input data generation and output documentation.

Before presenting the practical realization of an optimization procedure, we should define some of the terms used in structural optimization (see information given in Fig. 1).

Optimization algorithm	: mathematical procedure for constrained/unconstrained optimization (optimality-criteria procedures, mathematical programming methods).
Optimization procedure	: total concept for solving an optimization problem.
Optimization strategy	: approach used to reduce complex optimization problems to simplified substitute problems resp. smaller subsystems.
Preference function	: transformation of several objective functions into one scalar substitute objective function.
Structural model	: mathematical description of the structural behaviour (mathematical-physical model).
Structural parameters	: parameters of the structural model.
Analysis variables	: structural parameters which can be varied during optimization computations.
State variables	: response of the structural model.
Design variables	: design quantities to be varied.
Initial design	: initial values of the design variables.
Design model	: relationship between design and analysis variables.
Transformation variables	: transformed design varables for approximation concepts.
Evaluation model	: mathematical link between state variables and the objective function and constraint values under consideration of optimization strategies.
Optimization model	: comprehensive term for the design and evaluation model(s).

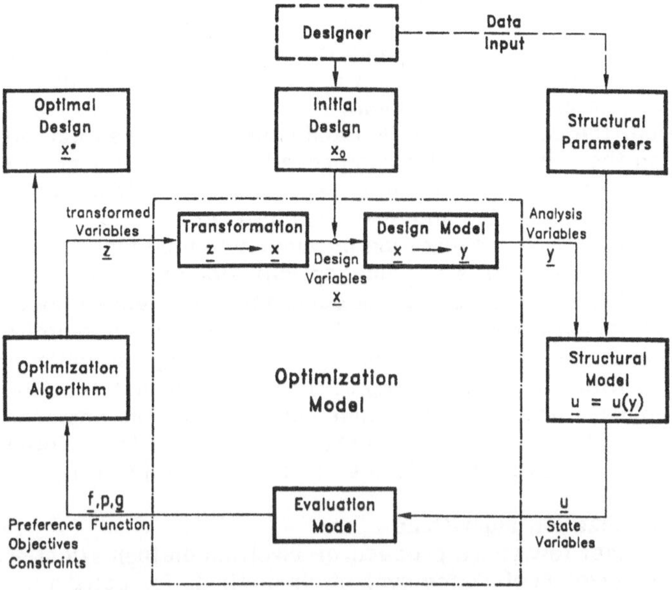

Fig. 1: The structure of an optimization loop

When dealing with an optimization problem, it is recommendable to proceed following the Three-Columns-Concept [1]. The first step is the theoretical formulation of the optimization problem, taking all of the relevant demands on the structure into consideration. An optimization algorithm is selected from the third column and combined with the structural model and the optimization model to form an optimization procedure. The columns are described in detail in the following:

Column 1: Structural model

An essential requirement of structural optimization is the formulation of the structural behaviour in mathematical terms. In the case of mechanical systems this refers to the typical structural response of static and dynamic loads, such as deformations, stresses, eigenvalues, buckling loads etc.

Here, all of the state variables required for formulating the objective function and constraints are provided. The structural calculation is carried out using efficient analysis procedures such as the finite element method or transfer matrices procedures. In order to ensure the largest possible field of application, it should be possible to use several structural analysis methods.

Column 2: Optimization modelling

From an engineer's point of view, this is the most important column of the optimization procedure. First of all, the quantities which are to be changed during the optimization process, i.e. the analysis variables, are selected from the structural parameters. The design model (variable linking, variable fixing, approach functions etc.) provides a mathematical link between the analysis variables and the design variables. In order to increase efficiency and to improve the convergence of the optimization computation, the optimization problem is adapted to meet the special requirements of the optimization algorithm by transforming the design variables into transformation variables. Using this approach it is possible to almost linearize the stress constraints of a sizing problem.

In addition, the objective function and constraint values can be determined by using the relevant evaluation procedures. When formulating the optimization model the engineer must take the demands from the fields of design, production, assembly and operation into account.

The treatment of an optimization problem with multiple criteria (Vector Optimization Problem VOP) is a typical optimization strategy. Characteristic of vector optimization problems is the occurence of objective conflicts which means that no design variable vector allows the simultaneous optimal fulfilment of all objectives. A VOP is solved by transforming the initial problem into scalar substitute problems using so-called preference functions. These substitute problems provide compromise solutions and the engineer selects the most preferable one of these. A detailed account of the solution strategies for vector optimization problems can be found in [1, 8-11].

Column 3: Optimization algorithms

During the last few years procedures involving mathematical programming have been given preference over other methods for solving nonlinear constrained optimization problems. These algorithms are iterative procedures, which proceeding with an initial design $\underline{x}_0$ generally provide improved design variables $\underline{x}_k$ as a result of each iteration k. The optimization calculation is terminated if a breaking-off criterion responds during an iteration. Numerous studies have demonstrated that the selection of the optimization algorithm has to ensue depending on the problem. This is particularly important for a reliable optimization and a high level of efficiency (computing times, rate of convergence). If, for example, all of the iteration results have to be within the feasible domain, a feasible working procedure, such as the Generalized Reduced Gradients (GRG), should be applied [13].

3. OPTIMIZATION PROCEDURE SAPOP

The software system SAPOP (Structural Analysis Program and Optimization Procedure, Fig. 2) was developed from the basic solution philosophy of the Three-Columns-Concept and from the requirements mentioned above [7, 12]. It is composed of three mutually independent blocks which communicate with each other via a Data-Management-System . Each of these blocks is divided into individual subsystems. Between these subsystems there exist standardized interfaces to ensure the largest possible modularity. Each subsystem contains a number of interchangeable program modules. When carrying out an optimization computation only those modules are taken which are actually needed from each subsystem.

The actual optimization computation is carried out by the SAPOP main module MAIN. First of all, an initialization phase is run and subsequently the optimization is started via the ONE-SYSTEM module of the DECOMP block. For the optimization of large structures and the partial optimization of

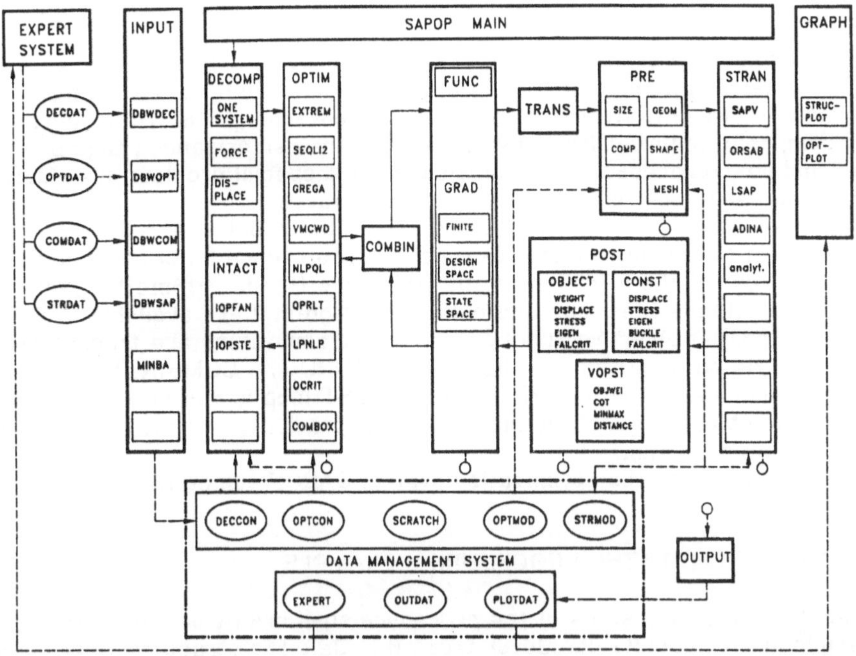

Fig. 2: Block diagram of the software system SAPOP

substructures, it is possible to apply two special decomposition strategies, the cutting force method FORCE and deformation method DISPLACE, [13]. The DECOMP modules have access to a number of different optimization algorithms (seven mathematical programming methods, stress ratio method).

For an optimization computation, the respective actual objective function and constraint values and, for most of the algorithms, their gradients, are required for each iteration. Via the interface module COMBIN, the control program FUNC is called for functional evaluations or the control programs for the sensitivity analysis (gradient calculation).

The module TRANS transfers the transformation variables into physically interpretable design variables. The subsequently called pre-processor contains different design models and determines the analysis variables from the design variables. The design model SIZE includes variable linking and fixing for cross-section optimization (sizing). The SHAPE, GEOM and MESH modules can be used for form and geometry optimization tasks. As far as composite optimization is concerned, COMP converts the design variables into layer thicknesses and ply angles of a fiber composite material and the respective material characteristics (stiffnesses and hygrothermic coefficients) are calculated [12].

The structural analysis is carried out using the updated analysis variables. These are part of the structural parameters of the mathematical-mechanical model, which describes the physical behaviour of the actual component. The equation and differential equation systems are solved by using efficient numerical methods. As mentioned before, the user can link his own structural analysis modules to SAPOP. Thus, it is possible to deal with structures using analytical calculations or with any examples from interdisciplinary fields.

The computed state variables are transferred to the post-processor in order to determine the objective function and constraint values. Modules are available for weight, stress, deformation, eigenvalues, local truss buckling and composite failure criteria evaluations. If the performance range of the pre- and post-processor is not sufficient for a special problem, user defined programs can be linked via standardized interfaces.

Vector optimization problems are solved by transforming the objective functions into a scalar substitute function (preference function). For these purposes four of the most important strategies (trade-off-method, min-max strategy, distance functions and objective weighting) are available [1,8-11].

The updated objective function and constraint values are transferred to the optimization algorithm via the modules FUNC and COMBIN. In the event of a gradient evaluation the PRE-STRAN-POST loop is run several times before the objective function and constraint gradients are available for the optimization algorithm.

More details on the optimization system SAPOP can be found in [7, 12].

4. APPLICATION OF THE PROCEDURE - EXAMPLE

Some definitions and steps of the procedure shall briefly be illustrated by the example of a composite cantilever truss (Fig.3a). The structural analysis has been carried out by finite elements. The structural parameters are the geometric quantities, the cross sections and the material data. As design variables the wall thicknesses of the tubes and the horizontal and vertical distances between the nodal points have been chosen. For a better solution behaviour of the optimization procedure the cross-section variables are transformed into reciprocal transformation variables.

The structural analysis provides the state variables $\underline{u}$ (forces S_i, nodal displacements u_{xi}; u_{yi}). For the evaluation model the objective functions and constraints are formulated as functions of the state variables and structural parameters. The objective function vector reads as follows:

$$\underline{f}(\underline{x}) = \begin{bmatrix} f_1(\underline{x}) := W = \sum_{i=1}^{15} \rho\, l_i\, A_i \\ f_2(\underline{x}) := u_{x_{max}} = u_{x_8} \end{bmatrix}$$

The evaluation model includes the transformation of the vector optimization problem into a scalar substitute problem using the trade-off method as well as the definition of the stress, buckling and failure constraints. The optimization algorithm uses the preference function and constraint values to calculate the new design. Various tests have shown that the VMCWD and the QPRLT [7] procedures provide the best solution behaviour for this application.

Fig. 7 presents various functional-efficient (or p-efficient) solution curves for the design problem of a steel and a composite cantilever. Functional-efficient solutions are compromise solutions. They are characterized by the fact that if one objective function is reduced, then the other one will increase in order to keep within the feasible domain of the optimization problem [1].

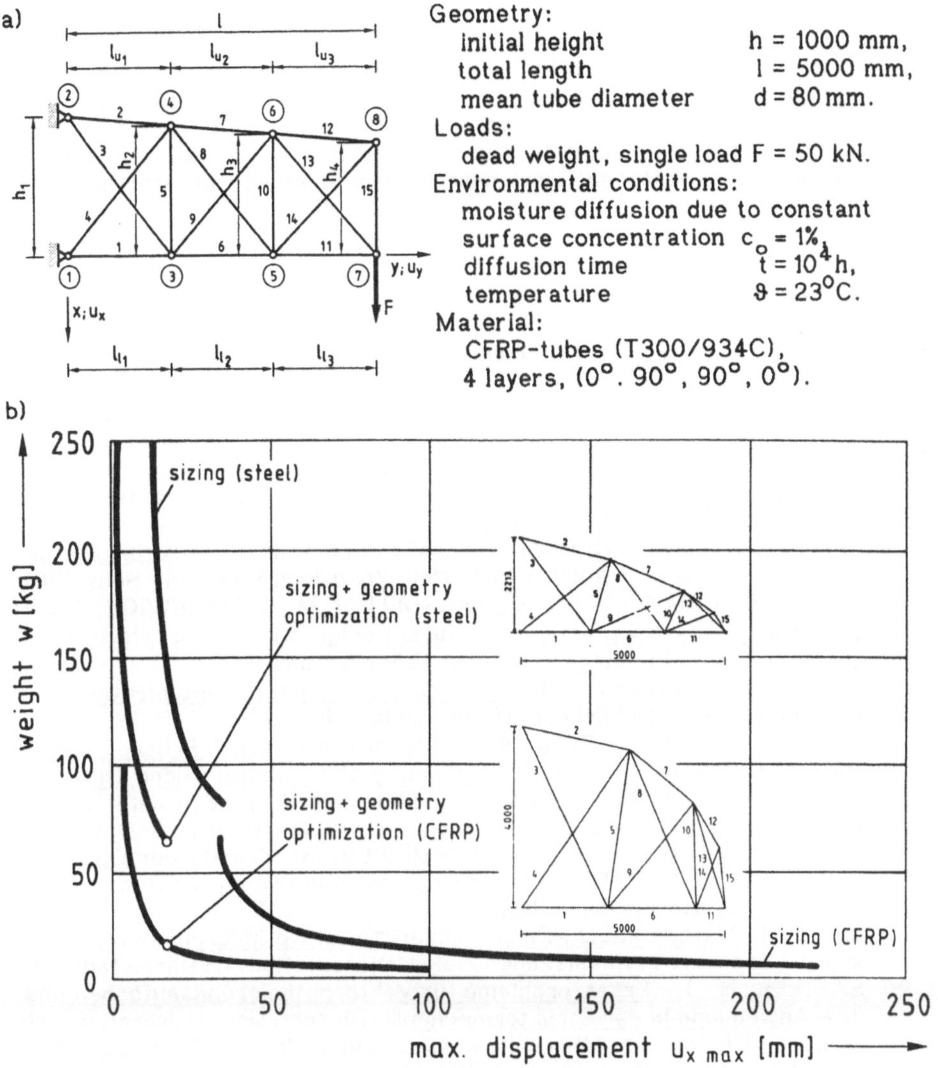

Fig. 3: Cantilever truss and functional-efficient boundaries

5. CONCLUSION

This paper introduces an optimization procedure as a concept for solving optimization problems in the field of structural optimization. Starting from the Three-Columns-Concept with the columns Structural Model, Optimization Algorithms and Optimization Model and taking other demands into consideration, the software system SAPOP (Structural Analysis Program and Optimization Procedure) was developed. For special optimization problems some optimization strategies are implemented (e.g. vector optimization, shape functions). The procedure is applied to a sizing and a geometrical geometrical layout of a composite and steel cantilever.

The performance of SAPOP is demonstrated by many successfully solved optimization problems in various fields of application [1, 7, 12, 13]. Until now problems with up to 100 design variables and 300 constraints have been treated. Because of its modularity and an efficient data management it is possible to implement the software system even on small workstations and personal computers.

REFERENCES

[1] ESCHENAUER, H.: Rechnerische und experimentelle Untersuchungen zur Strukturoptimierung von Bauweisen. DFG-Forschungsbericht, Universität GH-Siegen, Technische Informationsbibliothek TIB Hannover 1985 (in english and german)

[2] HAUG, E.J.; ARORA, J.S.: Applied Optimal Design: Mechanical and Structural Systems. Chichester, New York, Brisbane, Toronto, Singapore: John Wiley & Sons 1979

[3] REKLAITIS, G.V.; RAVINDRAN, A.; RAGSDELL, K.M.: Engineering Optimization: Methods and Applications. New York: John Wiley and Sons, 1983.

[4] ATREK, E.; GALLAGHER, R.N.; RAGSDELL, K.M.; ZIENKIEWICZ, O.C.: New Directions in Optimum Structural Design. Chichester, New York, Brisbane, Toronto, Singapore: John Wiley & Sons 1984

[5] HAFTKA, T.R.; KAMAT, M.P.: Elements of Structural Optimization. Maritinus Nijhoff Publishers, Netherlands 1985

[6] HÖRNLEIN, H.R.E.M.: 'Take Off in Structural Design.' Proceedings of the NATO/NASA/NSF/USAF Conference of Computer Aided Optimal Design, Vol III. Troja, Portugal, June 29 - July 11, 1986, Center of Mechanics and Materials of the Technical University of Lisboa, 176-199

[7] ESCHENAUER, H.; POST, P.U.; BREMICKER, M: 'Einsatz der Optimierungsprozedur SAPOP zur Auslegung von Bauteilkomponenten.' Bauingenieur (to appear 1988)

[8] BAIER, H.: Mathematische Programmierung zur Optimierung von Tragwerken insbesondere bei mehrfachen Zielen. Dissertation. TH Darmstadt 1978

[9] SATTLER, H.-J.: Ersatzprobleme für Vektoroptimierungsaufgaben und ihre Anwendung in der Strukturmechanik. Dissertation. Universität-GH-Siegen, VDI-Fortschrittbericht, Reihe 1, Düsseldorf: VDI-Verlag 1982

[10] KOSKI, J,: Truss Optimization with Vector Criterion. Tampere University of Technology, Department of Mechanical Engineering, Tampere, Finland, Publication No. 6, 1979

[11] OSYCZKA, A.: Multicriterion Optimization in Engineering. New York: John Wiley & Sons 1984

[12] POST, P.U.: Optimierung von Verbundbauweisen unter Berücksichtigung des Langzeitverhaltens. Dissertation. Universität-GH Siegen, Institut für Mechanik und Regelungstechnik (to appear 1988)

[13] BREMICKER, M.: Gestaltsoptimierung dreidimensionaler Strukturen unter Verwendung von Dekompositionsstrategien. Dissertation. Universität-GH-Siegen, Institut für Mechanik und Regelungstechnik, (to appear 1988)

STRUCTURAL SHAPE OPTIMIZATION USING OASIS

B.Esping & D.Holm
The Royal Institute of Technology
Department of Aeronautical Structures and Materials
S-100 44 Stockholm
Sweden

ABSTRACT. OASIS is a code for structural optimization. This paper demonstrates the capibilities in shape optimization. The CAD system ALADDIN is used for shape description and generation of FE-meshes.

What is structural optimization about ?

Let us consider a connecting rod (fig.1).
Its function is to transfer a force from the piston to the crankshaft. What should it look like ? In our approach, we give a proposal, which will be improved after optimization.

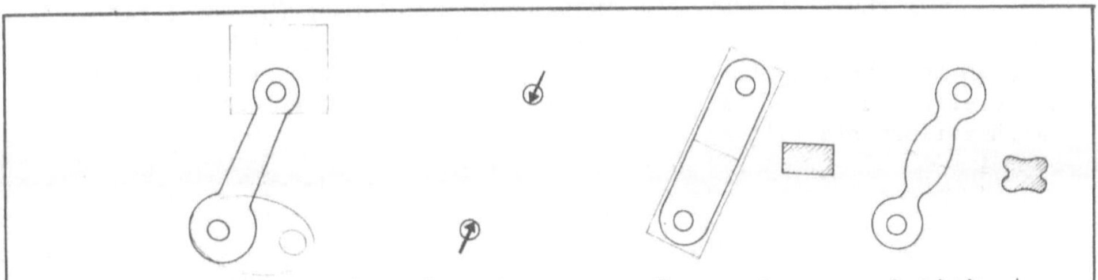

Fig. 1 Connecting rod Function Proposal Optimized

CAD for shape description and design variables.

Let us consider an arch bridge. Its shape can be described by a number of control points, which are connected by various types of lines (fig.2).

G. I. N. Rozvany and B. L. Karihaloo (eds.), Structural Optimization, 93–100.

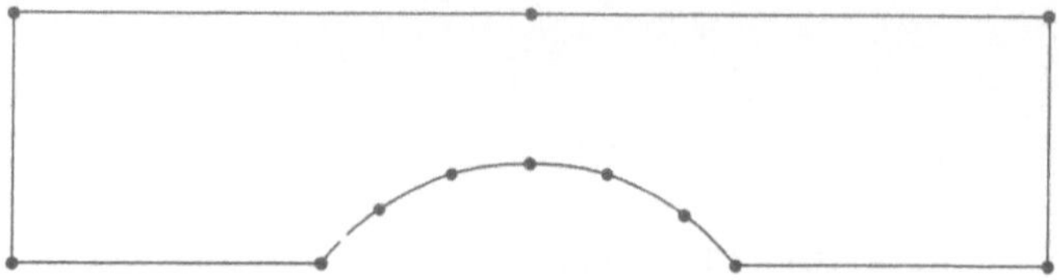

Fig. 2 CAD-model for an arch bridge

The bridge consists of what is inscribed by the lines. Let us say that our problem is to minimize the weight of the bridge, when it is subjected to some loadsystems. Let's further say that we are only allowed to change the shape of the arch, which is described by a spline. Next we simply add design variables to the points on the arch, (fig 3) (ref 1).

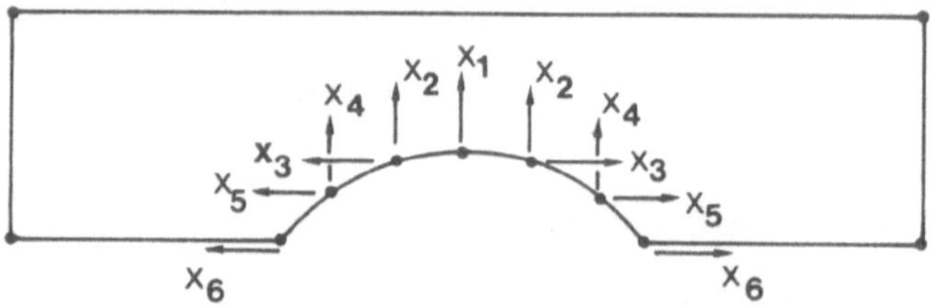

Fig. 3 Design variables

Analysis

We can use the finite element (FE), boundary element (BE) or finite difference (FD) method for the response analysis. In all cases we need a mesh for the computations. We use parametric mapping functions to to create the mesh from the CAD-model. It is then very easy to make mesh refinements (fig 4).

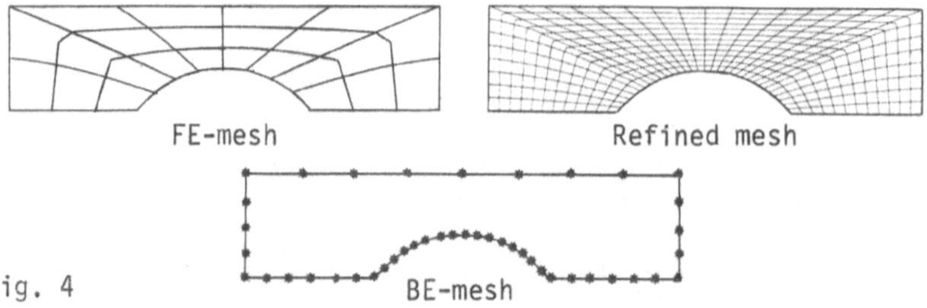

Fig. 4

Sensivities

Quasi numerical derivatives

For FE-structures we use quasi numerical derivatives.
Let us take a displacement d as an example. Its derivative with respect to a design variable x can be expressed as:

$$\frac{\partial d}{\partial x} = -\sum_e u_e^T \frac{\partial K_e}{\partial x} v_e$$

u_e is the element displacement vector corresponding to one of the load cases. v_e is the virtual displacement vector corresponding to the displacement d. K_e is the element stiffness matrix. Its derivative is derived using forward finite differences, (fig 5):

$$\frac{\partial K_e}{\partial x} = \frac{K_e(x+\Delta x) - K_e(x)}{\Delta x}$$

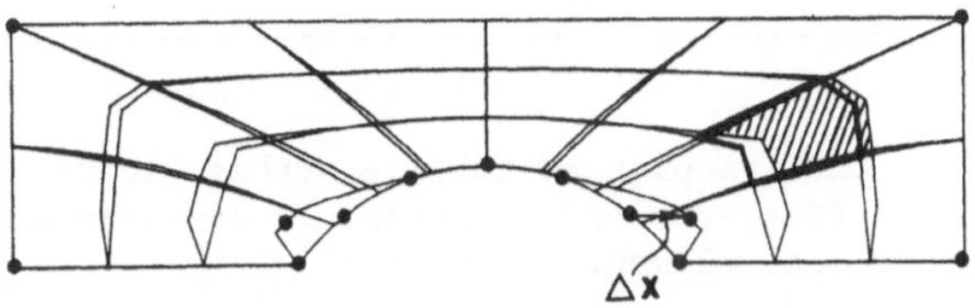

Fig. 5 Elements with and without variation of x

This technique can be used for stresses, displacements, eigenfrequences etc.

Finite difference derivatives
Let r(x) be a general response in our system. The forward finite difference derivative is:

$$\frac{\partial r}{\partial x} = \frac{r(x+\Delta x) - r(x)}{\Delta x}$$

This technique is more flexible, but also more costly. It can be used for explicit constraints, production constraints etc. in which cases it can be difficult to find expressions for the derivatives, or when the cost of an analysis is low.

Analytical derivatives

It is possible to give analytical expressions for the derivatives of stress, displacements etc. but the risk of making mistakes is big and the computational savings are small. The flexibility is also lost.

The OASIS system

The OASIS system (fig 6) (ref 2) consists of a number of exchangable moduls.

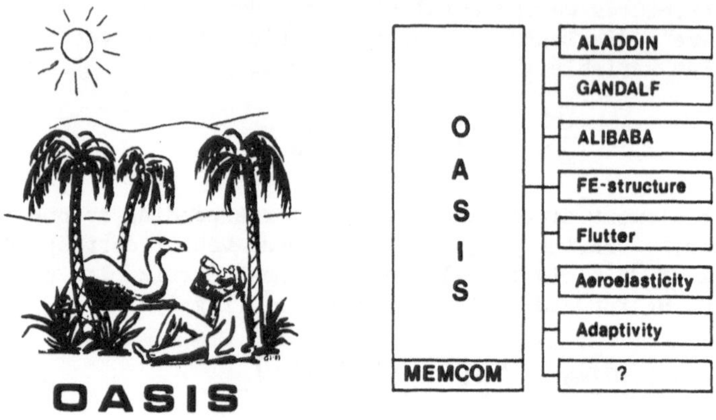

Fig. 6

ALADDIN

ALADDIN (ref 3) is a CAD system used to give meshes, loads, boundary conditions etc. to the FE-system and to OASIS. It can also be used as a postprocessor for stresses, element errors etc.

GANDALF

GANDALF is a monitor for the optimization process, but also a postprocessor for diagrams on objective and constraints functions versus number of iterations. GANDALF - ALADDIN are used to inspect intermediate results, modify design variables or constraints, extend the number of variables or constraints. ALADDIN can also be used for adaptive mesh modifications.

ALIBABA

ALIBABA is the mathematical optimization package and involves the linear approximation concept (LA), inverse approximation (IA), hybrid approximation (HA) and the method of moving a asymtots (MMA) for continuous optimization and Degedewe method for discrete or mixed continuous - discrete optimization.
It also involves methods for min-max optimization and relaxation for strongly infeasible problems.

FE-structure

FE-structure stands for a FE code for structural analysis. As the sensivity analysis is based on the quasi numerical approach it is very easy to exchange one FE-code to another. So far, OASIS has been interfaced with two FE-codes; BASIS and ASKA.
The SAAB company works on an interface with ABAQUS.

Flutter and aeroelasticity

Constraints on flutter has been implemented by SAAB (ref 4) and aeroelasticity by the Aeronautical Research Institute of Sweden (FFA). (ref 5)

Adaptivity

Error estimation and adaptive mesh refinement is under development in a collaboration with the University of Michigan, USA (ref 6).

Problems

OASIS can be used to solve problems, where stresses, displacements, weight, eigenfrequencies and local buckling either are constraints or selected as the objective function. OASIS can also be used to solve min-max problems.

Variables

Other kinds of variables are thicknesses in mono or multilayered membranes or shells, cross secton shape of beams and fibre orientation.

Heurestic and explicit functions

Some constraints or objective functions can be expressed explicity or heuresticly. This is often the case with production constraints.

Auxiliary

The flexible technique that we have used allows for other types of response constraints on: acoustics, magnetic fields, electrical fields, fluid mechanics, production etc. It is also very suitable for parallell processing.

Examples

In the following we briefly give a number of examples taken from structural shape optimization.

Hydraulic cylinder for Volvo BM

The cylinder consists of a cylindric part and an end plate. The two parts are welded together (fig 7). The joint is the weak point. A radial crack is assumed. The lifetime of the cylinder is mainly depending on how the crack grows, which can be calculated from the stress intensity factor K_I.
Our goal is to minimize K_I and also the maximum equivalent stress σ_{eff}. This is a min-max problem:

$$\min(\max(K_I, \sigma_{eff}))$$

K_I, σ_{eff} are here normalized with respect to their initial values.

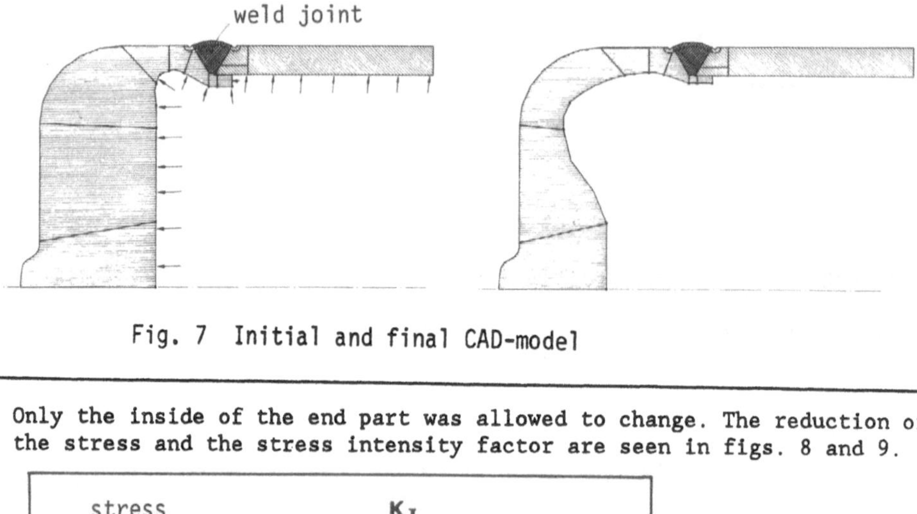

Fig. 7 Initial and final CAD-model

Only the inside of the end part was allowed to change. The reduction of the stress and the stress intensity factor are seen in figs. 8 and 9.

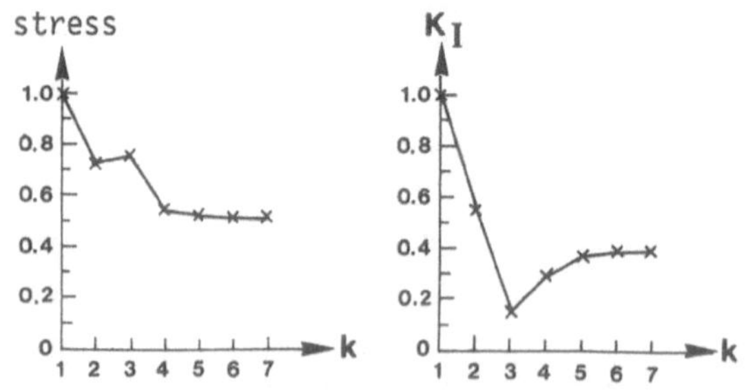

Fig. 8 Stress and K_I histories

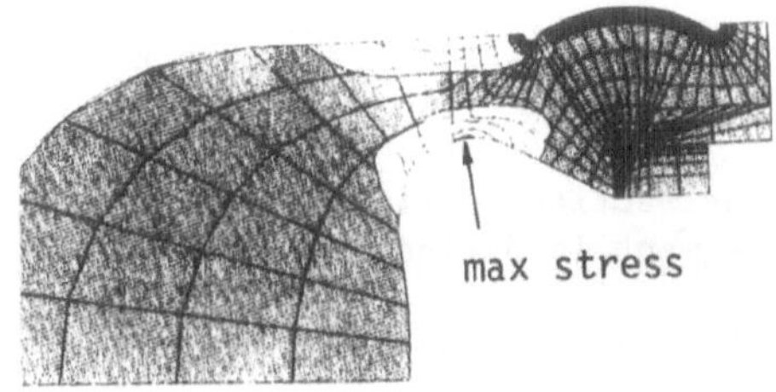

Fig. 9 Stress distribution before and after optimization

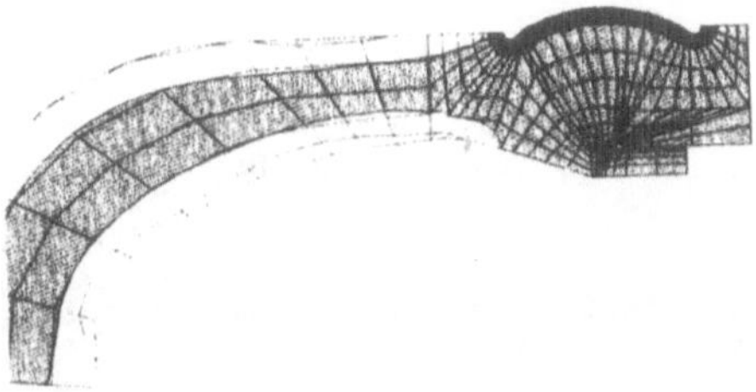

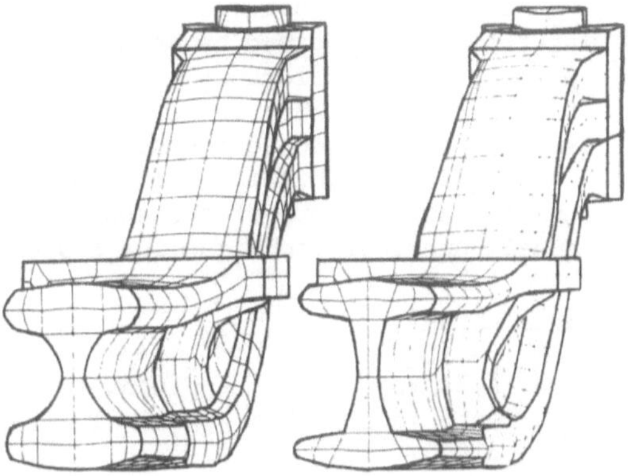

Fig. 10 Initial and final FE-model

Truck front axle beam for Scania

The beam (fig 10) (ref 7) is optimized for three load cases of which two correspond to fatigue. The forged beam has constraints on positive slopes of all surfaces and also on the change of the camber angle and maximum stress. Neither the camber angle nor the stress level may increase compared to the situation on already existing beams. Still the weight decrease by 15% which is 13 kg per beam. Altogether 20 000 beams are produced every year. The material cost alone is more than \$ 2 per kg.

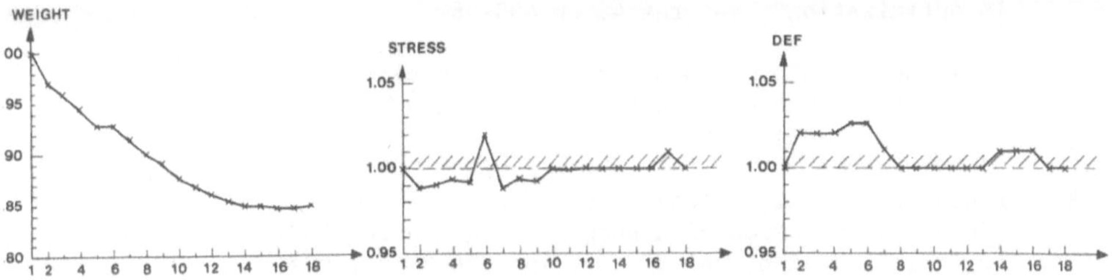

Fig. 11 Weight, stress and displacement histories

There were 60 design variables, 15000 degrees of fredom and 10000 constraints.

Auxiliary

We have a number of other examples. Only one additional approach to shape optimization will be mentioned here, namely shape optimization of cross sections of beams in frame works.
We have studied thin walled opened cross sections defined by cross section points joined by segments. A segment is defined by its two points and the thickness at its two ends (ref 9)
Design variables can be attached to both the coordinates of the cross section points and to the segment thicknesses.
The easiness with which this capacity has been implemented is due to the flexible nature of the code, which comes out as a result of the quasi numerical technique for response derivatives.

References

1. B.Esping, "A CAD approach to the minimum weight design problem", Int. journal for numerical methods in eng, Vol 21, pp 1049-1066 (1985)

2. B.Esping, "The OASIS structural optimization system", Computers & Structures, Vol 23, no 3, pp 365-377, (1986)

3. D.Holm,"ALADDIN - user's manual", Dept. of Aeronautical Structures and Materials, The Royal Institute of Technology, Stockholm, Sweden.

4. T.Brämå, "Weight optimization of aircraft structures", Computer Aided Optimal Design: Structural and Mech Systems, editor C.A.M Soares, Springer, pp 971-985, (1987)

5. O.Johansson, "Weight optimization under structural and aeroelastic constraints of a wing for a general aviation aircraft", FFA, TN 1985-3, FFA, Stockholm, Sweden (1985)

6. N.Kikuchi, "Adaptive grid design for finite element analysis in optimization", see ref 4, pp 493-562.

7. B.Esping, D.Holm, J.Lönnqvist, K-A Olsson, "Shape optimization of a truck front axle beam", report 87-7 see ref 3 (1987)

8. B.Esping, D.Holm, R.Isby, M.Larsson, "Shape optimization of a suspension arm using OASIS", Int. journal of vehicle design, vol 8, no 3, pp 326-334, (1987)

9. O.Hultgren, "Optimum design of an arbitrary thin walled open cross section", diploma work, see ref 3, (1987)

OPTIMAL DESIGN OF R.C. FRAMES BASED ON IMPROVED INELASTIC ANALYSIS METHOD

C.S. GURUJEE
Civil Engineering Department
Indian Institute of Technology
P.O. I.I.T. Powai
Bombay 400076
India

and

S.N. AGASHE
Structural Engineering Department
Victoria Jubilee Technical Institute
P.O. Matunga
Bombay 400018
India

ABSTRACT. A general method for minimum cost design of R.C. frames has been presented. Constraints based on improved inelastic analysis method which can account for local unloading are used in framing optimization problem for minimizing the cost of material and the formwork used for a frame. The optimization problem is solved by extended interior penalty function technique using variable metric method of unconstrained minimization. A single dimension general computer program has been prepared based on the method and the procedure is demonstrated by solving four illustrative examples.

1. Introduction

Any analysis or design solution of an R.C. frame problem must account for its inelastic behaviour. Extension of plastic analysis methods as used for steel frames for the analysis of R.C. frames is possible but it must account for non-linear moment-curvature relationship and limited rotation capacity of an R.C.section. The concept of a mechanism in case of a steel frame when sufficient number of plastic hinges are formed is used for calculating the maximum load that can be sustained by a frame. The load which transforms a structure into a mechanism is called ultimate load. A design based on a safety factor with respect to the ultimate load is more rational

G. I. N. Rozvany and B. L. Karihaloo (eds.), Structural Optimization, 101–108.

and can be easily adopted for frames made of ductile material such as steel. Even under the action of ultimate load the frame still remains in elastic region and the plastic behaviour is assumed to be concentrated only at a few points called plastic hinges. For an R.C. frame the same concept can be used for calculation of rotation at plastic hinges and limiting them to the required level for design purposes. This was proposed by Baker[1] who assumed elasto-plastic behaviour for an R.C. section. While Baker's work accounted for limited ductility of an R.C. section, it did not take into account non-linear moment-curvature relationship of an R.C. section. Baker used the expressions for rotations at a plastic hinge position as the one obtained under external loads for the reduced determinate frame obtained after introducing number of hinges equal to the statical indeterminacy of the structure plus those obtained by introducing plastic moments at the hinged points. Corradi, DeDonato and Maier[2] used the expressions for moments at a section as the superposed moments obtained by complete elastic behaviour of frame and those obtained due to inelastic rotations at hinged sections. Inelastic rotations were then expressed as the non-linear or equivalent piecewise linear relationship with the moment at that section. An analysis problem posed in this maner in terms of inelastic rotations is identified as Linear Complimentarity Problem (hereafter referred to as LCP) and is ideally suited for working on a computer. By reformulating this problem Kaneko[3] reduced computer storage requirements for an analysis and this approach has been converted into a general computer program using single array dimensioning by Gurujee and Agashe[4].

All the approaches mentioned so far assume that no local unloading occurs at a section during proportional increase of load. This phenomenon does occur in several cases. If the previous approaches are to be used manual intervention is required in getting solution to such cases and is not desirable in the interest of preparing a general purpose program. The present work proposes a method which can allow for local unloading. This is achieved by expressing the moment-rotation relationship at a section as a hexalinear relationship with three linear pieces in first quadrant and three linear pieces in third quadrant as against trilinear relationship in case of Kaneko's formulation. This improved method of inelastic analysis is explained in the next section of this paper.

The purpose of any structural analysis is to design and to obtain optimal design if possible. Optimal design of R.C. frames with inelastic analysis method was tried earlier by Krishnamoorthy and Munro[5]. The objective function used in this work was the amount of steel reinforcement and the problem of optimal design was solved as a sequence of linear programming problem. Solution of a nonlinear programming problem is obtained more efficiently as sequence of unconstrained optimization problems. This technique was used by Krishnamoorthy and Mosi[6] but their presentation lacked generality as they considered minimum cost of the total structure equal to the sum of minimized cost of separate parts of the structure (referred to as groups) obtained by considering constraints concerned only with that part. As a result, most of their results are seen to violate some of the constraints at the end of solution.

The present work has considered total cost of a frame as the objective function. The cost includes cost of material and cost of formwork. Constraints are based on the improved inelastic analysis procedure explained in the next section and the usual constraints on dimensions of a section and steel percentages. Total cost optimization process which consists of analysis-optimization process is explained further in this paper. The resulting optimizaton problem is solved by extended interior penalty function technique as proposed by Cassis and Schmit[7] by using variable metric method. A general computer program based on the optimization procedure has been prepared and its application to four problems is explained further.

2. Improved Inelastic Analysis

The problem considered here is as follows. Given geometrical details of a frame and material properties it is desired to find the moment field and displacements for a given set of external loads which can be expressed as a single vector $\{F\}$. By introduction of a scalar $\alpha \leq 1$, it is possible to express any intermediate loading stage leading upto $\alpha = 1$ representing the load at which response is desired. By inspection of loading pattern it is possible to identify locations in the frame where moment is likely to be critical. These sections, numbering n, are possible hinge or discontinuity locations. Once the moments at these locations are known, moment and axial and shear forces at any other location in the frame can be found by means of statical considerations. The assumptions in this method are as follows: i) Effects of geometric changes are not significant ii) Inelastic rotations (discontinuities) are concentrated at critical sections. iii) Moment-rotation relationship at each critical section can be expressed as a hexalinear relation.

Due to the first assumption it is possible to write equilibrium equations based on original geometry of the frame and due to the second assumption it is possible to express the moment at n critical sections in the form of following matrix equation.

$$\{\mu\} = \alpha\{\mu_E\} + [Z]\{\theta\} \tag{1}$$

where vector $\{\mu\}$ represents the moments at n critical sections at the load level indicated by α, $\{\mu_E\}$ represents the moments at n critical sections when there are no discontinuities at any of the n critical sections, $\{\theta\}$ represents the rotation discontinuities at critical sections at the load level α and $[Z]$ represents influence coefficient matrix. An element Z_{ij} represents moment at i th critical section when a unit discontinuity is introduced at j with all other discontinuities being equal to zero. If $\{\theta\}$ is known the moment field is defined throughout the frame. However elements of $\{\theta\}$ are related to the moment at that point by a nonlinear relation. This relationship is idealised by a series of six straight line segments is shown in Fig.1. This relationship called hexalinear moment-rotation relationship expresses three distinct stages of behaviour in either of the moment and rotation directions. The first stage

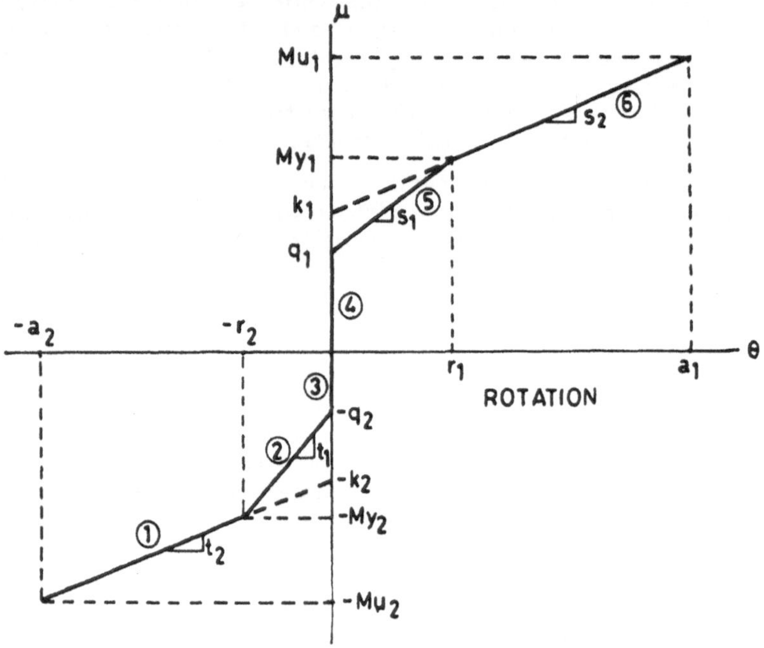

Figure 1. Hexalinear moment-rotation law.

0 to q_1 (or q_2) denotes the stage upto the first crack at the section, the second stage q_1 to M_{y1} (or q_2 to M_{y2}) denotes the stage from cracking and appearance of discontinuity upto the yielding of steel and the third from M_{y1} to M_{u1} (or M_{y2} to M_{u2}) denotes the stage from yielding upto the exhaustion of rotation capacity of the section. Denoting maximum rotations at the end of second and third stage as r_1 and a_1 (or $-r_1$ and $-a_2$) and noting that the section has to be in any one of the six segments at any given stage of loading, the general moment rotation relationship can be expressed as follows.

$$\mu_i = q_1 - w_1 + s_1 v_1 + s_2 x_1 - q_2 - w_2 - t_1 v_2 - t_2 x_2 \qquad (2)$$

$$w_1,\ v_1,\ x_1,\ w_2,\ v_2,\ x_2 \geq 0 \qquad (3)$$

$$v_1 \leq r_1 \quad \text{and} \quad v_2 \leq r_2 \qquad (4)$$

$$r_1 + x_1 \leq a_1 \quad \text{and} \quad r_2 + x_2 \leq a_2 \tag{5}$$

$$(q_1 - w_1)\,(q_2 - w_2) = 0 \tag{6}$$

$$w_1 v_1 = 0 \quad \text{and} \quad w_2 v_2 = 0 \tag{7}$$

$$(r_1 - v_1)\,x_1 = 0 \quad \text{and} \quad (r_2 - v_2)\,x_2 = 0 \tag{8}$$

$$\text{and} \quad \theta_i = v_1 + x_1 - v_2 - x_2 \tag{9}$$

In the above expressions suffix i denoting the critical section number has been dropped for clarity from all expressions (except on l.h.s. of (2) and (9)). s_1, s_2, t_1 and t_2 are slopes in various segments of moment-rotation law; x_1, v_1, x_2, v_2 are components into which rotation θ_i can be broken as expressed in relation (9) and w_1, w_2 denote the shortfall in moment as compared to cracking moment should that section be loaded at less than cracking moment. Relations (3) to (8) ensure that the point is in any one of the six segments. Now collecting such expressions for all the critical sections and denoting the expressions in matrix form by making use of capital letters while doing so, the expression (2) can be written as

$$\{\mu\} = \{Q_1\} - \{W_1\} + [S_1]\{V_1\} + [S_2]\{X_1\}$$
$$-\{Q_2\} + \{W_2\} - [T_1]\{V_2\} - [T_2]\{X_2\} \tag{10}$$

and equation (9) gets transformed to

$$\{\theta\} = \{V_1\} + \{X_1\} - \{V_2\} - \{X_2\} \tag{11}$$

Substituting equation (11) in equation (1) and then equating this with (10), the problem of finding response gets transformed to finding vectors $\{W_1\}$, $\{W_2\}$, $\{V_1\}$, $\{X_1\}$, $\{V_2\}$ and $\{X_2\}$ such that the following matrix equations are satisfied.

$$\{Q_1\} - \{Q_2\} - \{W_1\} + \{W_2\} + [M_1]\{V_1\} + [N_1]\{X_1\}$$
$$+ [M_2]\{V_2\} + [N_2]\{X_2\} = \{0\}, \tag{12}$$

and

$$\{W_1\}, \{V_1\}, \{X_1\}, \{W_2\}, \{V_2\}, \{X_2\} \geq \{0\} \tag{13}$$

$$\{V_1\} \leq \{R_1\} \quad \text{and} \quad \{V_2\} \leq \{R_2\} \tag{14}$$

$$\{X_1\} + \{R_1\} \leq \{A_1\} \quad \text{and} \quad \{X_2\} + \{R_2\} \leq \{A_2\} \tag{15}$$

$$[\{Q_1\} - \{W_1\}]^T\,[\{Q_2\} - \{W_2\}] = 0 \tag{16}$$

$$\{W_1\}^T\{V_1\} = 0 \quad \text{and} \quad \{W_2\}^T\{V_2\} = 0 \tag{17}$$

$$[\{R_1\} - \{V_1\}]^T \{X_1\} = 0 \quad \text{and} \quad [\{R_2\} - \{V_2\}]^T \{X_2\} = 0 \tag{18}$$

where $[M_1] = [S_1] - [Z]$, $[M_2] = [Z] - [T_1]$,

$$[N_1] = [S_2] - [Z] \quad \text{and} \quad [N_2] = [Z] - [T_2] \tag{19}$$

This problem is known as an LCP and can be solved by Greaves'[8] algorithm.

3. Formulation of the optimization problem

The previous section has explained the means of obtaining response for any given set of loads. The process of optimal design of R.C. frames is seen here as a series of analysis-optimization cycles. In analysis stage all the properties obtained from most recent optimization stage are assumed to be known and the moment field is found in this stage. In optimization phase the moment field is assumed to remain constant and optimal design is found with this moment field. It is seen that mostly in two or three such cycles convergence is reached. The optimization problem is framed as follows.

3.1. FORMULATION OF THE OBJECTIVE FUNCTION

The objective function denoting the cost of material and cost of form work is formulated in the same manner as done by Krishnamoorthy and Mosi[6] and the details of the assumptions done in this regard and their justification can be found in that reference. There are four design variables associated with a column which are breadth (b), effective depth (d), half of cross-sectional steel requirement at top (Ast) and at bottom (Asb) and five design variables associated with a beam which are breadth (b), effective depth (d), steel requirement at the two ends (As1 and As2) and steel requirement at the maximum bending moment position in the middle (Asm). Further denoting u_c as the cost of concrete per unit volume, u_s as the cost of steel per unit weight, γ_s as the weight of steel per unit volume, u_f as the cost of formwork per unit area, d_c as cover and L as the length of a member the total cost of a frame can be denoted as

$$W = \sum_{\text{Columns}} [b(d+d_c)Lu_c+(Ast+Asb)L\gamma_s u_s+2(b+d+d_c)Lu_f]$$
$$+\sum_{\text{Beams}} [b(d+d_c)Lu_c+((As1+As2)/4+Asm/2)L\gamma_s u_s+2(b+d+d_c)Lu_f] \tag{20}$$

3.2. FORMULATION OF THE CONSTRAINTS

Constraints for the beam consist of neutral axis limitation to ensure tensile failure, constraint on moment capacity and other constraints such as those on maximum and minimum reinforcements, depth constraints and non-negativity constraints making a total of 20 constraints for 5 variables for each beam member. Constraints for column consist of moment capacity constraint obtained by equating axial force obtained from the inelastic analysis to the axial force obtained from limiting conditions and other usual constraints making totally 13 constraints in terms of 4 variables for a column. Details of this can be found in reference 9.

3.3. SOLUTION OF THE OPTIMIZATION PROBLEM

Optimization design problem of an RCC frame is thus a typical non-linear programming problem. The number of variables is 4 times the number of columns plus 5 times the number of beams. The number of constraints is equal to 13 times the number of columns plus 20 times the number of beams. All the variables are scaled to lie between zero and one and constraints are also scaled to have more or less same values at the starting point. The problem is solved by extended penalty function technique of Cassis and Schmit[7].

4. Computer Programming and Numerical Examples

A general computer program in FORTRAN IV based on the optimization procedure described in the previous section has been prepared. The program consists of the main program which reads all the input data consisting of initial trial design, loads, material properties, moment-rotation relations for different sections, cost parameters and various tolerances used in optimization phase and controls flow between inelastic analysis and optimization routines. The output also is written by the main program. There are three major subroutines, one for formulation and solution of inelastic analysis problem as an LCP by Graves'[8] algorithm, second for formulation of optimization problem and scaling the variables and constraints and the third one for solution of an NLP by extended interior penalty function technique. The entire program is written by using two one dimensional arrays. One array is used for storing all the floating point quantities and the other for storing all the integer quantities. This has been done because on some computers different default field widths are used for fixed point and floating point numbers. It is thus possible to use this program on any computer. All the problems have been solved on DEC-20 computer as well on an Indian made micro-computer ESPL Super Micro which makes use of MC 68000 chip. Four illustrative problems are as follows; Example 1 - A two-bay two-storey frame; Example 2 - A one-bay six-storey frame; Example 3 - A two-bay six-storey frame and Example 4 - A three-bay four-storey frame. The details of these problems are availabe in references 6 and 9. Some significant data items used are as follows. Cost of concrete per cubic meter:Rs.400.00; Cost of reinforcing steel per tonne:Rs.4000.00; Cost of formwork per square metre: Rs.15.00; Cube strength of concrete (28 days): 200 kg/sq.cm.; Yield strength of steel (0.2 percent proof): 4250 kg/sq.cm. All the remaining data have been used from prevailing Indian Standard. Summary of the results for all the four problems has been shown in Table I. (Details in ref.9)

TABLE I: Results of Numerical Examples of Optimal Design

Sr. No.	Description of Parameter	Example Numbers			
		1	2	3	4
1.	Number of members	10	18	30	28
2.	Number of critical sections	24	42	72	68
3.	Number of design variables	44	78	132	124
4.	Number of constraints	158	276	474	448
5.	Initial cost in Rupees	5956	9213	13629	13781
6.	Optimum cost in Rupees	3254	5205	10676	13432
7.	Approx CPU time for LCP solution in seconds (DEC-20)	20	60	200	185
8.	CPU time for total optimization process in seconds (DEC-20)	260	714	946	911

5. Conclusion

The paper presents a method for optimal design of R.C.C. frames making use of a more general inelastic analysis method, more rational formulation of optimization problem and using a better optimization technique demonstrated by several examples.

6. References

1. Baker, A.L.L. 'The ultimate load thoery', Conc.Pubs, London (1956).
2. Corradi, L, Donato, O. De and Maier, G. 'Inelastic analysis of r.c. frames', ASCE, Str. Divn. **ST-100**, 1925-1942 (1974).
3. Kaneko, I. 'A mathematical programming method for inelastic anlysis of reinfroced concrete frame, 'IJNME, **11**, 1137-1154 (1977).
4. Gurujee, C.S. and Agashe, S.N, 'INACOF-A computer program for inelastic analysis of r.c. framed structures' IJNME, **21**, 330-363(1985)
5. Krishnamoorthy, C.S. and Munro, J. 'Linear program for optimal design of r.c. frames', IABSE, **33-I**, 119-141 (1973).
6. Krishnamoorthy, C.S. and Mosi, D.R. 'Optimal design of reinfroced concrete frames based o;n inelastic analysis, Engg.Optn, **5**, 151-167 (1981).
7. Cassis, J.H. and Schmit, L.A. 'On implementation of the extended interior penalty function', IJNME, **10**, 3-23 (1976).
8. Graves, R.L., 'A principal pivoting simplex algorithm for linear and quadratic programming', Op. Res, **15**, 482-494 (1967).
9. Agashe, S.N. 'Optimal design of reinforced concrete frames based on improved inelastic analysis method', Ph.D. thesis submitted to I.I.T. Bombay, 1986.

BOUNDARY ELEMENT METHODS IN OPTIMAL
SHAPE DESIGN - AN INTEGRATED APPROACH

Prabhat Hajela & Junhaur Jih
Department of Engineering Sciences
University of Florida, Gainesville, Florida
USA

ABSTRACT. The present paper describes an implementation of the boundary element method (BEM) in optimal shape design. The computational advantages of the boundary element technique over the more traditionally used finite element method of analysis, are examined in context of the shape synthesis problem. An integrated analysis and optimization approach is also proposed for this problem. Approximate strategies that can be used in conjunction with boundary element techniques are briefly discussed. Numerical results are presented for shape design of torsional shafts and of structural components that can be characterized by a state of plane stress.

1. INTRODUCTION.

Recent studies in structural optimization have extended the traditional structural member sizing problem to the determination of optimal shapes for desired structural response. The finite element method forms the basis for analysis in most of these developments. This method is conceptually a whole body discretization and often results in a very large number of elements, and consequently, large systems of linear algebraic equations. Boundary stress information obtained in this approach is often imprecise, and must therefore be used with caution in boundary shape modification. This problem is further compounded when shape change results in a distortion of the original finite element mesh, mandating the need for an adaptive mesh generation routine that interfaces with the optimization.

The boundary element method [1] provides an alternative solution strategy to a given set of differential equations by transforming into an equivalent set of integral equations, where the solution of the latter typically requires only the boundary of the domain be discretized. The order of the resulting linear system of equations is considerably lower than in the finite element method, but the sparseness of the coefficient matrix is generally lost. The use of this method in the shape synthesis of structures has been described in recent publications [2,3].

G. I. N. Rozvany and B. L. Karihaloo (eds.), Structural Optimization, 109–116.

The present paper reports preliminary efforts in the adaptation of boundary elements for optimal shape design, with emphasis on examining approximation concepts that would influence the computational requirements. An integrated optimization approach, in which the linear algebraic system of equations from the BEM are treated as equality constraints, allows the solution to the analysis and the optimization to proceed simultaneously. The selection of design variables and constraints is also examined.

2. BEM IN SHAPE DESIGN

Two distinct classes of shape design problems were attempted in the present effort. These were the determination of the boundaries for singly- and multiply-connected prismatic bars for prescribed torsional rigidity, and for structural components in a state of plane stress. The BEM analysis equations for these problems are summarized here for completeness.

For a solid shaft in torsion, the stress function u satisfies the Poisson state equation in the domain Ω, and is zero on the boundary of the domain Γ.

$$\nabla^2 u = -2 \qquad \text{in } \Omega \tag{1}$$

$$u = 0 \qquad \text{on } \Gamma \tag{2}$$

The fundamental solution of the above problem is of the form,

$$w = -\frac{1}{2\pi} \ln \left(\frac{1}{r}\right) \qquad r = |x - x_o| \tag{3}$$

and is obtained as a solution of

$$\nabla^2 w = -\delta(x, x_o) \tag{4}$$

where, $\delta(x,x_o)$ is the delta function. Using Green's identity and making use of the fundamental solution allows one to write an expression for u at a point x_p as follows.

$$U(x_p) = \int_\Gamma \left(\frac{\partial u}{\partial n} w - \frac{\partial w}{\partial n} u\right) d\Gamma + 2 \int_\Omega d\Omega \tag{5}$$

A numerical solution of (5) is facilitated by discretizing the boundary into N segments, and replacing the integral expressions by summations.

$$C_p u_i(x_p) + \sum_{j=1}^{n} u_j H_{ij} = \sum_{j=1}^{m} q_j G_{ij} + B \tag{6}$$

where,

$$C_p = \frac{1}{2} \qquad \text{for smooth boundary}$$

$$H_{ij} = \int_\Gamma \frac{\partial w}{\partial n} d\Gamma \qquad G_{ij} = \int_\Gamma w \, d\Gamma \tag{7}$$

$$q_j = \left(\frac{\partial u}{\partial n}\right)_j \qquad B = 2\int_\Omega d\Omega$$

The above equation may be expressed in an equivalent matrix notation as follows.

$$[H]\{u\} = [G]\{q\}+\{B\} \tag{8}$$

When a prismatic bar with a multiply connected domain (Figure 1) is examined for torsional stress distribution, the boundary conditions change as follows,

$$u = 0 \text{ on } \Gamma, \quad u = u_o \text{ on } \Gamma_o \tag{9}$$

$$\int_{S_o} \partial u/\partial n \, ds = 2\, D_o \text{ on } \Gamma_o$$

where, D_o is the area of the internal hole. The torsional rigidity of the section as obtained as follows.

$$K = 2\int_\Omega u \, d\Omega + 2\, u_o D_o \tag{10}$$

Note that the multiply-connected domain introduces one additional unknown u_o and hence also increases the dimensionality of (8) by one.

The boundary element method for elasticity problems is based on Somigliana's identity, and is given by

$$C_{ij}\, u_j(x_o) = \int_\Gamma (p(x)\, G_{ij}(x,x_o) - F_{ij}(x,x_o).u_j(x)]d\Gamma$$
$$+ \int_\Omega b.\, G_{ij}(x,x_o)\, d\Omega \tag{11}$$

where, C_{ij} is a coefficient that depends on the geometry of the boundary at point x_o; $u_j(x_o)$ is the displacement at point x_o in the j direction p,b are the actual state of traction and body force; $G_{ij}(x,x_o)$ and F_{ij} represent the fundamental Kelvin solution for displacement and traction, respectively.

The fundamental solutions are written as follows.

$$G_{ij} = \frac{1}{8\pi G(1-\nu)} \left[(3-4\nu)\ln\frac{1}{r}\delta_{ij} + \frac{\partial r}{\partial x_i}\frac{\partial r}{\partial x_j} \right]$$

$$F_{ij} = -\frac{1}{4\pi G(1-\nu)r} \left[\frac{\partial r}{\partial n} \{(1-2\nu)\delta_{ij} + 2\frac{\partial r}{\partial x_i}\frac{\partial r}{\partial x_j}\} - (1-2\nu)\left(\frac{\partial r}{\partial x_i} n_j - \frac{\partial r}{\partial x_j} n_i\right) \right] \tag{12}$$

The analysis described above was combined with nonlinear programming (NLP) based optimization algorithms such as the feasible-usable search directions approach and an exterior penalty function strategy, to solve several shape design problems. The general optimization problem can be stated as a requirement to

$$\text{Minimize } F(\bar{d})$$

$$\text{Subject to, } g_j(\bar{d}) < 0, \quad h_k(\bar{d}) = 0, \text{ and, } \quad \bar{d}_i^L \leq \bar{d}_i \leq \bar{d}_i^u$$

where $F(\bar{d})$ is a scalar objective function, $g_j(\bar{d})$ and $h_k(\bar{d})$ are inequality and equality constraints on the response quantities, and $\bar{d}_i^L$ and $\bar{d}_i^u$ are prescribed lower and upper bounds on the design variables, respectively. Two distinct strategies were implemented in the present effort. The first was a traditional approach where the analysis equations were solved for the response, and the latter used by the optimizer to produce a new design. A full nonlinear treatment of the constraints and a piecewise linear approximation was attempted. The second approach, for which only preliminary results are available, solved the analysis and optimization problems simultaneously, treating the analysis equations as additional equality constraints and response quantities as additional design variables. Approximation concepts were used extensively in this implementation, and are discussed in subsequent sections.

2.1. Approximation Strategies

The use of approximation concepts was explored in the context of the shape design problems. Such approximations are pertinent to both the BEM analysis and to the optimization problem, and these are inherently coupled. The selection of appropriate design variables is perhaps most critical in the shape design problem. The choice of specific node coordinates for this task may appear attractive for ease in implementation, but is not computationally viable. This is linked to the high cost of determining design sensitivity with respect to these variables. Further, experience clearly shows that allowing this degree of design freedom produces shapes that are unacceptable from a manufacturing standpoint. Two strategies were implemented in this exercise. The first was based on allowing individual nodes to move along predetermined radial lines, with the distance from a reference point chosen as the design variable. The other approach consisted of selecting suitable shape functions to describe the boundaries, and the choice of coefficients for such curves were the designated design variables in the problem. Both these approaches had the effect of reducing the overall number of variables, and hence the total computational effort.

When considering the shape design problem for allowable stress limits, the number of constraints were of the order of number of points where the analysis was performed. Further, when using the integrated approach in which the analysis equations are considered as additional equality constraints, the number of constraints were increased by a number equal to twice the number of equality constraints. To render this problem more computationally manageable, a cumulative constraint that represents a large number of constraints by a single measure [4], was adopted for the present task.

Finally, when BEM is used in an iterative manner as in the optimization problem, careful attention must be paid to the structure of the system matrices. In particular, it is relatively easy to identify those sections of the system matrix that must be modified when a shape variable is changed. This is illustrated as follows

$$\begin{bmatrix} H_{11} & H_{12} & H_{1k-1} & H_{1k} & H_{1k} & \cdots & H_{1n} \\ \vdots & \vdots & \cdot & \cdot & \cdot & & \\ H_{k11} & H_{k-12} & & & & & \\ H_{k1} & H_{k2} & \cdot & \cdot & \cdot & \cdots & \\ H_{k+11} & H_{k+12} & & & & & \\ \vdots & & \cdot & \cdot & \cdot & & \\ H_{n1} & \cdots & H_{nk-1} & H_{nk} & H_{nk+1} & \cdots & H_{nn} \end{bmatrix}$$

The above matrix for the solution of a torsion problem with constant elements clearly shows the banded portion of the matrix that is changed due to the perturbation of the boundary at the k-th element. Furthermore, the choice of shape variables should be based on the nature of terms involved in the boundary integrals. If, the G and H integrals can be written in terms of constants or combinations of constants that define the interpolating polynomial, then a proper choice of an intermediate variable would allow the system matrices to be extrapolated during the design process, and would avoid the computational cost of reformulation. This is explained better by considering a term from the G matrix and noting its linear dependence on ln (1/r), where r could be considered a shape variable.

$$G = \sum_{4} \ln \left(\frac{1}{r}\right) \; w_1 \; \frac{(\text{length of element})}{2}$$

The choice of ln (1/r) as an intermediate design variable improves the numerical efficiency of the optimization problem.

3. DISCUSSION OF NUMERICAL RESULTS

The first problem attempted in the optimum design study was to configure the boundary of a prismatic bar for maximum torsional rigidity and subject to a constraint of a fixed cross sectional area of 50 units, and an additional requirement that the bending inertia about the horizontal axis be greater than 350 units. The definition of design variables, geometry descriptors, and the initial and final designs for this shaft are shown in Figures 2 & 3. Six design variables were used to obtain the optimum design, using both a linear piecewise approximation to the constraint and objective functions, and a fully nonlinear treatment of these functions. The stress function was evaluated at 64 points. The CPU time requirements for the two approaches were comparable.

A multiply-connected domain problem was introduced in the form of a torsional shaft with a hole, as shown in Figure 4. There were fourteen design variables in this problem. The torsional rigidity and bending inertia requirements were similar to the solid shaft with additional requirements to constrain the total and inner hole section areas to 40 and 20 units, respectively. A total of 48 nodes was used in the analysis, with a converged result obtained in 10 iterations. The final design is shown in Figure 4.

Two plane stress problems were also attempted in this effort. A torque arm shown in Figure 5, was sized for minimum weight and subject to maximum allowable von Mises stress of 62,000 N/cm^2. The maximum stresses in this structure occur at the boundaries. A total of 133 linear boundary elements were used in the analysis, with six design variables used to define the geometry. Three additional geometry constraints were imposed to prevent boundary intersections during redesign. The optimization problem was solved using an exterior penalty function approach, and the evolution of the design is shown in Figure 5.

The second plane stress problem attempted in this work is somewhat of a classic, and requires a minimum area fillet to sustain a uniaxial distritibuted load. The initial and final design for this structure are shown in Figures 6 and 7, and represent two different set of design variables. The first set of results (Figure 6) was obtained for a single shape variable which defined the location at which a parabola from point A would meet the horizontal line OB with zero slope. The second set (Figure 7) was to determine the location of points A and B on the vertical and horizontal axis, which could be connected by an elliptical curve, and satisfy the von Mises stress requirements. This problem, was also attempted by defining design variables as radial locations from a fixed reference point. An optimum design could not be obtained for this set because of severe oscillations in the domain boundary, as shown in Figure 8. This problem also reinforced the need to use an adaptive node generation capability in the BEM based optimum design.

4. ACKNOWLEDGEMENTS

This work was partly supported by the United States Air Force Office of Scientific Research, under research grant No. S-760-6MG-002.

5. REFERENCES

[1]. P.K. Bannerjee and R. Butterfield, Boundary Element Methods in Engineering Science, McGraw Hill, New York, 1981.

[2]. C.A. Mota Soares, H.C. Rodrigues, L.M. Oliveira Faria and E.J. Haug, 'Optimization of Geometry of Shafts Using Boundary Elements', J. Mech. Transm. Autom. Des., 106, 1984.

[3]. T. Burczynski and T. Adamczyk, 'The Boundary Element Formulation for Multiparameter Structure Shape Optimization', App. Math Modelling, 9, 1985.

[4]. P. Hajela, Techniques in Optimum Structural Synthesis With Static and Dynamic Constraints, Ph. D. Thesis, Stanford University, 1982.

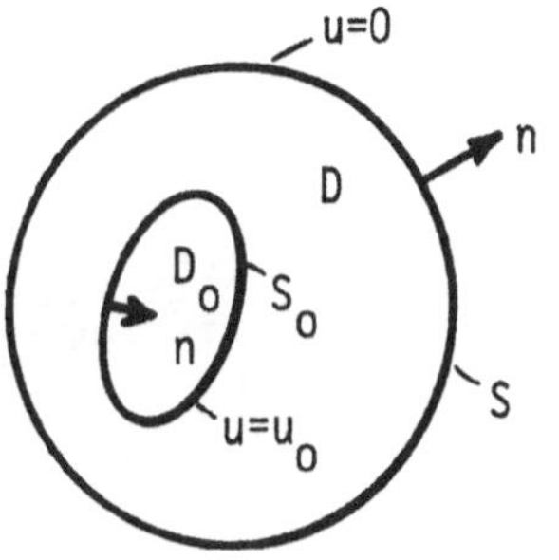

Figure 1. A multiply-connected domain

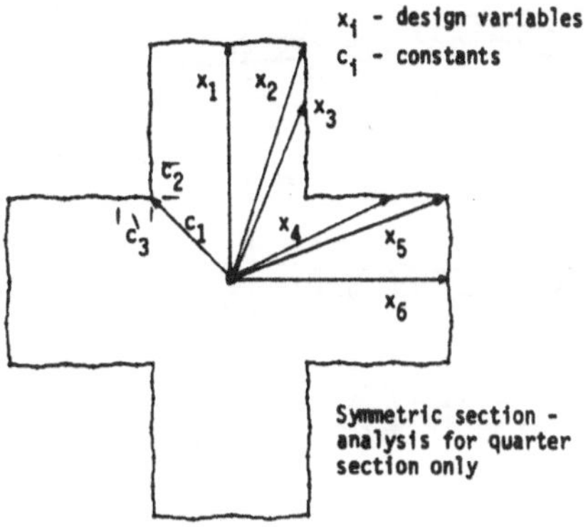

Figure 2. Initial design and design variable definition for torsional shaft

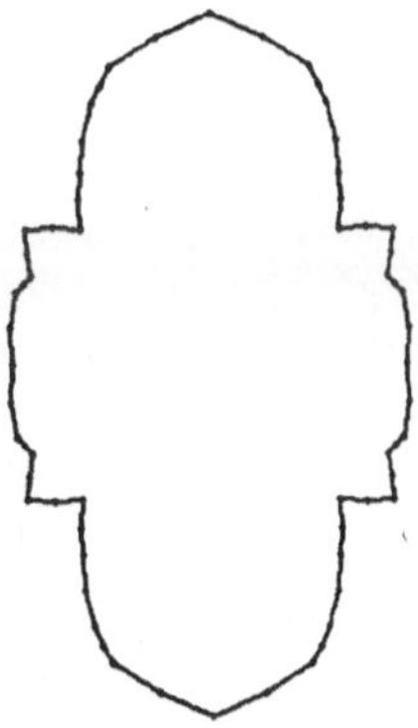

Figure 3. Final design for torsional shaft

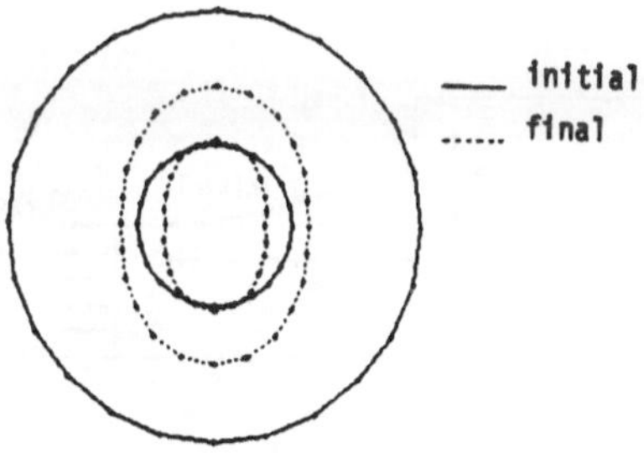

Figure 4. Initial and final design for torsional shaft

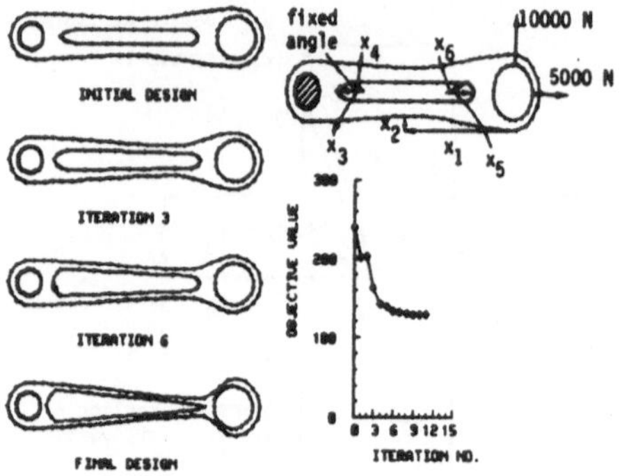

Figure 5. Design variable definition and iteration history for torque arm

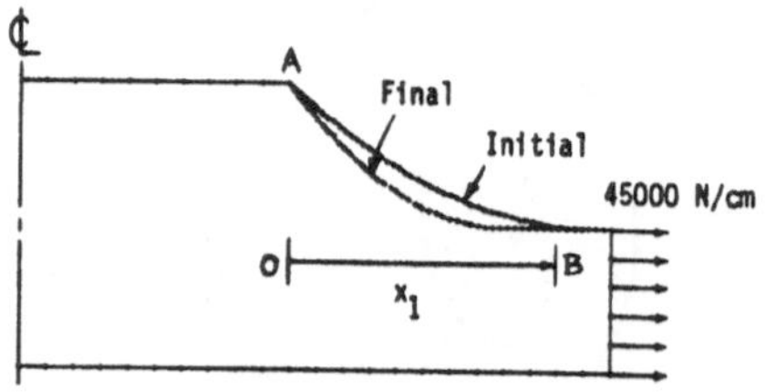

Figure 6. Initial and final design of fillet (parabolic interpolation)

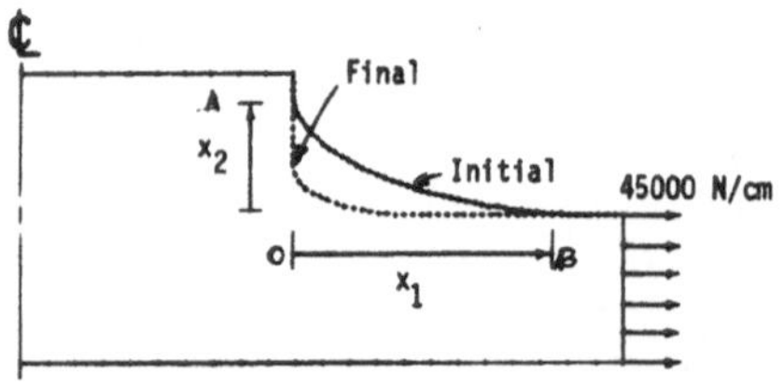

Figure 7. Initial and final design of fillet (elliptic interpolation)

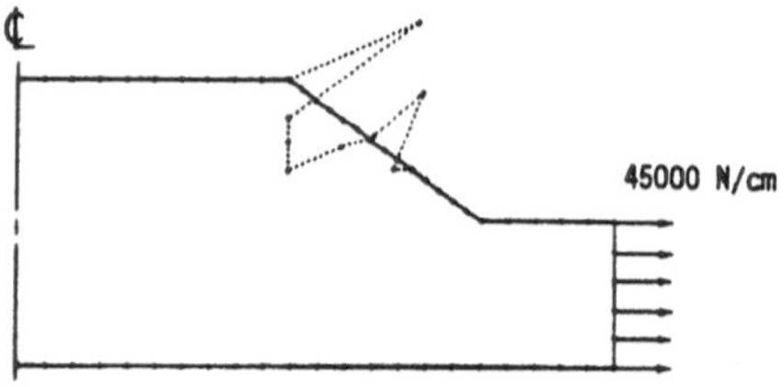

Figure 8. Design generated when node locations are used as design variables

A MICHELL TYPE CRITERION FOR SHELLS

W.S. Hemp

ABSTRACT. A lower bound, depending upon a restricted virtual displacement system, is found for the volume of material required by thin shells, which carry given loads, including hydrostatic pressure, to given supports. The loads are carried by membrane stress-resultants and Tresca's safety criterion is satisfied. Sufficient conditions for the attainment of the bound lead to optimum designs in some simple cases, namely those for which the optimum structure is indeed a thin shell and not a three-dimentional continuum.

1. OPTIMUM DESIGN OF SHELLS

A shell is determined by its middle surface and the distribution of its thickness t. Its field of membrane stress-resultants is determined by the given forces applied to it and the shape of its middle surface.† The thickness distribution necessary to ensure safety follows from a strength criterion, such as that of Tresca, which will be used here. The volume of material required V can then be expressed as a functional of the shape of the middle surface and that shape can be determined for the optimum shell by making V a minimum.

This direct approach leads to a problem in the calculus of variations and is indeed a perfectly feasible method for the solution of optimum shell problems. The present paper, however, attacks the problem in the manner of Michell by seeking a lower bound to V_{min} and formulating sufficient criteria for this bound to be equal to V_{min}. Means are thus provided for the identification of optimum shells in certain cases.

The bound is a known multiple of the virtual work W of the external forces, acting on the competing shell designs, resulting from a virtual deformation of the space in which the shells must lie. This work must be the same for all the shells considered and this requirement imposes restrictions on the virtual deformation to be used (section 2).

The virtual deformation must also be restricted in magnitude. If ε_i (i=1,2,3) are the principal strains, then, for the whole space considered,

† The contribution of 'edge effects' to the 'cost' V is neglected.

G. I. N. Rozvany and B. L. Karihaloo (eds.), Structural Optimization, 117–123.

$$|\varepsilon_i| + |\varepsilon_j| + |\varepsilon_i + \varepsilon_j| \leq 2\varepsilon, \ (i,j=1,2,3;\ i \neq j). \tag{1.1}$$

On the basis of these assumptions, it will be shown that, for all feasible shells,

$$W/\sigma\varepsilon \leq V \tag{1.2}$$

where σ is the allowable tensile stress used in the Tresca criterion. It follows that

$$W/\sigma\varepsilon \leq V_{min} \tag{1.3}$$

for the optimum shell.

2. RESTRICTION OF THE VIRTUAL DISPLACEMENT

The virtual deformation must satisfy equation (1.1) as well as satisfying the kinematic conditions imposed upon the shells, like fixed edges, where the displacements must vanish.

If the external forces are given in position and direction, then W will be the same for all shells. However in the case of shells loaded by hydrostatic pressure p, which in general may vary through space, the condition that W is the same for all shell design gives, see Fig. 1, for any two shells (1) and (2),

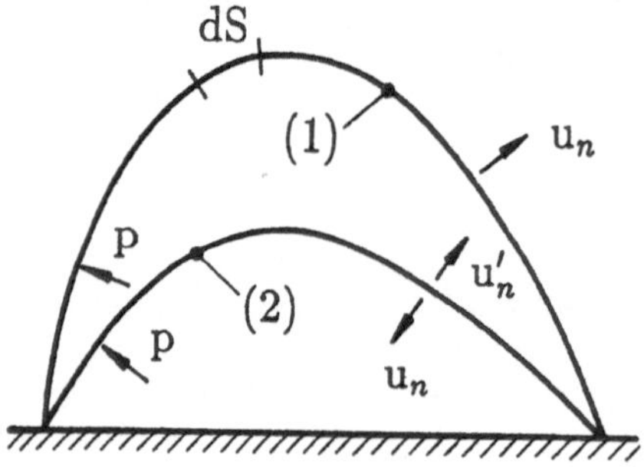

Fig. 1

$$W = \int_{(1)} pu_n dS = \int_{(2)} pu'_n dS$$

or

$$\int_{(1)+(2)} pu_n dS = 0 \tag{2.1}$$

where u_n and u'_n ($=-u_n$) are normal displacements at the shells. Equation (2.1) implies that

$$\int_{V_{12}} \operatorname{div}(p\underset{\sim}{u})\, dV = 0 \tag{2.2}$$

where $\underset{\sim}{u}$ is the vector displacement and dV an element of the volume V between shells (1), (2). Since the shells can take any shape, it follows that the virtual displacement must satisfy †

$$\operatorname{div}(p\underset{\sim}{u}) = 0. \tag{2.3}$$

3. PRINCIPAL MEMBRANE STRAINS IN THE SHELLS

The principal membrane strains e_1 and e_2 in a shell at any point of the

† In the case when the volume enclosed by the shell is prescribed $\operatorname{div}(p\underset{\sim}{u})$ = constant.

middle surface with a unit normal (ℓ,m,n) refered to directions of ε_i (i=1,2,3) at that point are determined by

$$\underset{L,M,N}{\text{Stat.}}\ e = \varepsilon_1 L^2 + \varepsilon_2 M^2 + \varepsilon_3 N^2 , \tag{3.1}$$

subject to,

$$L\ell + Mm + Nn = 0,\ L^2 + M^2 + N^2 = 1 , \tag{3.2}$$

where (L,M,N) is the direction of the direct strain e. Introducing Langranian multipliers λ,μ gives the equivalent problem

$$\text{Stat.}\ e = \varepsilon_1 L^2 + \varepsilon_2 M^2 + \varepsilon_3 N^2 - 2\lambda(L\ell + Mm + Nn) - \\ - \mu(L^2 + M^2 + N^2 - 1), \tag{3.3}$$

which yields

$$(\varepsilon_1-\mu)\ L = \lambda\ell,\ (\varepsilon_2-\mu)\ M = \lambda m,\ (\varepsilon_3-\mu)\ N = \lambda n \tag{3.4}$$

and, by (3.2) and (3.1),

$$\mu = e,\ \lambda = \varepsilon_1 L\ell + \varepsilon_2 Mm + \varepsilon_3 Nn . \tag{3.5}$$

The determinant of the equations (3.4) and the first of (3.2) for L, M, N and λ must vanish since L = M = N = 0 is impossible by the second of (3.2). This yields an equation for $\mu = e$, with roots e_1 and e_2, namely

$$e^2 -\{\ell^2(\varepsilon_2 + \varepsilon_3) + m^2(\varepsilon_3 + \varepsilon_1) + n^2(\varepsilon_1 + \varepsilon_2)\}\ e + \\ + \ell^2\varepsilon_2\varepsilon_3 + m^2\varepsilon_3\varepsilon_1 + n^2\varepsilon_1\varepsilon_2 = 0 . \tag{3.6}$$

Equation (1.1) can be written in the equivalent form

$$|\varepsilon_i| \le \varepsilon\ (i = 1,2,3),\quad |\varepsilon_i + \varepsilon_j| \le \varepsilon\ (i,j = 1,2,3;\ i\neq j) \tag{3.7}$$

as may be seen by the geometrical representation of Fig. 2, which shows the hexagonal region defined by (1.1).

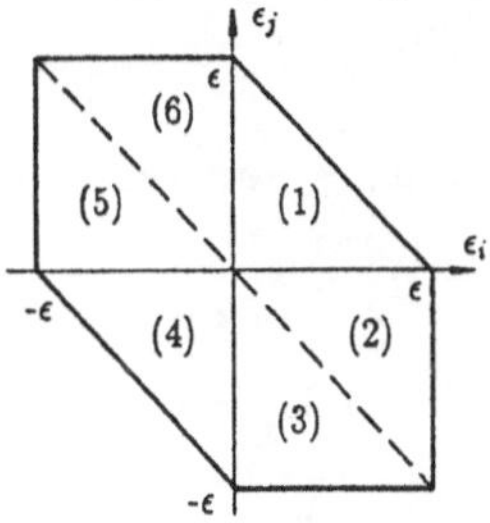

Fig. 2

Equations (3.1) and the second of (3.2)together with (3.7), shows that any direct strain e satisfies

$$|e| \le |\varepsilon_1|L^2 + |\varepsilon_2|M^2 + |\varepsilon_3|N^2 \le \varepsilon$$

and hence that

$$|e_1| \leq \varepsilon, \; |e_2| \leq \varepsilon \tag{3.8}$$

Equation (3.6) gives

$$e_1 + e_2 = \ell^2(\varepsilon_2 + \varepsilon_3) + m^2(\varepsilon_3 + \varepsilon_1) + n^2(\varepsilon_1 + \varepsilon_2)$$

and so

$$|e_1 + e_2| \leq \ell^2|\varepsilon_2 + \varepsilon_3| + m^2|\varepsilon_3 + \varepsilon_1| + n^2|\varepsilon_1 + \varepsilon_2|$$

which by (3.7) gives

$$|e_1 + e_2| \leq \varepsilon \, . \tag{3.9}$$

Finally, by the arguments used to derive (3.7), (3.8,9) are equivalent to

$$|e_1| + |e_2| + |e_1 + e_2| \leq 2\varepsilon \, . \tag{3.10}$$

The condition of (3.10) may also be written as

$$e_1 = \lambda_1 - \mu_1 + \lambda_3 - \mu_3, \; e_2 = \lambda_2 - \mu_2 - \lambda_3 + \mu_3, \tag{3.11}$$

subject to

$$\sum_{i=1}^{3} (\lambda_i + \mu_i) \leq \varepsilon \; ; \; \lambda_i, \mu_i \geq 0 \; (i=1,2,3) \, . \tag{3.12}$$

It is clear by direct substitution from (3.11), using (3.12), that

$$|e_1| + |e_2| + |e_1 + e_2| \leq 2\sum_{i=1}^{3} (\lambda_i + \mu_i) \leq 2\varepsilon \, . \tag{3.13}$$

Conversely, given (3.10), values of λ_i, μ_i (i=1,2,3) can be found for each of the regions (1) to (6), shown on Fig. 2, such that (3.11,12) are satisfied. These values are given in the following table:

	Case	Values of non-zero λ_i, μ_i (i=1,2,3)
(1)	$e_1 \geq 0, e_2 \geq 0, e_1 + e_2 \leq \varepsilon$	$\lambda_1 = e_1$, $\lambda_2 = e_2$
(2)	$e_1 \leq \varepsilon, e_2 \leq 0, e_1 + e_2 \geq 0$	$\lambda_1 = e_1 + e_2$, $\lambda_3 = -e_2$
(3)	$e_1 \geq 0, e_2 \geq -\varepsilon, e_1 + e_2 \leq 0$	$\lambda_3 = e_1$, $\mu_2 = -e_1 - e_2$
(4)	$e_1 \leq 0, e_2 \leq 0, e_1 + e_2 \geq -\varepsilon$	$\mu_1 = -e_1$, $\mu_2 = -e_2$
(5)	$e_1 \geq -\varepsilon, e_2 \geq 0, e_1 + e_2 \leq 0$	$\mu_1 = -e_1 - e_2$, $\mu_3 = e_2$
(6)	$e_1 \leq 0, e_2 \leq \varepsilon, e_1 + e_2 \geq 0$	$\lambda_2 = e_1 + e_2$, $\mu_3 = -e_1$

4. TRESCA'S CRITERION OF SAFETY

Let the membrane stress-resultants in a shell, refered to the orthogonal coordinates defined by the lines of the principal strains e_1 and e_2, be T_1, T_2, and S. The maximum shear stress must be not greater than $\sigma/2$ in absolute value and so

$$(T_1^2/4 + S^2/2)^{1/2} \leq \sigma t/2, \quad (T_2^2/4 + S^2/2)^{1/2} \leq \sigma t/2,$$
$$\{(T_1 - T_2)^2/4 + S^2\}^{1/2} \leq \sigma t/2 , \tag{4.1}$$

where the first two impose Tresca's condition on the shear stress in planes at $\pi/4$ to the middle surface and the last does the same for the greatest membrane shear stress. It thus follows that

$$|T_1| \leq \sigma t, \quad |T_2| \leq \sigma t, \quad |T_1 - T_2| \leq \sigma t \tag{4.2}$$

or alternatively,

$$\begin{aligned} T_1 + p_1 &= \sigma t , & -T_1 + q_1 &= \sigma t , \\ T_2 + p_2 &= \sigma t , & -T_2 + q_2 &= \sigma t , \\ T_1 + T_2 + p_3 &= \sigma t , & -T_1 + T_2 + q_3 &= \sigma t , \end{aligned}$$
$$\text{where } p_i,\ q_i \geq 0 \ (i=1,2,3) \tag{4.3}$$

Equations (4.2) or (4.3) mean that (T_1,T_2) must lie inside or on the well known Tresca hexagon. (Fig. 3)

5. THE LOWER BOUND

Imposing the virtual strain on any shell, which is safely in equilibrium with the given external forces, gives, since the virtual displacement vanishes at the supports, the following expression for the common virtual work W:

$$W = \int (T_1 e_1 + T_2 e_2) dS , \tag{5.1}$$

where the integral is taken over the whole middle surface. Substituting from (3.11) for the strains e_1 and e_2 and from (4.3) for the stress-resultants gives

$$W = \int \{\lambda_1 T_1 + \mu_1(-T_1) + \lambda_2 T_2 + \mu_2(-T_2) + \lambda_3(T_1 - T_2) + \mu_3(-T_1 + T_2)\} dS$$
$$= \int \{\sigma t \sum_{i=1}^{3} (\lambda_i + \mu_i) - \sum_{i=1}^{3} (\lambda_i p_i + \mu_i q_i)\} dS \tag{5.2}$$

and so by (3.12) and (4.3)

$$W \leq \sigma\varepsilon \int t dS = \sigma\varepsilon V , \tag{5.3}$$

which is (1.2). Since this applies to any safe shell, including the optimum shell, the lower bound $W/\sigma\varepsilon$ of (1.3) has been established.

6. SUFFICIENT CONDITIONS FOR OPTIMUM SHELLS

Conditions for an optimum can now be formulated. These are conditions which insure equality in (5.3). A sufficient set are as follows:

(1) One of the inequalities of (1.1) for the principal virtual strains must be satisfied as an equality on the middle surface of the shell. For definiteness take that for i=1, j=2.

(2) The principal planes corresponding to the strains ε_1 and ε_2 must be tangent planes at all points on the middle surface. This means by (3.6), with ℓ=m=0, n=1, that $e_1 = \varepsilon_1$ and $e_2 = \varepsilon_2$ and so, by (1), (3.10) is satisfied as an equality.

(3) Equation (5.2) shows that the conditions

$$\sum_{i=1}^{3} (\lambda_i + \mu_i) = \varepsilon \ , \quad \lambda_i p_i = \mu_i q_i = 0 \ (i=1,2,3) \tag{6.1}$$

must be imposed, since all λ_i, μ_i, p_i, q_i are ≥ 0. Since $p_i > 0$, $q_i > 0$, (i=1,2,3) gives $\lambda_1 = \mu_1 = 0$ (i=1,2,3), which contradicts the first of (6.1), it follows that t must be chosen so that (T_1, T_2) satisfies at least one of (4.3) and so lies on the Tresca hexagon. This means by (4.1) that S=0 and so the principal directions of the stress-resultants must coincide with the principal directions of the virtual strains i.e. the directions of e_1 and e_2. It also follows, as in plasticity theory, that the virtual strains must be 'normal' to the Tresca hexagon at the point (T_1, T_2). For example on the side p_1=0 (see Fig. 3), (6.1)gives $\lambda_1 = \varepsilon$ and so $e_1 = \varepsilon$, e_2=0, a direction normal to this side. On the side p_3=0 it follows that $\lambda_3 = \varepsilon$ and $e_1 = \varepsilon$, $e_2 = -\varepsilon$ which is again normal. At a corner $p_1 = p_2 = 0$, $\lambda_1 + \lambda_2 = \varepsilon$ and $e_1 = \lambda_1$, $e_2 = \lambda_2$ and so any direction between $(\varepsilon, 0)$ and $(0, \varepsilon)$ is allowed. Similarly at a corner $p_1 = p_3 = 0$, $\lambda_1 + \lambda_3 = \varepsilon$ and $e_1 = \varepsilon$, $e_2 = -\lambda_3$ and so any direction between $(\varepsilon, 0)$ and $(\varepsilon, -\varepsilon)$ can occur. Other sides and corners give similar results.

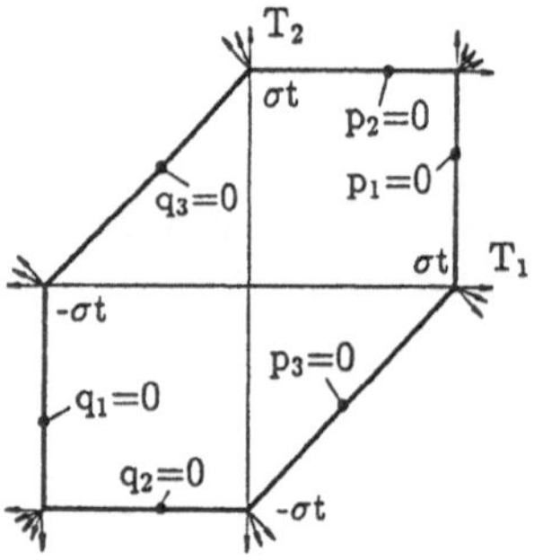

Fig. 3

7. EXAMPLES

Examples of simple shell designs, which can be shown to be optimum shells, using the criteria of equation (1.1) and sections 2 and 6 are: (1) the closed pressure vessel of spherical form, (2) Michell's shell for torsion (Ref. 1), which is part of a sphere, (3) a spherical dome, with base angle $\pi/3$, fixed to a circular base and under uniform pressure, (4) a conical shell with semi-angle less than $\pi/4$, carrying a vertical load to a circular base and (5) a conical roof, of semi-angle $\pi/4$, carrying uniform snow loading, after a 'vertical fall' of snow.

In all these cases, a displacement field can be found, which satisfies (1.1); which, when appropriate, satisfies a condition of the type considered in section 2 and which, together with the membrane solution for these loaded shells, satisfies the sufficient conditions of section 6.

8. GENERALISATION

In all the examples of section 7 a virtual displacement field can be found such that $W/\sigma\varepsilon = V_{min}$ and so optimality is established for the assumed shell. The generality of this method must now be considered. Take a virtual displacement field, satisfying (1.1) and a restriction of the kind imposed in section 2, but otherwise completely general and seek the maximum of the virtual work W for some problem of shell design. The theory developed above shows that

$$W_{max}/\sigma\varepsilon \leq V_{min}, \tag{8.1}$$

assuming the existence of $V_{min.}$, but there is no guarantee of equality. To establish equality it must be shown that max. W and min. V are dual principles. However the dual of max. W, which is subject to conditions in three dimensions, like (1.1) and (2.3), must lead in general to a principle min. V for structures made up of material distributed through finite regions of space, with pressure loading distributed as body forces and not necessarily giving minima with material concentrated in shells.

The analogous case of Michell frameworks can be cited (Ref. 2 pp. 86-8). The problem (4) of section 7, for height/base radius $\geq$ 0.765, is solved by a continuum of compression members lying on a cone. However for height/base radius < 0.765, the optimum solution involves both a cone filled with tension members, as well as a conical-cum-spherical shell. A section of the structure for height = 0 is shown in Fig. 4.

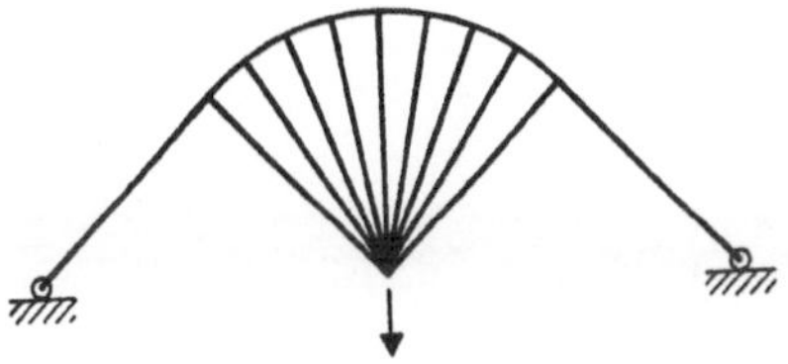

Fig. 4

It would seem therefore that a generalisation to a max. W principle will only yield an optimum shell, when it is not possible to find a structure with smaller V, made up using, in part, a finite distribution of material. The lower bound is of course valid in all cases.

9. REFERENCES

1. Michell, A.G.M., (1904). 'The limit of economy of material in frame structures'. Phil. Mag. 8, (4).

2. Hemp, W.S., (1973). Optimum Structures. Clarendon Press, Oxford.

SHAPE OPTIMIZATION OF THE CROSS-SECTIONS OF THINWALLED BEAMS SUBJECTED TO BENDING AND SHEAR

M. Hýča
National Research Institute of Machine Design (SVÚSS)
Dept. of Loading Capacity and Service Life of Structures
Husova 8
110 00 Praha 1
Czechoslovakia

ABSTRACT. This paper presents the general condition for zero warping of symmetrical cross-sections of thinwalled beams under bending and shear. The condition derived within the linear elastic analysis is represented by a nonlinear functional equation coupling the function of the wall-thickness distribution with the Cartesian coordinates of points of the cross-sectional center line which are parallel to the plane of the bending moment. The solution of the corresponding functional problem, sought with the use of the discrete contracting mapping, makes it possible to determine the optimum cross-sections of beam-like structural elements which do not warp under bending and shear, the beam thus remaining free from shear-lag irrespective of the beam end conditions and the transverse loading variation.

1. INTRODUCTION

With the bending and shear of thinwalled beam-like structural elements reinforced by diaphragms the secondary axial membrane stresses are generated due to the restraint of warping of cross-sections. These secondary stresses, representing self-balanced internal force system, are superimposed on the primary axial membrane stresses, corresponding to the fictious displacement field arising at free warping of cross-sections (these primary membrane stresses being determinable with the help of the classical bending theory, founded on the Bernoulli hypothesis of plane sections). This phenomenon, known as shear-lag or shear diffusion, and often interpreted as the effective width problem of wide beam flanges is most marked in built-in cross-sections and in cross-sections in the region of abrupt changes of the loading or internal shear force, particularly with beams at low values of the slenderness ratio or if the load is acting at a relatively small distance from a supported or built-in cross-section compared with the length of a beam.

Since the loading capacity of considered beams may decrease due to

G. I. N. Rozvany and B. L. Karihaloo (eds.), Structural Optimization, 125–134.

the shear-lag phenomenon (especially as far as the fatigue behaviour of welded beams and girders of notch sensitive material exhibiting high crack propagation rate is concerned |e.g. 1,2,3| or in structures of low brittle fracture strength) the nonwarpable cross-sections may sometimes be preferrable to the warpable ones. Thus the problem of theoretical determination of the cross-sections warpless in bending becomes important in practice.

To the author's knowledge the problem of the optimization of thin-walled beam cross-sections with regard to the shear-lag effect in bending, which consists in minimization of maximum axial membrane stress by means of the elimination of bending warpability of cross-sections, has not yet been solved in sufficiently general way, the only warpless-in--bending thinwalled cross-section known till now being represented by an annular cross-section |4|. The condition for oval cross-section to be warpless in bending which uses a varying stringer-pitch in fuselage-like sections was presented by Williams in |5|. A possibility of reducing the unfavourable effect of the shear-lag by enlarging the flange thickness in the areas adjacent to the vertical webs of rectangular box-section girders was investigated by Křístek et al in |6|.

2. FUNCTIONAL CONDITION FOR ZERO WARPING AND ITS SOLUTION

The theory for determining symmetrical cross-sections nonwarping in bending, not necessitating the initial limitation on the cross-sectional shape and assuming the walls fully effective in transmitting both axial membrane stress and shear flow, has been proposed by the author in |8,9|.

2.1. Assumptions and Basic Relations

Consider a thinwalled beam of a symmetrical, open or closed cross-section subjected to plane bending and shear due to a transverse loading, p_z, with arbitrary discontinuities and under various possible end conditions as shown in fig. 1. Adopt the linear elastic analysis, limited to beams with nonbuckling cover, and assume that (a) the cross-sections are undeformable in their planes, the deformation of an arbitrary cross--section thus consisting of a rigid translation and rotation of the plane of section and a warping, i.e., displacement perpendicular to the plane of the cross-section; (b) the bending stiffness of the thin wall may be neglected; (c) the beam wall-thickness is small compared to cross-sectional dimensions and it may vary round the cross-sectional perimeter but is constant along the beam; (d) the material of the beam is homogeneous and linearly elastic. The necessary and sufficient conditions that must be met in order to obtain the cross-section which is nonwarpable in bending regardless of the transverse loading and end conditions (the beam thus remaining free from shear-lag) can then be derived in the form of a nonlinear functional problem (1), (2) in which S denotes the cross-sectional perimeter; s is the length of the segment of cross--sectional center line which determines the position of points on the center line; $t = t(s)$ is the wall-thickness at the center line point in distance s; $z = z(s)$ stands for the Cartesian coordinate, parallel to the plane of the bending moment, of the same point; and $\dot{z} = dz/ds$.

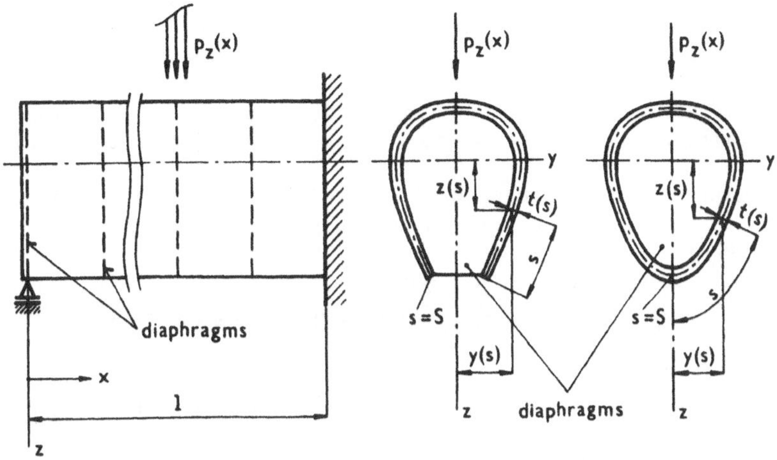

Figure 1. Thinwalled beam with symmetrical open or closed cross-section subjected to arbitrary end conditions and arbitrary transverse loading, p_z, in the x-z plane of symmetry.

$$\dot{z}\,t\int_0^{S/2}\left(\int_0^{\xi} z\,t\,d\nu\right)^2\frac{d\xi}{t} + \int_0^{s} z\,t\,d\nu \cdot \int_0^{S/2} z^2 t\,d\nu = 0, \qquad s \in \langle 0\,;S/2 \rangle \tag{1}$$

$$\int_0^{S/2} z\,t\,ds = 0\,, \tag{2}$$

Given the function t(s), the z-coordinate of the points of the nonwarpable cross-section center line can be determined by solving the nonlinear functional problem (1), (2) or, conversely, the function z(s) being given, the wall-thickness t(s) of the nonwarpable cross-section can be found analogously |10|.

Since the inequality $t(s) > 0$ is always valid, the equation (1) implies the following permissibility conditions for the shape of the center line of the symmetrical nonwarping cross-sections |10|

$$\dot{z}(s=0) = \dot{z}(s=S/2) = 0\,, \qquad \dot{z}(s \in (0\,;S/2)) < 0\,, \tag{3}$$

which are valid for any z(0). Thus the solution $t(s) > 0$ of the functional problem (1), (2) exists only in the class of the center lines whose coordinates z(s) meet the conditions (3). Some examples of the permissible center lines of bisymmetrical cross-sections nonwarpable under bending and shear in the x-z plane of the beam are shown in |10|. It then holds that the cross-sections, whose center lines do not meet the conditions (3), will always warp under bending and shear in the x-z plane of the beam, irrespective of the pattern of the wall-thickness, t(s). The I-, T-, rectangular channel- or closed box-section may serve as an example. Some other shapes of cross-sections strongly warpable under bending and shear are in |10,11|. The conditions (3) represent, therefore, the necessary conditions for symmetrical cross-sections to be nonwarpable under bending and shear in the x-z plane of the beam, thus providing the fundamental rules of their optimization |11,13|.

2.2. Existence and Uniqueness of the Solution to the Functional Problem (1), (2)

The existence and uniqueness of the solution to the functional problem (1), (2) can easily be proved for only $t(s) \equiv$ const. The equation (1) then reads

$$\dot{z}.\int_0^{S/2}\left(\int_0^{\xi} z\,d\nu\right)^2 d\xi + \int_0^{s} z\,d\nu.\int_0^{S/2} z^2 d\nu = 0, \qquad s \in \langle 0;S/2\rangle . \tag{4}$$

The unknown function, $z(s)$, may be assumed to have the form of

$$z(s) = \frac{z_0}{2} + \sum_{i=1,2,\dots}^{n} z_i.\cos 2\pi i \frac{s}{S} . \tag{5}$$

The condition (2) then yields

$$z_0 = 0 . \tag{6}$$

The remaining coefficients, $z_1, z_2, \dots, z_n$, in eq. (5), determining the hitherto unknown shape of the center line of the cross-section having $t(s) \equiv$ const., will be determined by eq. (4) which, after substituting from eqs. (5) and (6), yields

$$\sum_{i=1,2,..}^{n} \frac{z_i}{i} \sum_{j=1,2,..}^{n} z_j^2 \sin 2\pi i \frac{s}{S} = \sum_{i=1,2,..}^{n} i z_i \sum_{j=1,2,..}^{n} \left(\frac{z_j}{j}\right)^2 \sin 2\pi i \frac{s}{S} . \tag{7}$$

The condition of equality of these two truncated series is therefore

$$\sum_{j=1,2,..}^{n} z_j^2 - i^2 \sum_{j=1,2,..}^{n} \left(\frac{z_j}{j}\right)^2 = 0, \qquad i = \overline{1,n} . \tag{8}$$

However, eq. (8) can only be met if $i = 1$, i.e. for $n = 1$, the magnitude of coefficient z_1 becoming arbitrary. Considering the identity (6), the relation (5) then implies that the functional equation (4) will hold true for any $s \in \langle 0;S/2\rangle$ only if

$$z(s) = z_1.\cos 2\pi \frac{s}{S} , \qquad s \in \langle 0;S/2\rangle \tag{9}$$

which represents the exact and unique solution of the functional problem (1), (2) for $t(s) \equiv$ const. The simplest example of the center line of the corresponding nonwarping cross-section thus may be represented by a circle.

Given $t(s) \not\equiv$ const. or if the permissible pattern of the z-coordinate is prescribed, the functional problem (1), (2) can only be solved numerically. The existence and uniqueness of the continuous solution z(s)or t(s) then may only be deduced from the existence and uniqueness of the discretized solution z_i or t_i which has been determined with properly chosen initial approximation of the function to be found.

It is worth noting that uniqueness of the solution, z(s), of the functional problem (1), (2), irrespective of the prescribed variation of t(s) > 0, does not imply the uniqueness of the shape of center line of the nonwarping cross-section. This is a consequence of the fact that the functional problem in question does not contain y-coordinates of points of the cross-sectional center line. Therefore there are still at least three degrees of freedom, namely the magnitude of the integration constant, y(0), the cross-sectional perimeter, $S \geqslant S_{min}$ (see conditions (25)), and the sign in eq. (22) of the y-coordinate of points of the cross-sectional center line at least of one point in both the symmetrical halves of the center line . The nonwarpable cross-section, determined by the functions z(s) and t(s) which do meet the functional problem (1), (2), can thus have at least ∞^3 shapes of center lines |10|.

2.3. Solution of the Functional Problem (1), (2)

Given the wall-thickness variation, t(s) > 0, s ε <0 S/2>, the z-coordinate of the theoretically nonwarping cross-section can be determined with the help of the mapping |11|

$$z(s) = \frac{\int_0^{S/2} z^2 t\,d\nu}{\int_0^{S/2}\left(\int_0^{\nu} z\,t\,d\xi\right)^2 \frac{d\nu}{t}} \left(\frac{\int_0^{S/2} t\left[\int_0^{\nu}\left(\int_0^{\mu} z\,t\,d\xi\right)\frac{d\mu}{t}\right]d\nu}{\int_0^{S/2} t\,d\nu} - \int_0^{s}\left(\int_0^{\nu} z\,t\,d\xi\right)\frac{d\nu}{t} \right) \qquad (10)$$

valid for s ε <0 S/2>. The mapping has been derived by integrating eq. (1) from 0 to s and by meeting the orthogonality condition (2). Provided that the mapping (10) is contracting, the z-coordinate of the center line of nonwarping cross-section can thus be determined by successive approximations starting with an arbitrary initial approximation which must comply with the conditions (3), e.g.

$$z(s) = z(0).\cos 2\pi\frac{s}{S}\,, \qquad s \in \langle 0\,; S/2\rangle\,. \qquad (11)$$

Discretizing the interval <0;S/2>, the given wall-thickness variation, t(s), and z-coordinate variation, z(s), and assuming some suitable interpolation or approximation rule, various forms of the mapping (10) may be derived in order to determine the discrete pattern of the z-coordinate.

Introduce the dimensionless z-coordinate, ζ, dimensionless wall--thickness, ϑ, and dimensionless circumferential coordinate, $\bar{s}$, according to (12) and assume a nonuniform discretization (13), (14):

$$\zeta(\bar{s}) = \frac{z(\bar{s})}{z(0)}\,, \qquad \vartheta(\bar{s}) = \frac{t(\bar{s})}{t(0)}\,, \qquad \bar{s} = \frac{s}{S/2} \in \langle 0\,;1\rangle\,, \qquad (12)$$

$$\langle 0;1\rangle = \langle \bar{s}_0;\bar{s}_n\rangle = \bigcup_{i=1}^{n}\langle \bar{s}_{i-1};\bar{s}_i\rangle, \qquad \bar{s}_i - \bar{s}_{i-1} = \delta_i ,$$

$$\bar{s}_0 = 0, \qquad \bar{s}_{i=\overline{1,n}} = \sum_{j=1}^{i}\delta_j , \qquad \bar{s}_n = \sum_{j=1}^{n}\delta_j = 1, \tag{13}$$

$$\zeta_i = \frac{z(\bar{s}_i)}{z(0)}, \qquad \vartheta_i = \frac{t(\bar{s}_i)}{t(0)} . \tag{14}$$

Adopting piecewise linear interpolation, the following discretized iterative procedure can be derived on the basis of the mapping (10):

$$\zeta_i = \frac{\sum_{j=0}^{n-1}\delta_{j+1}\left(\zeta_j^2\vartheta_j + \zeta_{j+1}^2\vartheta_{j+1}\right)}{\sum_{j=0}^{n-1}\delta_{j+1}\left(\frac{\psi_j^2}{\vartheta_j} + \frac{\psi_{j+1}^2}{\vartheta_{j+1}}\right)}\left(\frac{\sum_{j=0}^{n-1}\delta_{j+1}\left(\phi_j\vartheta_j + \phi_{j+1}\vartheta_{j+1}\right)}{\sum_{j=0}^{n-1}\delta_{j+1}\left(\vartheta_j + \vartheta_{j+1}\right)} - \phi_i\right),$$

$$i = \overline{0,n} \tag{15}$$

in which

$$\phi_j = \begin{cases} 0 & \dots \; j = 0, \\ \frac{1}{2}\sum_{k=0}^{j-1}\delta_{k+1}\left(\frac{\psi_k}{\vartheta_k} + \frac{\psi_{k+1}}{\vartheta_{k+1}}\right) & \dots \; j = \overline{1,n}, \end{cases} \tag{16}$$

$$\psi_j = \begin{cases} 0 & \dots \; j = 0, \\ \frac{1}{2}\sum_{k=0}^{j-1}\delta_{k+1}\left(\zeta_k\vartheta_k + \zeta_{k+1}\vartheta_{k+1}\right) & \dots \; j = \overline{1,n}. \end{cases} \tag{17}$$

Assuming the discrete initial approximation of the ζ-coordinate in the form of

$$\zeta_i = \cos \pi\bar{s}_i , \qquad i = \overline{0,n} , \tag{18}$$

as follows from (11), the discrete mapping (15) makes it possible to determine the discrete pattern of the dimensionless coordinate ζ.

Similar approach can be adopted for determining the discrete wall-thickness variation round the perimeter of the theoretically nonwarping cross-section with the prescribed center line which is compatible with the conditions of permissibility (3) as shown in |10,12|.

3. COMPUTING THE y-COORDINATES OF POINTS OF THE CENTER LINE OF A NONWARPABLE CROSS-SECTION WITH PRESCRIBED VARIATION OF t(s)

In order to fully determine the shape of the center line of a nonwarpable cross-section, the y-coordinates of points of the center line are to be added. Introducing the dimensionless y-coordinate

$$\eta(\bar{s}) = \frac{y(\bar{s})}{z(0)}, \qquad \bar{s} = \frac{s}{S/2} \in \langle 0; 1 \rangle, \tag{19}$$

and using Pythagorean theorem we get

$$\eta(\bar{s}) = \eta(0) \pm \int_0^{\bar{s}} \sqrt{\varrho^2 - \dot{\zeta}^2(\bar{\nu})}.d\bar{\nu}, \qquad \bar{s} \in \langle 0; 1 \rangle \tag{20}$$

in which

$$\varrho = \frac{S/2}{z(0)}, \qquad \dot{\zeta}(\bar{s}) = \frac{d\zeta}{d\bar{s}} \tag{21}$$

and $\eta(0)$ is an optional integration constant. Adopting the discretization as in eqs. (13) and using the piecewise linear interpolation again, the relation (20) yields

$$\eta_i = \eta_{i-1} \pm \frac{\delta_i}{2}\left(\sqrt{\varrho^2 - \dot{\zeta}_{i-1}^2} + \sqrt{\varrho^2 - \dot{\zeta}_i^2}\right), \qquad i = \overline{1, n}, \tag{22}$$

in which

$$\dot{\zeta}_0 = \frac{\zeta_1 - \zeta_0}{\delta_1}, \quad \dot{\zeta}_n = \frac{\zeta_n - \zeta_{n-1}}{\delta_n}, \quad \dot{\zeta}_j = \frac{\zeta_{j+1} - \zeta_{j-1}}{\delta_{j+1} + \delta_j} \quad (j = \overline{1, n-1}). \tag{23}$$

The relation (22) then implies that optional quantity, ρ, must meet the condition

$$|\varrho| \geq \max_{i=\overline{0,n}} |\dot{\zeta}_i| = |\varrho|_{min}, \tag{24}$$

which yields

$$S \geq 2|z(0)|.|\varrho|_{min} = S_{min} \text{ or } |z(0)| \leq \frac{S}{2|\varrho|_{min}} = |z(0)|_{max}. \tag{25}$$

Thus with the chosen quantity $z(0)$ the minimum possible perimeter of the center line, S_{min}, is determined, or, similarly, given S, the maximum magnitude of the quantity $|z(0)|$ is limited as well.

4. COMPARISON WITH TEST RESULTS

In order to demonstrate that the high axial membrane stresses at the corners of usual box-sections of girders with low slenderness ratio can be eliminated by using the above nonwarping cross-sections, two relatively short, simply supported plastic beams under transverse load at the mid span were investigated experimentally using strain gauges |7,13|.

Consider the flat, rectangular bisymmetrical reference box-section of 200 x 100 mm with flange thickness of 8 mm and web thickness of 4 mm, and the corresponding theoretically nonwarpable bisymmetrical cross-section of the same wall-thickness distribution and the same perimeter (i.e. with the same cross-sectional area), computed for $S = S_{min} = 600$ mm, $|z(0)| = |z(0)|_{max}$, $y(0) = 0$, using plus sign in eq. (22), one half of the perimeter having been subdivided into 60 equal intervals. The quadrants of both these cross-sections can be seen in fig. 2. Each of the beams of length 500 mm was reinforced by five equally spaced plastic diaphragms 5 mm thick, the transverse load acting through the plastic support of the same thickness. Young's modulus, E, and Poisson's ratio, ν, of the materials varied from 3294 to 3383 MPa and from 0.34 to 0.35, respectively. The wall-thickness of the cross-sections was limited by the material available for manufacturing of the beams.

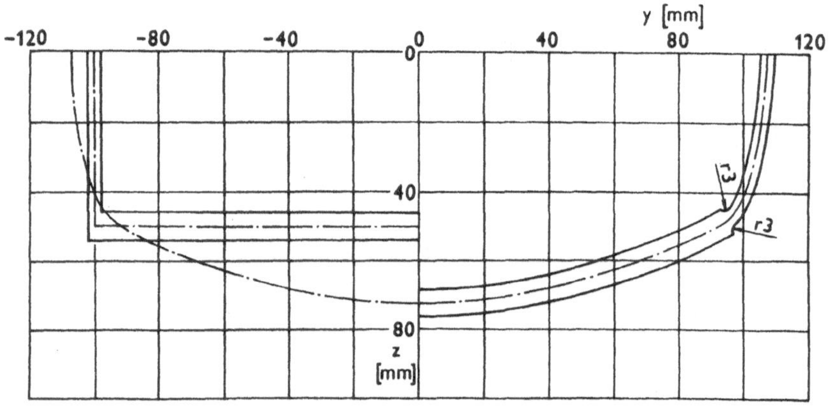

Figure 2. The quadrant of the reference, bisymmetrical rectangular warpable cross-section (left) and the quadrant of the corresponding, bisymmetrical theoretically nonwarpable cross-section with the same area and wall-thickness distribution, determined with the help of equations (15) and (22) (right).

The average axial stress, $\sigma_x = (\sigma_x^o + \sigma_x^i)/2$, where σ_x^o and σ_x^i are the axial stresses on the outer and inner surface of the beam walls, respectively, were determined in the vicinity of the loaded sections ($x = 0.98(l/2)$) using strain gauges. Fig. 3 shows that the maximum magnitude of the average axial stress, σ_x, in the nonwarping cross-section is less than 40 % of the corresponding stress in the beam with the reference warping box-section of the same area and the same wall-thickness distribution. The same conclusion is valid for all the theoretically nonwarpable cross-sections stiffened by diaphragms obtained using the same wall-thickness distribution and S but varying the magnitude of constant y(0) and the sign in eq. (22).

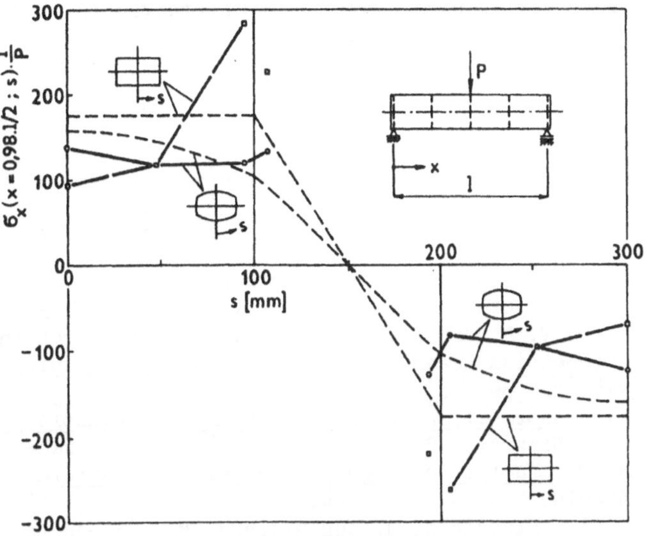

Figure 3. ——— and — — — : Linearly interpolated variation of average axial stress, $\sigma_x.l^2/P$, determined experimentally with the help of the relation $\sigma_x = (\sigma_x^o + \sigma_x^i)/2$, in beams with cross-sections shown in fig. 2, at the distance $x = 0.98(l/2)$. – – – : Axial membrane stress, $\sigma_x.l^2/P$, computed from the classical bending theory of beams which corresponds to perfectly nonwarpable cross-section

5. DISCUSSION

The results show that the unfavourable effect of shear-lag in a simply supported relatively short thinwalled beam-like structural element with the rectangular box-section, subjected to transverse load in the half span can be eliminated by using theoretically nonwarpable cross-section. The same conclusion holds true irrespective of the beam end conditions, the loading distribution along the beam axis and the geometry of the reference warping box-section.

The zero warping functional condition presented here may thus be applied to optimize the cross-sections of relatively short thinwalled beam-like structural elements with acceptable accuracy regardless of the beam end conditions and the transverse loading distribution. This is especially desirable in welded structural elements of notch sensitive material exhibiting high crack propagation rate or in structures of low brittle fracture strength where the use of the nonwarpable cross-sections may result in increased loading capacity. Alternatively, theoretically nonwarpable cross-sections can provide sufficient loading capacity at a lower weight compared with structures having the usual box--sections. The nonwarpable cross-sections requiring no extra material or allowing a decrease in weight of the structure will therefore be more convenient than the warpable ones, providing the material consumption or possible increase of loading capacity are preferred to production costs.

REFERENCES

1 'Trends in the Design of Steel Box Girder Bridges', J. of the Structural Division, ASCE, Vol. 93, No ST3, June 1967

2 Albrecht, P. and Fisher, J. W.: 'An Engineering Analysis of Crack Growth at Transverse Stiffeners', Mém. AIPC, 35-I, 1975, pp. 1 - 22

3 Gorpichenko, V. M. and Kulkova, N. N.: 'Fatigue Strength of Thin-walled Box-Section Beams' (in Russian), Vestnik Mashinostroeniya, No 3, 1976, pp. 30 - 33

4 Panc, Vl.: Theory of Thinwalled Structures on Elastic Foundation (in Czech), Rozpravy ČSAV, Vol. 75, No 4, NČSAV, Praha 1965, 137 p.

5 Williams, D.: An Introduction to the Theory of Aircraft Structures, Edward Arnold, London, 1960, 448 p.

6 Křístek, Vl., Studnička, J. and Škaloud, M.: 'Shear-Lag in Wide Flanges of Steel Bridges', Acta Techn. ČSAV, No 4, 1981, pp. 464-488

7 Jaroš, P. and Matoušek, B.: Experimental Research of the Theoretically Nonwarpable Cross-Sections of Thinwalled Beams under Bending, (in Czech), Res. Rept. SVÚSS No 84-02118, 38 p.

8 Hýča, M.: 'An Optimization Criterion for Cross-Sections of Short Thinwalled Beams under Bending' (in Czech), in Optimalizačné metódy navrhovanija ocelových konštrukcií, ČSVTS, Vysoké Tatry - Stará Lesná, Dec. 1978, pp. 75 - 84

9 Hýča, M.: 'A Functional Condition for Zero Warping of Cross-Sections of Thinwalled Beams under Bending', Acta Techn. ČSAV, No 2, 1984, pp. 151 - 176

10 Hýča, M.: 'Optimization of Thinwalled Prismatic Beams under Bending' (in Czech), Strojnícky Čas., Vol. XXVI, No 3, 1975, pp. 265 - 280

11 Hýča, M.: 'Profiloptimierung der dünnwandigen Stäbe', Wissenschaftliches Kolloquium für Mechanik am Mechanik-Zentrum der Techn. Univ. Braunschweig, June 1986, 16 p.

12 Hýča, M.: 'Numerical Treatment of the Problem of Optimization of Thinwalled Beam Cross-Sections under Bending', Wiss. Z. der Hochsch. für Archit. und Bauwesen Weimar, 23 J., 1976, H.3, pp. 261 - 264

13 Hýča, M.: 'Relation Between the Wall-Thickness and the Coordinates of the Points of the Center Lines of the Symmetrical Warpless-in--Bending Cross-Sections of Thinwalled Beams' (in Czech), Stavebnícky Čas., Vol. 34, No 7, 1986, pp. 557 - 576

OPTIMIZATION OF STRUCTURES USING THE FINITE ELEMENT METHOD

Ir. L.F. Jansen
University of Twente
P.O. Box 217
7500 AE Enschede
The Netherlands

ABSTRACT. The optimization code presented in this paper takes fullest advantage of the finite element modelling of structures. The link between the finite element model and the optimization problem is maintained by so-called design elements, which give this code a wide applicability and flexibility. Furthermore, the cost function and the constraints are regarded as a part of the finite element definition, rather than to be part of the optimization code. Therefore different cost functions and constraints can be used on structures, in order to meet the specific design requirements of the problem at hand. This is achieved by means of user-supplied routines on finite element level only. The gradient of the cost function and constraints are determined analytically, avoiding numerical differentiation. As a result a general purpose-structural optimization program, called Optisys, was obtained.
As an example, simulation of bone remodelling will be discussed in this paper. This example can be regarded as a combined shape and material-property optimization problem, since bone remodelling is the change of shape and property of natural bone due to a change of external loading.

Introduction

Finite element calculations often assist researchers in the optimization of structures. The present paper presents a structural optimization program which takes fullest advantage of the finite element modelling of structures. The integration of the optimization code and the finite element code is such that any structure, of which a finite element model can be made, can be optimized with respect to any parameter of the code's data-base (material properties etc.) and with respect to any calculated quantity (stresses, displacements etc.). This is achieved by defining the cost function and the constraints on finite element level.

Furthermore the link between the optmization problem and the finite element model is maintained by design elements. Design elements were first introduced by Imam [1] for the purpose of shape optimization. The design elements used in the present code have been further developed and

G. I. N. Rozvany and B. L. Karihaloo (eds.), Structural Optimization, 135–141.

can be used in any type of optimization problem, allowing a reduction of independent variables.

The optimization is defined as

$$\begin{aligned} &\min.\ F(\mathbf{u}, \omega, \mathbf{a}) \\ &\quad C_i(\mathbf{u}, \omega, \mathbf{a}) < 0 \end{aligned} \tag{1}$$

Where F represents the costfunction and C_i represent the constraints.

The cost function and the constraints are a function of the design variables **a**, which can be expressed in terms of finite element data, and the state variables **u** and ω, i.e. the nodal displacements and the eigenfrequenties, respectively.

Design Elements

In conjunction with design elements, optimization variables are introduced. The optimization variables are associated with the nodes of a design element mesh which is defined for the structure. The values of the design variables are obtained by interpolation of the optimization variables. The interpolation functions can be taken identical to the displacement functions known from finite element theory, e.g.

$$a(\mathbf{z}_{el}) = \sum x_i L_i(Z_d, \mathbf{z}_{el}) \tag{2}$$

Where $\mathbf{z}_{el}$, $a(\mathbf{z}_{el})$, x_i, L_i, Z_d, denote the coordinates of the nodal points of a finite element, the design variable related to a finite element, the optimization variable related to the nodes of the design element, the interpolation functions and the coordinates of the nodal points of the design element, respectively.

As such, the FE (finite element) model is a function of the optimization variables. Also the cost function and constraints are a function of the optimization variables and instead of eq. (1) we write

$$\begin{aligned} &\min.\ F(\mathbf{u}, \omega, \mathbf{x}) \\ &\quad C_i(\mathbf{u}, \omega, \mathbf{x}) < 0 \end{aligned} \tag{3}$$

The number of optimization variables can be chosen independently of the number of finite elements, whereas the number of design variables is usually related to the number of finite elements. Thus the number of independent variables can be reduced.

For shape optimization, design elements have been developed which map the original mesh onto a new mesh, avoiding elaborate mesh generation procedures.

Definition of Cost Function and Constraints

The cost function and the constraints define the actual optimization problem. The succes of a general purpose optimization code depends on the way these functions can be defined by the user, such that design requirements are met.

In general, the cost function can be regarded as an integral of a function which is defined on (a part of) the structure, such as its volume or its weight. Hence the cost function is defined as the sum of the contributions of finite elements

$$F(\mathbf{u},\omega,\mathbf{x}) = \sum_{FN} f_{el}(\,\mathbf{u}_{el},\ \omega,\ \mathbf{a}_{el}\,) \tag{4}$$

Where FN denotes the set of finite elements which contribute to the cost function. The subscript el indicates that variables are local with respect to finite elements. The constraints are defined in the same way, i.e.

$$C_i(\mathbf{u},\omega,\mathbf{x}) = \sum_{NCi} c_{el}(\,\mathbf{u}_{el},\ \omega,\ \mathbf{a}_{el}) \tag{5}$$

Where NCi denotes the sub set of finite elements contributing to the constraint C_i. These contributions to the cost function and the constraints are independent of a FE model and will, in general, only depend on the design requirements and the finite element type.

The gradient of the cost function can also be expresed as the sum of finite element contributions, i.e.

$$\nabla F(\mathbf{u},\partial\mathbf{u}/\partial\mathbf{x},\omega,\partial\omega/\partial\mathbf{x},\mathbf{x}) =$$

$$\sum_{NF} \nabla f_{el}(\mathbf{u}_{el},\partial\mathbf{u}_{el}/\partial\mathbf{x},\omega,\partial\omega/\partial\mathbf{x},\mathbf{a}_{el},\partial\mathbf{a}_{el}/\partial\mathbf{x}) \tag{6}$$

For the gradients of the constraints a similar expression can be derived. The derivatives $\partial\mathbf{a}/\partial\mathbf{x}$ are readily obtained from eq. (2). The gradients $\partial\mathbf{u}/\partial\mathbf{x}$ are derived from the equation of equilibrium which, after differentiation with respect to x_k, yields

$$\mathbf{K}\ \partial\mathbf{u}/\partial x_k = \sum_{Nel} [\ \partial\mathbf{p}_{el}/\partial\mathbf{a}_{el} - \partial\mathbf{K}_{el}/\partial\mathbf{a}_{el}\mathbf{u}_{el}\,]\ \partial\mathbf{a}_{el}/\partial x_k \tag{7}$$

Where **K** denotes the stiffness matrix and **p** the load vector. The right-hand side vector of eq. (7) can be composed of finite element contributions. The gradients $\partial\omega/\partial\mathbf{x}$ are derived from the eigenvalue problem which after differentiation with respect to x_k and some manipulation yields

$$\partial\omega/\partial x_k\ 2\omega\mathbf{v}^T\mathbf{M}\mathbf{v} =$$

$$\sum_{Nel} [\ \mathbf{v}_{el}^T\ \partial\mathbf{K}_{el}/\partial\mathbf{a}_{el}\mathbf{v}_{el} - \mathbf{v}_{el}^T\partial\mathbf{M}_{el}/\partial\mathbf{a}_{el}\mathbf{v}_{el}\ ]\ \partial\mathbf{a}_{el}/\partial x_k \tag{8}$$

Where **M** denotes the mass matrix, **v** the eigenvector and ω the corresponding eigenfrequency. The right-hand side of eq. (8) can also be regarded as the sum of finite element contributions.

Summarizing it is found that the optimization problem, i.e. the cost function and constraints, can be defined on finite element level, independently of a FE model. The contributions of a finite element to the functions and gradients metioned before, only depend on the finite element type and the design requirements. The present code is based on this dependency. Routines that return these contributions for a certain FE type, may be linked to the main program, allowing the user to define the desired optimization problem.

Simulation of Bone Remodelling

Natural bone cannot be regarded as an ordinary engineering material. Bone is an addaptive material, with the capacity to adjust itself to new circumstances. The bones of healthy individuals are more or less stable. However, when, due to some cause, the loading on the bone changes permanently, the adaptive capability of bone becomes evident. Such dramatic changes in loading occur when foreign material is attached to the bone, such as artificial joint implants. Due to this adaptive capability, loosening of these implants eventually may occur.

The mechanical properties of bone, such as Young's modulus and yield stress, are related to the so-called apparent density ρ (g/cm^3). The apparent density is defined as the weight of the bone-mass of a specimen, devided by the volume of that specimen and is related linear to the porosity of the material. Klever's [5] experiments on specimens taken from a human tibia, showed the following relations

$$E = 6.3\ 10^3\ \rho^{3.1}$$
$$\sigma_c = 0.012\ E^{0.97} \qquad (9)$$

Where E denotes Young's modulus (MPa) and σ_c denotes the yield stress (MPa). A cross section of the tibia shows that the apparent density is not constant throughout the bone, but changes dramatically. Several mathematical models have been developed over the last decade with the aim to predict the apparent density distribution of the human femur and tibia [6,7].

This paper presents a bone remodelling simulation of the tibia, taking both shape and apparent density distribution into account, as indicated by Wolff's law [8]. This law states that bones have a optimal weight design. Hence the optimization problem (O.P.1) is defined as

$$\begin{aligned} \text{min. } F &= \sum \rho_{el} Vol_{el} \\ C_i &= e_{el} / e_0 - 1 < 0 \end{aligned} \qquad (10)$$

Where e_{el} represents the strain energy density and e_0 represents the corresponding critical value according to eq. (9) ,assuming a constant safety factor, i.e.

$$e_0 = 0.454\ \rho^{2.9} \tag{11}$$

A cylinder of homogeneous material, subject to a distributed load, is used as an initial design for the upper part of the tibia. The finite element model constists of 135 isoparametric elements as shown in Fig. 1a.

Two design element meshes where used to link the optimization problem to the finite element model. The design elements for the apparent density optimization consists of 4-node isoparametric elements (see Fig. 1b). The interpolation functions used in this design element (see eq. (2)) are known from finite element theory [2], i.e.

$$\begin{aligned} L_1 &= 1/4\ (\ \xi-1\ )(\ \eta-1\ ) \qquad L_2 = 1/4\ (\ \xi+1\ )(\ \eta-1\ ) \\ L_3 &= 1/4\ (\ \xi+1\ )(\ \eta+1) \qquad L_4 = 1/4\ (\ \xi-1\ )(\ \eta+1\ ) \end{aligned} \tag{12}$$

Where ξ and η are the local coordinates of the centroid of a finite element with respect to the design element.

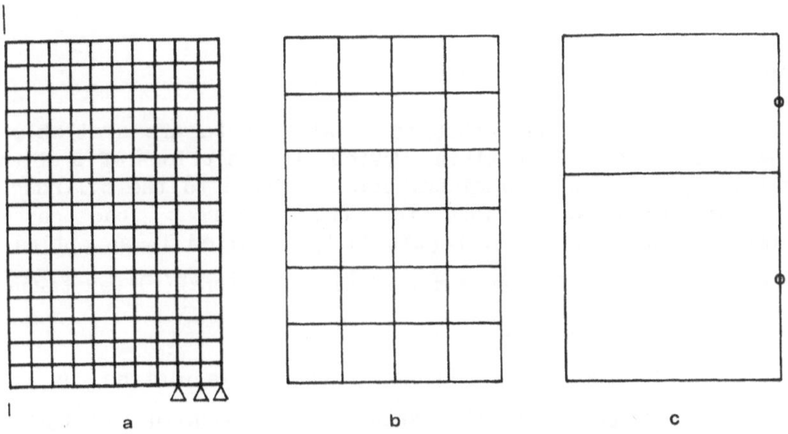

Figure 1. a. Finite element mesh. b. Apparent density design element mesh. c. Shape design element mesh.

For the shape optimization, two 3-node design elements were used, (see Fig. 1c) . These design elements map a rectangular mesh onto a new mesh. The interpolation functions are taken as follows

$$\begin{aligned} & r_{el}(\ r^0_{el},\ z^0_{el}) = r^0_{el} \sum\ x_i L_i(\zeta) \\ & L_1 = \zeta\ (\ \zeta-1\ )/2 \quad L_2 = (\ 1-\zeta\ )(\ 1+\zeta\ ) \quad L_3 = \zeta\ (\ \zeta+1\ )/2 \end{aligned} \tag{13}$$

Where r^0_{el} and z^0_{el} represent the mesh coordinates of the original FE mesh and r_{el} the coordinates of the new mesh , and ζ represents the local z_{el} coordinate with respect to the design element. The radius of the tibia-plateau is held constant during the optimization.

Optimization Method

The optimization problem is solved using a gradient projection method in conjunction with a Quasi-Newton method [3], implemented according to Powell [4].

Simultaneous optimization of shape and apparent density of O.P.1 converges to a local minimum due to poor scaling of the optimization problem. Therefore, the shape optimization and apparent density optimization were performed separately. A start design was obtained by an apparent density optimization, starting from a homogeneous apparent density distribution of $\rho = 2.0$, using $e_0/\rho^{2.9} = 0.1$. The optimization was continued in an incremental way for increasing values of $e_0/\rho^{2.9}$ [0.2, 0.25, 0.3, 0.4]. Each increment consisted of a shape optimization, which resulted in major reductions of the cost function, followed by an apparent density optimization. An average of 30 iterations was needed for an apparent density optimization, whereas the shape optimization took an average of 8 iterations, for each increment. This method converged to a near optimal design.

Results

It should be mentioned that no detailed data are available concerning the load acting on the tibia. Obviousely, solutions of problem O.P.1 are a function of the load distribution applied to the cylinder. Therefore problem O.P.1 was solved for two different load cases. A load distribution of constant magnitude P_0 resulted in an optimal design as shown in Fig. 2a. The second load distribution P(r) was taken according

$$P(r) = P_1 \sin(\pi r/R_0) \tag{14}$$

Where R_0 is the outer radius of the tibia-plateau, r is the polar coordinate, and P_1 was taken such that the magnitude of the resultant of the two load distribution were identical. Result of the ditribution of the apparent density and of the final shape of the bone is shown in Fig. 2b.

Both results can only be compared with experimental data in a qualitative way. The radii of the optimized cylinders show a reduction of about 40%, for both load cases, whereas in practise a reduction was found about of 55%. The apparent density at the tibia plateau, as reported by Klever, ranges from 0.25 to 0.6 g/cm^3. For the uniformely distributed load, a range from 0.7 to 0.8 g/cm^3 was found. For the second load case a range of 0.4 to 0.8 g/cm^3 was found.

The load distribution has a significant influence on the apparent density distribution, whereas the shapes of the two optimal cylinders are almost identical. This indicating that the apparent density distribution is more sensitive to changes of loading than the shape of the bone. This is confirmed by clinical results.

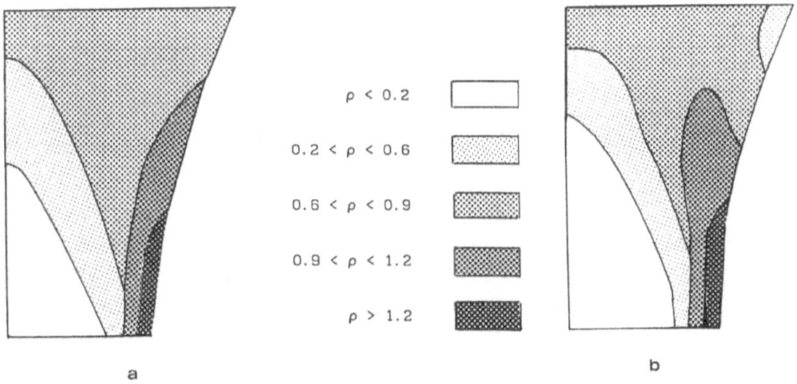

Figure 2. Apparent density distribution ρ (g/cm^3) for P_0= 10 Mpa (a) and P_1 = 16 Mpa (b)

References

[1] 'Three-Dimensional Shape Optimization.' M.H. Imam. *International Journal for Numerical Methods in Engineering* **18**, pp 661-673, 1982.

[2] *The Finite Element Method.* O.C. Zienkiewicz. Mc Graw Hill, New York 1977

[3] 'Quasi-Newton Methods, Motivation and Theory'. J.E. Dennis J.R. Moré, J.J. Moré. *SIAM Review* **19** no 1 Jan 1977

[4] 'A Fast Algorithme for Nonlinearly Constrained Optimization Calclulations'. M.J.D. Powell. *Lecture Notes in Mathematics.* A. Dold, B. Eckman (ed.) Procedings, Biannial Conferece Dundee 1977. Springer Verlag Berlin 1978

[5] *On the Mechanics of Failure of Artificial Knee Joints.* Frans J. Klever, Phd Thesis, University of Twente, 1984

[6] 'A Saturating Time Dependent Model for Bone Remodeling'.D.P. Fyhrie, D.R. Carter, G.S. Beaupré.34th Anual Meeting, Orthopaedic Research Society,Febr. 1988, Atlanta, Georgia.

[7] 'Adaptive Bone-Remodeling Theory Applied to Prosthetic-Design Analysis'. R. Huiskes, et. al. *J. Biomechanics.* **20**, Nr 12. Submitted Aug. 1987.

[8] 'Ueber die Bedeutung der Architektur der spongiösen Substanz'. J. Wolff. *Zentralblatt fuer die medizinische Wissenschaft.* **VI.** Jahrgang 223-234, 1869.*As cited by Huiskes*

COMPARISON OF NLP TECHNIQUES IN OPTIMUM STRUCTURAL FRAME DESIGN

B.L.KARIHALOO and S. KANAGASUNDARAM
School of Civil and Mining Engineering
The University of Sydney
Sydney, N.S.W. 2006
Australia

ABSTRACT.The aim of this paper is to study the efficiency of three non-linear programming techniques for the solution of optimization problems of plane structural frames under multiple load systems. The problem is one of minimizing the mass of the frame subject to constraints on normal and shear stresses, the maximum transverse deflection and buckling load. Numerical examples are presented to indicate the efficiency of the various NLP techniques.

1 Introduction

This paper studies the efficiency of three non-linear programming techniques for the solution of optimization problems of plane structural frames under multiple load systems. The three techniques are: 1)sequential convex programming with move-limits,SCP[1]; 2)sequential linear programming with move-limits,SLP[2]; and 3)sequential unconstrained minimization techniques, SUMT[3].

The optimization problem is one of minimizing the mass of a frame in such a way that the normal and shear stresses, the maximum transverse deflection anywhere in the frame, the critical load factor in buckling, and some sizing parameters do not violate respective prescribed limits under any load system. To obtain realistic designs the variation of the mass(stiffness) along the frame members is restricted to splines of order zero, one or two. Within these restricted classes of variation the mathematical optimization problem can be reduced to a non-linear programming problem(NLP) by the application of minimax variational approach to the differential constraints[4]. The solution of the resulting NLP is attempted by the three methods mentioned above using a new general purpose structural optimization package called ADS[5]. Several examples will be presented to demonstrate the merits and disadvantages of the various solution techniques.

2 Mathematical formulation of the problem

Consider an isolated elastic member of a plane frame, subjected to reactive forces from the joints as well as external forces $f(x)$ and self-weight $g(x)$. The external forces $f(x)$ consist of several individual force systems $\{f_\ell(x), \ell = 1, \ldots, NL\}$ each acting on the frame on separate occasions. It is assumed that the mass of a member m (which is proportional to

G. I. N. Rozvany and B. L. Karihaloo (eds.), Structural Optimization, 143–150.

its area $A_m(x)$) is related to its flexural stiffness (which is proportional to second moment of area $I_m(x)$) through

$$I_m(x) = cA_m^n(x) \tag{1}$$

in which c and n are constants determined by the cross-sectional shape[4]. Next, each member is divided into a given number of segments ($NSEG_m$) and the variation of the cross-sectional area is prescribed (NOR_m) using splines. For example, splines of order zero ($NOR_m = 0$) are used to represent constant area within a segment and of order one and two – linear and parabolic variations. Irrespective of the order of spline chosen for each segment, $A_m(x)$ will be of the form

$$A_m(x) = A_m(x, \mu_k^m); (k = 1, \ldots, N_m) \tag{2}$$

where μ_k^m denote the unknown design variables at the knots and N_m is the total number of unknowns: $N_m = NSEG_m$ if $NOR_m = 0$ and $N_m = NSEG_m + 1$ if $NOR_m = 1 or 2$. Using this simplification, the optimization problem can be written as[4]: Minimize mass

$$W = \sum_{m=1}^{NM} \gamma_m \int_0^{L_m} A_m(x, \mu_k^m) dx; (k = 1, \ldots, N_m) \tag{3}$$

subject to constraints on stresses, deflection, buckling load and geometry:

$$\Omega_i^*(x, \mu_k^m) \le 0; (i = 1, 2) \tag{4}$$

$$|(u_{max})_\ell|_m \le \Delta_0^m \tag{5}$$

$$(\lambda_{cr})_\ell / FS \ge 1; (\ell = 1, \ldots, NL) \tag{6}$$

$$G_j(A_m) = 0; (j = 1, \ldots, NC) \tag{7}$$

Here,γ_m is the mass density of the material,$(u_{max})_\ell$ and $(\lambda_{cr})_\ell$ are respectively the deflection within a member and the critical load factor in buckling of the frame under the load system $f_\ell(x)$; L_m is the length of the member, FS is factor of safety and Δ_0^m is the allowable deflection. The Equation(4) refers to constraints on both normal (i=1) and shear (i=2) stresses. The geometric constraints (Eq.7) may be imposed on the design such that, for instance, the optimum sizing parameter has the same value on either side of a joint other than a support.

3 Solution of the problem

The optimization problem was solved for two types of cross-sections: a) rectangular sections of constant depth and variable width($n = 1$), b) rectangular sections of constant width and variable depth($n = 3$). The iterative solution scheme is shown as a flow chart in Figure 1. The stress design was obtained by minimizing the mass of the frame subject only to the stress constraints [4].This involves solution of linear programs by an active set strategy[6]. Using the stress design $\tilde{A}$ as a lower bound the following non-linear programming problem was formulated: Minimize mass W (Eqn.3) subject to constraints on deflection(Eqn.5), buckling(Eqn.6), geometry(Eqn.7) and lower bounds on sizing parameters

$$A_m \ge \tilde{A}_m; (m = 1, \ldots, NM) \tag{8}$$

The solution of the NLP was attempted by the three methods[1–3]. It is not the intention here to describe these methods but to compare their effectiveness in solving the problem in hand. All the three methods are available in a new general purpose package called the ADS (Automated Design Synthesis)[5]. The critical load factor associated with the buckling of the frame under load system $f_\ell(x)$ was calculated by solving for the lowest eigenvalue of the characteristic equation[7]

$$\mathbf{K}(\lambda_\ell)\mathbf{D} = \mathbf{0} \tag{9}$$

where $\mathbf{K}(\lambda_\ell)$ is the overall stiffness matrix of the frame dependent on λ_ℓ and $\mathbf{D}$ is the displacement vector. The characteristic equation(9) was solved for all loading cases.

Several examples are presented in the next section.

4 Examples and Comparison of Solution Techniques

The results of several examples are shown in Fig. 2–8. The following material properties were assumed in the examples: $\sigma_0 = 5MPa$,$\tau_0 = 0.5MPa$,$E = 20GPa$,$\gamma = 2450kg/m^3$. With the exception of the frame shown in Fig.2, the frames were subjected to the following uniformly distributed loads on the beams in accordance with the recommendations of the Code of Practice[8]: Dead load(DL)=27.6kN/m(21.6kN/m at roof level);Live load(LL)=18kN/m(3kN/m at roof level). The wind loads are shown on the respective figures. The frames were designed to withstand the following load combinations: DL+LL, DL+LL+WL,DL+WL. The maximum deflection to span ratio for each member was limited to 1/350 (a very stringent value of 1/1000 was assumed for the frame shown in Fig.2) and the factor of safety(FS) against the buckling of the frame was assumed to be 3. The members of the frame in Fig.2 were designed to consist of one-metre long segments, while those of remaining frames consisted of only one segment. The stress constraints in each of the segments were calculated at sections spaced at 0.05m intervals, and the integrals were evaluated using Simpson's rule.

The solution of the non-linear programming problems by the SCP and SLP techniques provided quick convergence, and the computer times for solution were comparable (Table I). The SUMT, on the other hand took considerably longer for the solution of some of the examples (Figs. 4,7); it was therefore decided not to persist with this method.

Starting with the same uniform feasible design, all the three methods yielded practically identical optimum designs. The designs using zeroth and first order splines are shown in Figures 2–8 (the designs using the second order spline generally follow that of the first order and are not shown here). The designs with constant area segments(zeroth order spline) shown in the Figures are such that beams at any one level are of the same area; the columns at the same level are such that the outer columns have the same area, and the corresponding inner columns on either side of the line of symmetry at that level have the same area. The designs with linearly varying area(first order spline) on the other hand are such that all members meeting at any joint other than a support have the same value of the optimum sizing parameter. Under these geometric constraints the number of unknown design variables is reduced, because the simple equality constraints (Eq.7), can be omitted after adjusting the set of unknown variables. For instance, the total number of variables for the frame shown in Fig.2 reduces from 16 to 14 when $NOR = 0$, and from 19 to 17 when $NOR = 1$ (see Table I, column 3). A comparison of the volumes of optimum frames is presented in Table II along with the volumes of the corresponding frames of uniform design. It was found that in all the optimum designs, except those shown in Fig.2 and 3 the buckling and deflection constraints were inactive and the optimum designs were decided by the stress requirements alone. In the designs of Fig. 2 and 3 the deflection constraints were active for some of the members, along with the stress constraints, although again the buckling constraints were inactive. In order to verify the correctness of the numerical computations, all the optimum frames were analysed, and the stresses and maximum deflection were found to be within the prescribed limits.

The number of design variables and the number of non-linear constraints in each of the examples are shown in Table I. For the example frame shown in Fig.2, the number of design variables is greater than the number of constraints, but the converse applies to the remaining examples. Also shown in the Table are the number of constraint function evaluations for each of the methods. The CPU times(on microVAX) for each of the methods are also shown in the Table. It should be noted that SUMT required excessively large number of constraint function evaluations and hence, also the CPU time.In most of the cases, SCP required lesser number of evaluations and the solution was faster. At this point it is useful to mention that the gradients of the objective and constraint functions were calculated by finite differences and each set of evaluation of the constraint functions required analyses of the frame under all loading cases.

In conclusion, it may be said that of the three methods used in this study both the SCP and SLP were found suitable to optimize the frame designs under stress, deflection and buckling constraints, with the SCP slightly the better of the two.

5 Acknowledgement

The work reported in this paper was supported by grant F84/16078 to B.L.Karihaloo from the Australian research Grants Scheme.

6 References

1. **C. Fleury and V. Braibant**, Structural optimization:A new dual method using mixed variables, *Int. J. Num. Methods Eng.*,**23**,409–428 (1986)

2. **P. Pedersen**, The integrated approach to FEM–SLP for solving problems of optimal design, In *Optimization of Distributed Parameter systems*, Vol. I (Edited by **E. J. Haug** and **J. Cea**), Sijthoff & Noordhoff, The Netherlands, pp. 739–756 (1981)

3. **A. V. Fiacco and G. P. McCormick**, *Nonlinear Programming: Sequential Unconstrained Minimization techniques*, John Wiley, New York (1968)

4. **B. L. Karihaloo and S. Kanagasundaram**, Optimum design of statically indeterminate beams under multiple loads, *Comput. Struct.*,**26**,521–538 (1987). See also three further papers by the authors in this Journal in 1988

5. **G. N. Vanderplaats and H. Sugimoto**, A general-purpose optimization program for engineering design, *Comput. Struct.*, **21**,13–21 (1986)

6. **M. J. Best and K. Ritter**, *Linear Programming (Active Set Analysis and Computer Programs)*, Prentice-Hall, Englewood Cliffs, NJ (1985)

7. **W. P. Howson**, A compact method for computing the eigenvalues and eigenvectors of plane frames, In *Engineering Software*,(Edited by **R. A. Adey**, Pentech Press, London, pp. 281–300 (1979)

8. *Minimum Design Loads on Structures*, Australian Standard AS 1170–1981, Parts 1 and 2

Table I. Comparison of Solution Techniques

Fig	*NOR*	Number of variables	Number of non-linear constraints	Number of constraint function evaluations			CPU time, Seconds		
				SCP	*SLP*	*SUMT*	*SCP*	*SLP*	*SUMT*
2	0	14(16)	12	147	120	*	7110	5703	*
	1	17(19)	12	324	252	*	15604	12208	*
3	0	14(16)	12	133	225	*	6345	10530	*
	1	17(19)	12	504	666	*	23570	30817	*
4	0	5(8)	27	39	66	2230	1873	2775	83633
	1	8(16)	27	69	117	*	2865	5050	*
5	0	5(8)	27	39	54	*	1800	2398	*
	1	8(16)	27	60	108	*	2526	4339	*
7	0	8(15)	48	51	72	4176	4300	5799	305109
	1	13(30)	48	137	182	*	10536	13812	*
8	0	8(15)	48	57	72	*	5158	6331	*
	1	13(30)	48	177	266	*	14016	20594	*

* not available

Table II. Comparison of Volumes (m^3) of Optimum Frames

Figure	*NOR*	Optimum design		Uniform Design
		SCP	*SLP*	
$2(n=1)$	0	5.9301	5.8894	9.6801
	1	5.5276	5.5957	
$3(n=3)$	0	4.9906	5.0183	7.1719
	1	4.6745	4.7338	
$4(n=1)$	0	8.0956	8.0956	10.6391
	1	7.6560	7.6560	
$5(n=3)$	0	6.5610	6.5610	7.6328
	1	5.7610	5.7610	
$7(n=1)$	0	23.4970	23.4970	63.1290
	1	20.1335	20.1335	
$8(n=3)$	0	17.7634	17.7634	22.9505
	1	17.9826	17.9826	

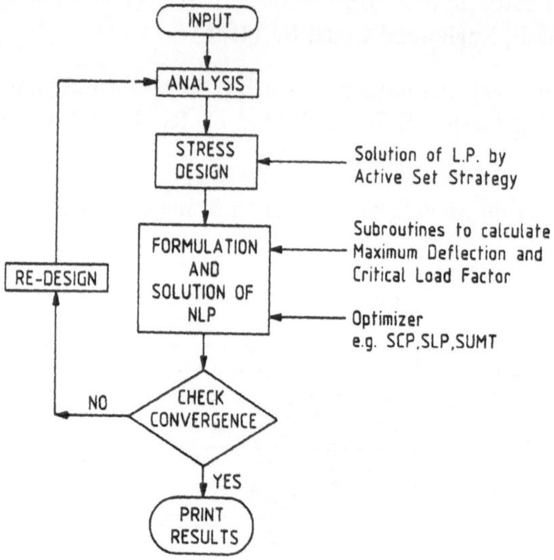

Figure 1. The iterative solution scheme for the optimization problem.

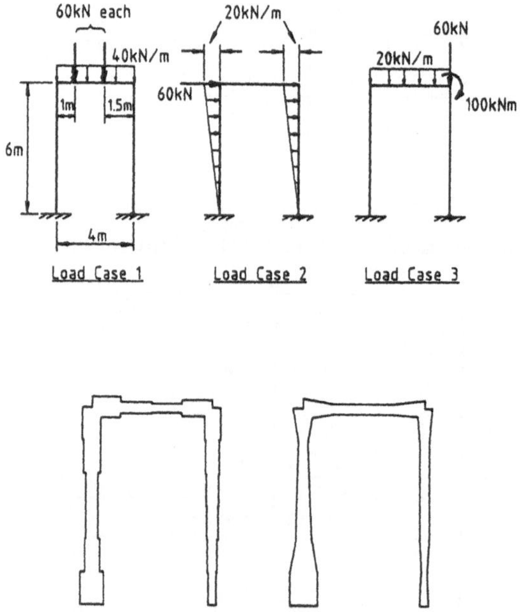

Figure 2. Minimum-weight design of a portal frame subject to multiple load systems as shown plus self-weight. All members are rectangular in cross-section with constant depth $h = 0.8m$ i.e.,$n = 1$.

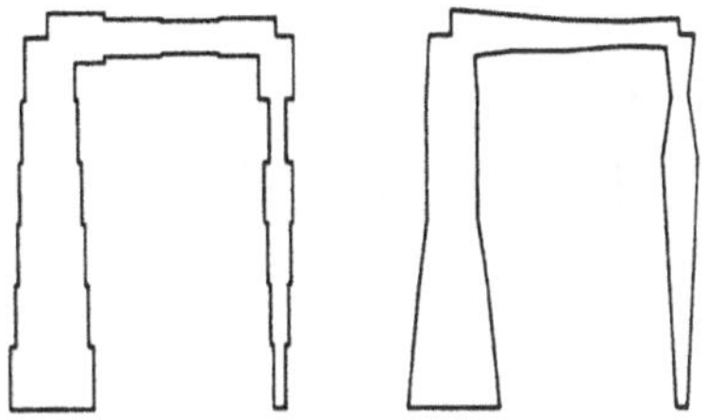

Figure 3. Minimum-weight design of the frame shown in Figure 2 but with members of constant width $b = 0.4m$ i.e.,$n = 3$.

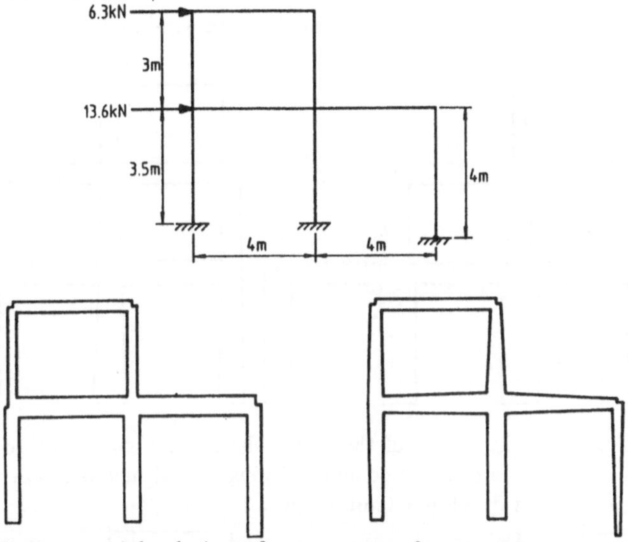

Figure 4. Minimum-weight design of a two-storey frame subject to multiple loads prescribed in the Code of Practice and self-weight. Wind loads are applied at floor levels. All members are rectangular in cross-section with constant depth $h = 0.6m$ i.e.,$n = 1$.

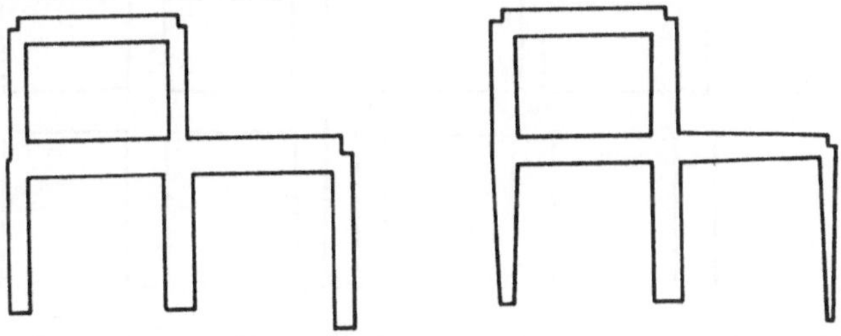

Figure 5. Minimum-weight design of the frame shown in Figure 4 but with members of constant width $b = 0.3m$ i.e., $n = 3$.

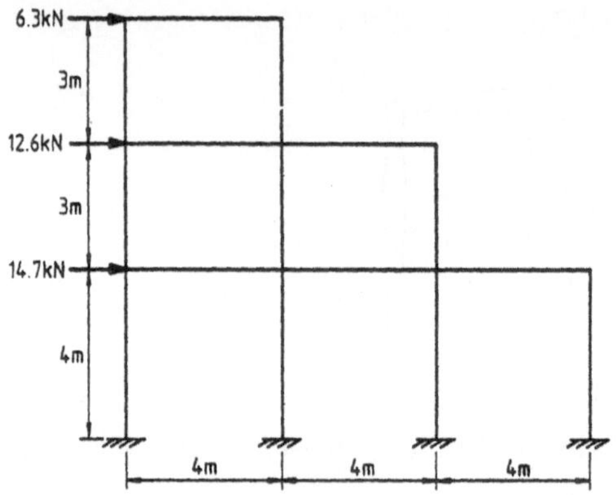

Figure 6. An example of a multi-storey frame subject to wind loads applied at floor levels.

Figure 7. Minimum-weight design of the frame shown in Figure 6 subject to multiple loads prescribed in the Code of Practice and self-weight. All members are rectangular in cross-section with constant depth $h = 0.6m$ i.e.,$n = 1$.

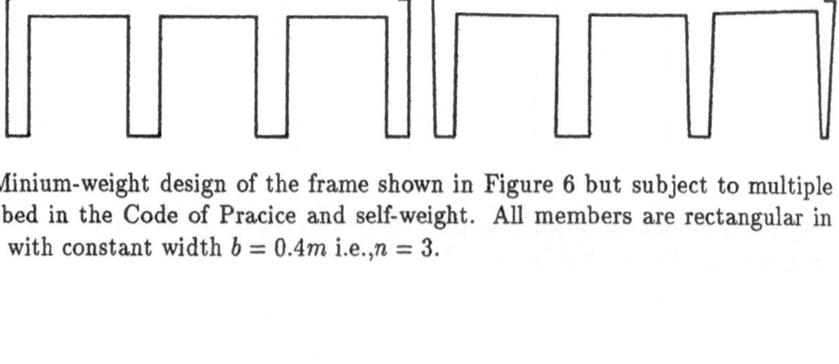

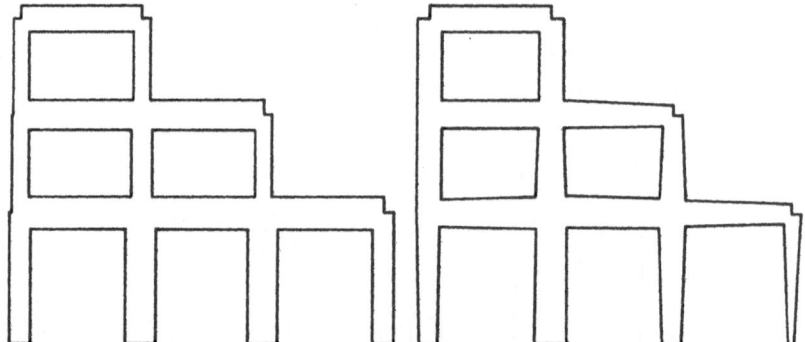

Figure 8. Minium-weight design of the frame shown in Figure 6 but subject to multiple loads prescribed in the Code of Pracice and self-weight. All members are rectangular in cross-section with constant width $b = 0.4m$ i.e.,$n = 3$.

STRUCTURAL AND CONTROL OPTIMIZATION WITH WEIGHT AND FROBENIUS NORM AS PERFORMANCE FUNCTIONS

N. S. Khot
Air Force Wright Aeronautical Laboratories (AFWAL/FIBR)
Wright-Patterson Air Force Base, Ohio 45433-6553 USA

R. V. Grandhi
Dept of Mechanical Systems Engineering
Wright State University, Dayton, Ohio 45435 USA

ABSTRACT. A simultaneous structural and control optimization approach of flexible structures is presented. The constraints are imposed on the closed-loop eigenvalue distribution and the damping parameters. The design variables are the cross-sectional areas of the members. The effect of minimizing the weight of the structure and the Frobenius norm of the control gains is investigated. An ACOSS-FOUR structure is designed for numerical comparison.

1. Introduction

Recently, there has been a great deal of research activity in large space structures design. Large flexible structures have ambitious performance requirements that must be met in the presence of a variety of disturbances. The natural frequency spectrum of such structures is typically quite dense, and damping levels are small. Therefore, these structures have to be actively controlled to achieve the required performance without increasing the structural mass. An efficient design of such structural systems is of fundamental interest to both the structural and control engineers. Systematic design approaches in structural optimization and optimal control theory are well developed. However, the problem of optimal structural design and that of optimal control design is solved with little or no interaction. Initially, the structure is designed subject to stress, displacement, and frequency requirements while ignoring the active control system, and subsequently the active control system is designed to minimize the control energy of the above specified structure. In the last three or four years[1,4] there has been a growing interest in the simultaneous integrated design of the structure and active control system to produce a truly optimum configuration which results in least weight and least control energy consumption.

This paper contains the results obtained by using the Frobenius norm and weight of the structure as the objective functions with constraints on the closed-loop eigenvalue distribution and the damping factors. The two problems are treated separately and their relationship has been investigated. In Ref. 4 it has been shown that for the two bar truss problem the two approaches give identical results. The present investigation indicates that this is not necessarily true for all structures. Multivari-

G. I. N. Rozvany and B. L. Karihaloo (eds.), Structural Optimization, 151–158.

able control theory is used with a linear quadratic regulator with constant feedback. The optimum structural design is obtained by modifying the cross-sectional areas of the members of the optimally controlled structure.

2. Preliminaries

The state space equation describing the dynamic behavior of a structure is given by

$$\{\dot{x}\} = [\mathrm{A}]\{\mathrm{x}\} + [\mathrm{B}]\{\mathrm{f}\} \tag{1}$$

where $\{x\}$ is the state variable vector and $\{f\}$ is the input vector. In Eq. 1, [A] and [B] are the plant and input matrices given by

$$[\mathrm{A}] = \left[\begin{array}{c|c} 0 & \nwarrow \mathrm{I} \searrow \\ \hline \nwarrow -\omega^2 \searrow & \nwarrow -2\varsigma\omega \searrow \end{array}\right] \tag{2}$$

$$[\mathrm{B}] = \left[\begin{array}{c} 0 \\ \hline \phi^{\mathrm{T}}\mathrm{D} \end{array}\right] \tag{3}$$

where $\{\omega^2\}$ is the vector of the square of the structural circular frequency, $\{\varsigma\}$ is the vector of the modal damping factor, [D] is the applied load distribution matrix and $[\phi]$ is the modal matrix. Eq. 1 is known as the state input equation and the state output equation is given by

$$\{y\} = [\ \mathrm{C}\ ]\{x\} \tag{4}$$

where $\{y\}$ is an output vector, and [C] is the output matrix which would be equal to $[\mathrm{B}]^T$ if the actuators and sensors are colocated. In order to design a linear quadratic regulator a performance index (PI) can be defined as

$$\mathrm{PI} = \int_0^t \left(\{x\}^{\mathrm{T}}[\mathrm{Q}]\{x\} + \{f\}^{\mathrm{T}}[\mathrm{R}]\{\mathrm{f}\}\right) \mathrm{dt} \tag{5}$$

where [Q] and [R] are the state and control weighting matrices which have to be positive semidefinite and positive definite respectively. The result of minimizing the quadratic performance index and satisfying the state input equation gives the state feedback control law

$$\{f\} = -[\tilde{\mathrm{G}}]\{x\} \tag{6}$$

where

$$[\tilde{\mathrm{G}}] = [\mathrm{R}]^{-1}[\mathrm{B}]^{\mathrm{T}}[\tilde{\mathrm{P}}] \tag{7}$$

In Eq. 7 $[\tilde{P}]$ is the Riccati matrix and is obtained by the solution of the algebraic equation

$$[\mathrm{A}]^{\mathrm{T}}[\tilde{\mathrm{P}}] - [\tilde{\mathrm{P}}][\mathrm{B}][\mathrm{R}]^{-1}[\mathrm{B}]^{\mathrm{T}}[\hat{\mathrm{P}}] + [\tilde{\mathrm{P}}][\mathrm{A}] + [\mathrm{Q}] = 0 \tag{8}$$

Using Eqs. (1) and (6) the governing equation of the optimum closed-loop system can be written as

$$\{\dot{x}\} = [\bar{\mathrm{A}}]\{x\} \tag{9}$$

where

$$[\bar{A}] = [A] - [B][\tilde{G}] \tag{10}$$

$$= [A] - [X][\tilde{P}] \tag{11}$$

and

$$[X] = [B][R]^{-1}[B]^T \tag{12}$$

The complex eigenvalues of the closed-loop matrix $[\bar{A}]$ can be written as

$$\lambda_i = \tilde{\sigma}_i \pm j\tilde{\omega}_i \tag{13}$$

and the damping factor ξ_i is given by

$$\xi_i = -\frac{\tilde{\sigma}_i}{(\tilde{\sigma}_i^2 + \tilde{\omega}_i^2)^{1/2}} \tag{14}$$

The Frobenius norm, S_G, of the gain matrix is given by

$$S_G = \text{Tr} \ \ [\tilde{G}]^T[\tilde{R}][\tilde{G}] \tag{15}$$

where Tr indicates the trace of a matrix and $[\tilde{R}]$ is the gain weighting matrix.

The sensitivity of the closed-loop eigenvalues with respect to the design variable A_ℓ, i.e, the cross-sectional area of member ℓ, is given by

$$\lambda_{i,l} = \{\beta\}_i^T[\bar{A}]_{,l}\{\alpha\}_i \tag{16}$$

where $\{\beta\}_i$ and $\{\alpha\}_i$ are the left-hand and right-hand eigenvectors of $[\bar{A}]$.

Using Eq. 7 and 15 the Frobenius norm can be written as

$$\text{Tr} \ \ S_G = \text{Tr} \ \ \left[[\tilde{P}]^T[B][F][B]^T[\tilde{P}]\right] \tag{17}$$

where

$$[F] = [R]^{-1^T}[\tilde{R}][R]^{-1} \tag{18}$$

The sensitivity of the Frobenius norm can be written by differentiating Eq. 17 with respect to the design variable A_ℓ.

3. Weight Optimization

The structural weight minimization problem can be stated as

$$\textit{minimize the weight } W = \sum_{i=1}^{m} \rho_i A_i l_i \tag{19}$$

with constraints

$$g_j(\xi_i) = \xi_i - \bar{\xi}_i \geq 0 \tag{20}$$

$$g_j(\tilde{\omega}_i) = \tilde{\omega}_i - \bar{\tilde{\omega}}_i \geq 0 \tag{21}$$

where ρ_i is the mass density, ℓ_i is the length of the element, $\bar{\xi}_i$ is the desired value of the closed-loop damping factor and $\bar{\tilde{\omega}}_i$ is the lower bound of the imaginary part of the closed-loop eigenvalue. In addition to the constraints in Eqs. 20 and 21 minimum size constraints are imposed on the design variable A_i. In weight optimization the objective function is linear and the constraints are nonlinear. The sensitivity of the objective function can be written by differentiating Eq. 19 with respect to the design variables while the gradients of the constraints can be obtained by using the procedure given in the previous section.

4. Frobenius Norm Optimization

In this case the problem can be stated as:

minimize the norm

$$S_G = Tr[\tilde{G}]^T[\tilde{R}][\tilde{G}] \tag{22}$$

with constraints as given in Eqs. 20 and 21. The main difference between this problem and the previous one is that the objective function is the Frobenius norm instead of the weight. In this case the objective function as well as the constraints are nonlinear. The optimum designs are obtained by using the NEWSUMT-A program which is based on an extended interior penalty function and a modified Newton's method.

5. Numerical Results

The finite element model of the tetrahedral truss is shown in Fig. 1. The edges of the tetrahedron are 10 units long. The structure has twelve degrees of freedom and four masses of 2 units each are attached at nodes 1 through 4. Young's modulus of the members is 1.0, and the density of the structural material is assumed to be 0.001. The dimensions of the structure are defined in unspecified consistent units. The coordinates of the node points are given in Ref. 3. The actuators are located in the six tripod legs. The sensor locations are assumed to coincide with the actuators. The weighting matrices [Q] and [R] are assumed to be identity matrices, and the passive damping (ς) in Eq. 2 is assumed to be zero. The initial design used for optimization was Design A whose areas are given in Table 1. The constraints imposed for this problem were $\tilde{\omega}_1 \geq 1.341$, $\tilde{\omega}_2 \geq 1.6$, and $\xi_1 \geq 0.15$

This structure was optimized with the weight as well as the Frobenius norm as the objective functions. For each case, at each iteration, the associated Frobenius norm or weight of the structure was calculated. Fig. 2 shows the iteration histories for (i) the weight as the objective function (Design I) and the associated Frobenius norm, and (ii) the Frobenius norm as the objective function (Design II) with the associated weight with an upper bound of 1000 on the design variables. For this structure it is seen that the weight minimization and the Frobenius norm minimization are not equivalent. The optimum designs for the two cases are different. The optimum cross-sectional areas, the structural frequencies, and the closed-loop damping parameters are given in Tables 1-4 for Designs I and II. Design II is driven by the upper limit value of the design variables. Some of the elements reach their upper limit values. It was found that if the upper bound was increased to 2000 in the design with the Frobenius norm as the objective function, some of the areas were also nearly equal to the upper bound. In the two-bar truss problem where the weight and the Frobenius norm minimization were found to be equivalent,[4] the constraints were imposed on all the damping parameters and the eigenvalues of the closed-loop system. In order to create similar design condition the tetrahedral truss was solved by imposing inequality constraints on all the damping parameters and the eigenvalues of the closed-loop system. The constraints are given in Table 2. Design III (weight as the objective function) and Design IV (Frobenius norm as the objective function) were obtained with the initial design the same as Design A. The iterative history for the two designs is shown in Fig. 2. The optimization design variables, the structural frequencies and the closed-loop damping parameters are given in Tables 1-4. In the case of twenty-four constraints the cross-sectional areas of Design IV did not reach the upper bound values.

6. Conclusions

The weight and Frobenius norm as the objective function give different optimum design variables and frequency distributions. When the structure was not completely constrained, some of the cross-sectional areas of the members for the Frobenius norm minimization were nearly equal to the upper bound limit. However, when constraints were imposed on all the damping parameter and frequencies, the optimum design was not governed by the upper bound on the areas of the members. It was found that different designs could be obtained satisfying all the constraints with a different distribution of variables depending on the initial design.

7. References

1. Haftka, R. T., Martinovic, Z. and Hallauer, Jr., W.L., 'Enhanced Vibration Controllability by Minor Structural Modifications,' *AIAA Journal*, Vol. **23**, No. 8, August 1985, pp. 1260-1266.
2. Bodden, D. S. and Junkins, J. L., 'Eigenvalue Optimization Algorithms for Structure/Controller Design Iterations,' *J. Guidance and Control*, Vol. **8**, 1985, pp. 697-706.
3. Khot, N. S., Eastep, F. E. and Venkayya, V. B., 'Optimal Structural Modifications to Enhance the Optimal Active Vibration Control of Large Flexible Structures,' *AIAA Journal*, Vol. **24**, No. 8, 1986, pp. 1368-1374.
4. Khot, N.S., Grandhi, R.V. and Venkayya, V.B., 'Structural and Control Optimization of Space Structures,' AIAA/ASME/ASCE/AHS 28th Structures, Structural Dynamics and Materials Conf., Monterey, CA, (87-0939), April, 1987.

Table 1. Cross-Sectional Areas of the Members

Element No.	Design A Nominal	Design I	Design II	Design III	Design IV
1 (1-2)	1000.	190.15	998.07	174.042	244.429
2 (2-3)	1000.	379.95	998.07	384.171	479.971
3 (1-3)	100.	386.62	996.58	369.825	424.907
4 (1-4)	100.	386.52	998.10	398.526	399.482
5 (2-4)	1000.	380.05	998.07	367.499	822.001
6 (3-4)	1000.	271.88	998.06	299.246	359.891
7 (2-5)	100.	253.20	921.64	251.667	263.864
8 (2-6)	100.	252.84	982.75	253.237	196.770
9 (3-7)	100.	34.19	36.93	31.341	32.260
10 (3-8)	100.	242.73	194.24	215.926	112.041
11 (4-9)	100.	34.11	30.09	39.421	38.966
12 (4-10)	100.	242.82	67.51	274.215	350.984
WEIGHT	43.69	22.950	66.086	22.948	30.1211
FROBENIUS NORM	17.170	20.014	21.969	21.168	20.263

Table 2. Constraints for Design III and IV

Constraints		Constraints	
$\xi_1 \geq 0.15$	$\omega_1 \geq 1.341$	$\xi_7 \geq 0.03$	$\omega_7 \geq 6.25$
$\xi_2 \geq 0.09$	$\omega_2 \geq 1.60$	$\xi_8 \geq 0.04$	$\omega_8 \geq 6.90$
$\xi_3 \geq 0.10$	$\omega_3 \geq 2.13$	$\xi_9 \geq 0.04$	$\omega_9 \geq 7.10$
$\xi_4 \geq 0.06$	$\omega_4 \geq 3.65$	$\xi_{10} \geq 0.04$	$\omega_{10} \geq 7.24$
$\xi_5 \geq 0.04$	$\omega_5 \geq 3.78$	$\xi_{11} \geq 0.03$	$\omega_{11} \geq 8.08$
$\xi_6 \geq 0.05$	$\omega_6 \geq 4.44$	$\xi_{12} \geq 0.02$	$\omega_{12} \geq 8.64$

Table 3. Structural Frequencies (ω_j^2) of Acoss-Four

Mode	Design A Nominal	Design I	Design II	Design III	Design IV
1	1.80	1.81	2.08	1.81	1.93
2	2.77	2.55	3.13	2.58	2.85
3	8.35	4.61	5.57	4.61	4.89
4	8.74	13.39	15.91	13.33	13.37
5	11.54	14.31	33.49	14.35	16.22
6	17.67	19.84	39.31	19.84	19.99
7	21.73	39.11	85.79	39.10	39.22
8	22.61	47.69	94.51	47.65	47.59
9	72.92	50.63	101.50	50.55	50.69
10	85.57	52.56	177.30	53.28	66.05
11	105.77	65.45	202.70	65.37	74.91
12	166.54	74.82	231.30	75.25	111.20

Table 4. Modal Damping Parameters of the Closed-Loop Eigenvalues

Mode	Design A Nominal	Design I	Design II	Design III	Design IV
1	0.0546	0.1500	0.1689	0.1509	0.1525
2	0.0653	0.1000	0.1436	0.0902	0.1054
3	.0.0737	0.1141	0.1159	0.1157	0.1107
4	0.0801	0.0674	0.0667	0.0650	0.0606
5	0.0839	0.0367	0.0475	0.0401	0.0661
6	0.0864	0.0567	0.0369	0.0568	0.0516
7	0.0760	0.0344	0.0255	0.0334	0.0478
8	0.0723	0.0377	0.0269	0.0444	0.0406
9	0.0341	0.0494	0.0179	0.0471	0.0414
10	0.0298	0.0484	0.0163	0.0452	0.0400
11	0.0207	0.0403	0.0239	0.0387	0.0304
12	0.0064	0.0252	0.0187	0.0266	0.0200

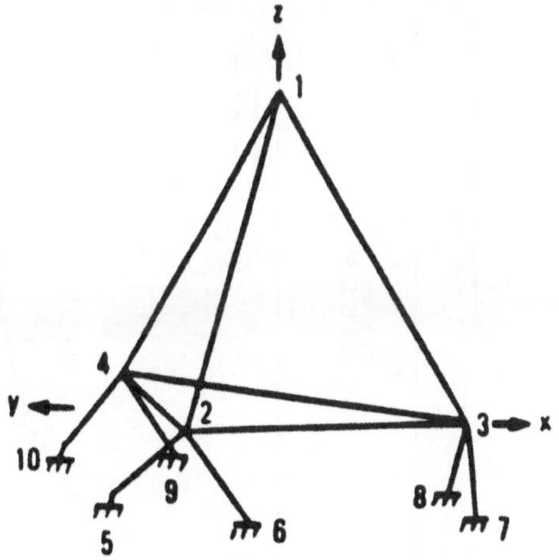

Fig. 1 ACOSS-FOUR Model

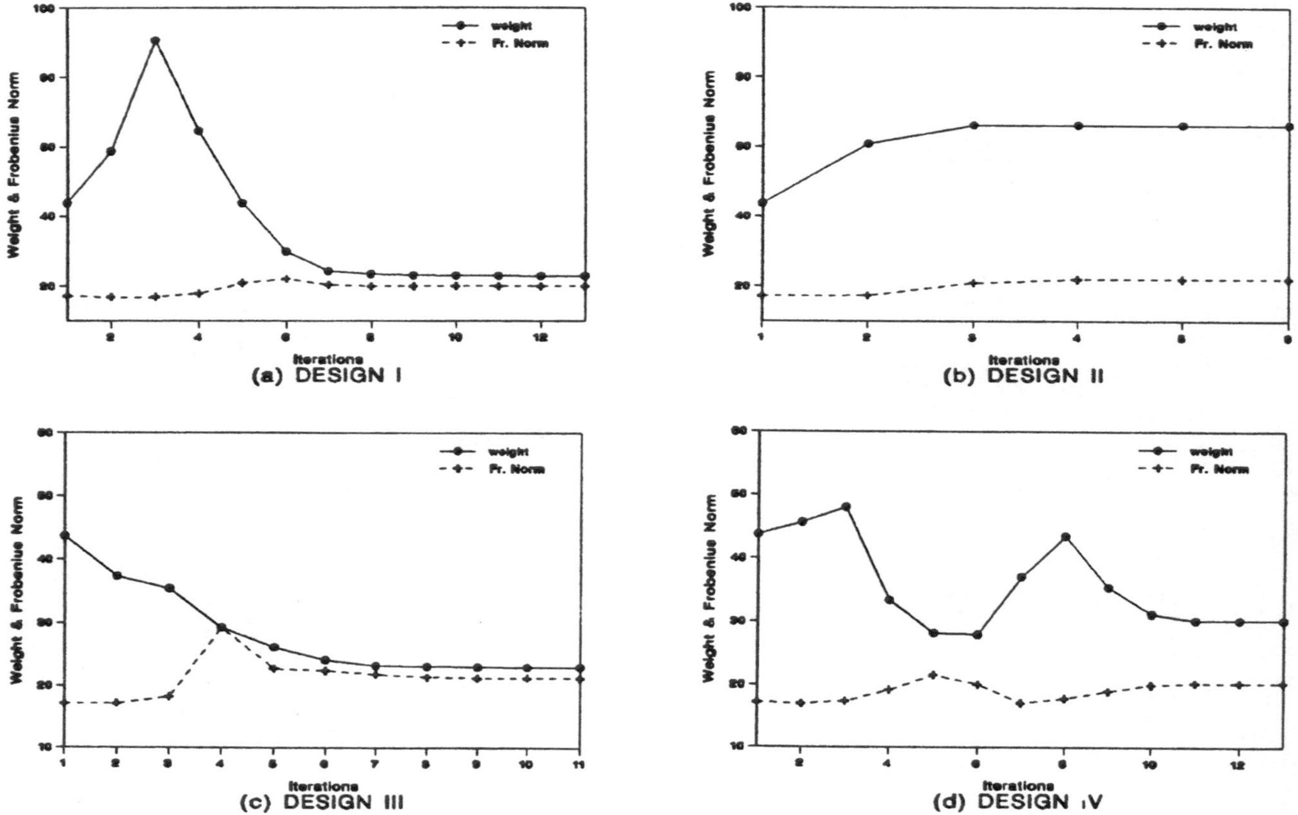

Fig. 2 Design Iteration History

MULTICRITERION PLATE OPTIMIZATION

J. Koski and R. Silvennoinen
University of Oulu and
Tampere Univ. of Technology
Linnanmaa
SF-90570 Oulu
Finland

M. Lawo
Kernforschungszentrum
Karlsruhe GmbH
Postfach 3640
D 7500 Karlsruhe 1
Bundesrepublik Deutschland

ABSTRACT. A multicriterion plate problem, where the material volume and the lowest natural frequency as well as some nodal displacements have been chosen as criteria, is formulated and methods for generating Pareto optima for it are discussed. A three-criterion plate example, where the static and dynamic analysis is based on the finite element method, is presented as an application.

1. INTRODUCTION

The first applications of the concept of Pareto optimality to structural optimization appeared in the scientific literature in the late 1970's and during the last decade a variety of articles based on the multicriterion approach has been published in the field of structural mechanics [1], [2]. Most of the applications treat a structural weight and some other quantities, usually associated with the stiffness of a structure, as the criteria to be minimized simultaneously. Generally, Pareto optima for these problems have been generated by different methods and some effort has been devoted to find the best Pareto optimal structure by applying an interactive approach.

In this paper the multicriterion optimization has been applied to plate design by formulating the problem and solving a few Pareto optima for it. The purpose here is to consider the possibilities of the multicriterion approach, in connection with the finite element analysis and optimization, to solve a plate bending design problem where in addition to the material volume also the static flexibility and the dynamic behaviour of the structure are used as criteria.

G. I. N. Rozvany and B. L. Karihaloo (eds.), Structural Optimization, 159–167.

2. MULTICRITERION PLATE PROBLEM

Several conflicting requirements may appear in plate design. Usually it should be a light-weight structure which can support given loads and which is stiff enough for the aimed purpose. This kind of plate is often sensitive to different dynamic perturbations. In cases where the frequency of the excitation is known the designer generally tries to avoid the situation where it is equal to some natural frequency of a structure. If the lowest of the natural frequencies can be raised over the dangerous value then also all the other natural frequencies are on the safe side. In practical applications, however, it is often difficult to impose exact limits for such quantities as natural frequencies or displacements. The safety of a structure can be increased by raising the lowest natural frequency from the imposed value but, on the other hand, the material volume of a structure will usually increase simultaneously. Even more clear is the conflict between the material volume and the displacements.
It seems reasonable to choose the lowest natural frequency and some important displacements Δ_i as criteria in addition to material volume V, instead of treating them as constraints. Accordingly, multicriterion plate problem P_m can be stated as

$$\min_{\underline{x}\in\Omega} [V \quad \omega \quad \Delta_1 \quad \Delta_2 \ \ldots \ \Delta_{m-2}]^T \tag{1}$$

where the vector objective function consists of m criteria. Second criterion $\omega^2 = -\min_i \mu_i$ where μ_i is the i-th eigenvalue of the natural vibration equation shown in (2). Vector of optimization variables $\underline{x} = [t_1 \ t_2 \ \ldots \ t_n \mid r_i \ \mu_i \ y_1^i \ \ldots y_N^i]^T$, i=1, 2,..., N, where plate thicknesses t_i are the design variables which determine the shape of a plate, and nodal displacements r_i as well as eigenvalues μ_i and the corresponding eigenvectors $\underline{y}^i$ are the state variables associated with the system equations. Consequently, the feasible set is defined by

$$\Omega = \{\underline{x}\in R^k \mid \underline{K}\ \underline{r} = \underline{R},\quad \underline{\sigma} = \underline{C}\ \underline{r},\ \sigma_i^l \le \sigma_i \le \sigma_i^u,\ t_i^l \le t_i \le t_i^u,$$
$$(\underline{K} - \mu_i\ \underline{M})\underline{y}^i = \underline{0}\ \} \tag{2}$$

It is possible to reduce the dimension of problem (1) from m to 2 by combining ω and all displacement criteria Δ_i linearly into another criterion in addition to V. This is motivated because the material volume is strongly conflicting with all the other criteria but these are usually not very competing. This results in reduced problem P_2 expressed as

$$\min_{\underline{x}\in\Omega} [\, V \quad \lambda_0\omega + \sum_{i=1}^{m-2} \lambda_i \Delta_i \,]^T, \qquad \lambda_i > 0, \quad \sum_{i=1}^{m-2} \lambda_i = 1, \tag{3}$$

which has the lowest possible dimension to be still a multicriterion problem. From computational viewpoint, it is more advantageous than problem P_m, because the number of constraints is decreased in solving Pareto optima by the constraint method. This formulation is also flexible for the designer because it has two strongly conflicting criteria but all the original criteria of problem P_m have been preserved using parameters λ_i by which the importance of the criteria can be controlled.

The minimum values of conflicting criteria in Ω cannot be obtained at the same design point. In multicriterion problems it is reasonable to apply the concept of Pareto optimality which for problem P_m is defined as:

A vector $\underline{x}^*\in\Omega$ is Pareto optimal for problem (1) if and only if there exists no $\underline{x}\in\Omega$ such $V(\underline{x})\leq V(\underline{x}^*)$, $\omega(\underline{x})\geq\omega(\underline{x}^*)$, $\Delta_i(\underline{x})\leq \Delta_i(\underline{x}^*)$ for $i=1,2,...,m-2$, and a strict inequality (<) occurs for at least one criterion.

Instead of one optimal solution usually a set of Pareto optima is characteristic to a multicriterion problem.
Corresponding vectors $\underline{Z}^*=[V(\underline{x}^*)\ \omega(\underline{x}^*)\ \Delta_1(\underline{x}^*)\ ...\Delta_{m-2}(\underline{x}^*)]^T$ in criterion space R^m are called minimal solutions. In the sequel, certain reduced problems and the constraint method are considered in solving Pareto optimal plates for problem P_m.

3. CHARACTERIZATION OF PARETO OPTIMA

In order to compute Pareto optima problem P_m is converted into a scalar optimization problem [3]. All Pareto optima can be found by minimizing V in the set Ω with additional constraints $\omega\geq\varepsilon_0$ and $\Delta_i\leq\varepsilon_i$ $(i=1,...,m-2)$ and varying upper limits $\varepsilon_j (j=0,1,...,m-2)$. When this constraint method is applied to the bicriterion problem P_2, the additional constraint is $\lambda_0\omega + \sum_{i=1}^{m-2}\lambda_i\ \Delta_i \leq \varepsilon$.

All the solutions of P_2 are also solutions to P_m, when the parameters λ_j are strictly positive. However, from the point of view of the designer, it makes no difference if small positive values for λ_j are used instead of $\lambda_j=0$. Thus it is possible to treat also those special cases of P_2 where some weights λ_j are zero.

In the sequel a three-criterion problem and its bicriterion version

$$\min_{\underline{x}\in\Omega} [V \;\; \omega \;\; \Delta]^T, \qquad \min_{\underline{x}\in\Omega} [V \;\; \lambda_0 \omega + \lambda_1 \Delta]^T \tag{4}$$

are considered.
Four different Pareto optima, which are common to both problems, can be found by the following natural approaches:

$$\text{(a)} \min_{\substack{\underline{x}\in\Omega \\ \omega\le\varepsilon_0}} V \qquad \text{(b)} \min_{\substack{\underline{x}\in\Omega \\ \Delta\le\varepsilon_1}} V \qquad \text{(c)} \min_{\underline{x}\in\Omega} V \qquad \text{(d)} \min_{\substack{\underline{x}\in\Omega \\ \omega\le\varepsilon_0 \\ \Delta\le\varepsilon_1}} V . \tag{5}$$

By choosing parameters λ_i properly it is possible to obtain a good overview of the minimal set by applying reduced problems (a), (b) and (c) only once. These can be interpreted as constraint method approaches for bicriterion problem (4) (cases $\lambda_1=0$, $\lambda_0=0$ or $\lambda_1=\lambda_0=0$).

4. NUMERICAL COMPUTATION OF PARETO OPTIMAL PLATE

4.1. Analysis

The analysis is based on the finite element method. Loads are applied dynamically, statical loading is treated as a special case.
In the following only structures idealized by plate finite elements are considered.
The displacement method is applied using a consistent mass matrix. Each node of the finite element mesh has three degrees of freedom: one displacement and two rotations. The equilibrium equations are

$$\underline{M}\,\ddot{\underline{r}} + \underline{K}\,\underline{r} = \underline{R}(t) \tag{6}$$

where $\underline{K}$ is the stiffness matrix, $\underline{M}$ is the mass matrix, $\underline{r}$ and $\ddot{\underline{r}}$ are the nodal displacements and accelerations and $\underline{R}(t)$ are the time dependent nodal loads.
A linear eigenvalue problem is solved, followed by a modal analysis. The loading is represented by means of response spectra. In this way loadings such as earthquakes, periodically running machines, impact etc. can be represented [4].
The analysis yields the natural frequencies, the nodal displacements and the maximum stresses at the design control points located on the top and bottom surface above and below

the centre of gravity of each element

$$\sigma_i = \sqrt{(\sigma_1^2 + \sigma_2^2)} \qquad (7)$$

with σ_1 and σ_2 as principal stresses.

4.2. Design and Optimization

Being independent from the finite element mesh, design variables and constraints are 'adjoined' to design groups. The finite elements belonging to the same design group all have their thickness expressed as design variable t_i.
The objective function is the scaled volume of the structuere. Therefor to reduce the risk of numerical defficiencies involved with digital computers the objective function is 1 at the beginning of each optimization process

$$\min\{V(\underline{t}) \equiv \sum_{i=1}^{n} t_i (\sum_{j=1}^{ni} A_j)/V^0\} \qquad (8)$$

where

$$V^0 = \sum_{i=1}^{n} t_i^0 (\sum_{j=1}^{ni} A_j)$$

and A_j is the area of a finite element, n is the number of design variables, ni is the number of finite elements of design group i, and V^0 is the unscaled objective function value at the beginning of the optimization process.
The stresses, the lowest natural frequency, important displacements and the design variables are constrained. The stress constraints are

$$G_i(\underline{t}) \equiv 1 - \sigma_i(\underline{t})/\sigma_i^o \geq 0 \qquad i=1,\ldots,n \qquad (9)$$

where $\sigma_i(\underline{t})$ is the maximum of all stresses at the control points within a design group.
By the frequency constraint a lower bound ε_0 for the natural frequencies is given by

$$G_{n+1}(\underline{t}) \equiv -\omega(\underline{t})/\varepsilon_0 - 1 \geq 0. \qquad (10)$$

The m-2 displacement constraints are used to restrict the deformation of the plate in m-2 predefined degrees of freedom

$$G_{n+1+j}(\underline{t}) \equiv 1 - |r_j(\underline{t})|/\Delta_j \geq 0, \qquad j=1,\ldots,m-2. \qquad (11)$$

Additionally are defined the side constraints in explicit form

$$t_i^l \leq t_i \leq t_i^u. \qquad (12)$$

5. THREE-CRITERION PLATE EXAMPLE

As a test example the clamped plate of fig. 1 is optimized. Two point loads R are assumed. Application of symmetry conditions reduces the size of the problem, hence one quarter of the plate is discretized by 50 elements. The finite element mesh is also given in fig. 1.

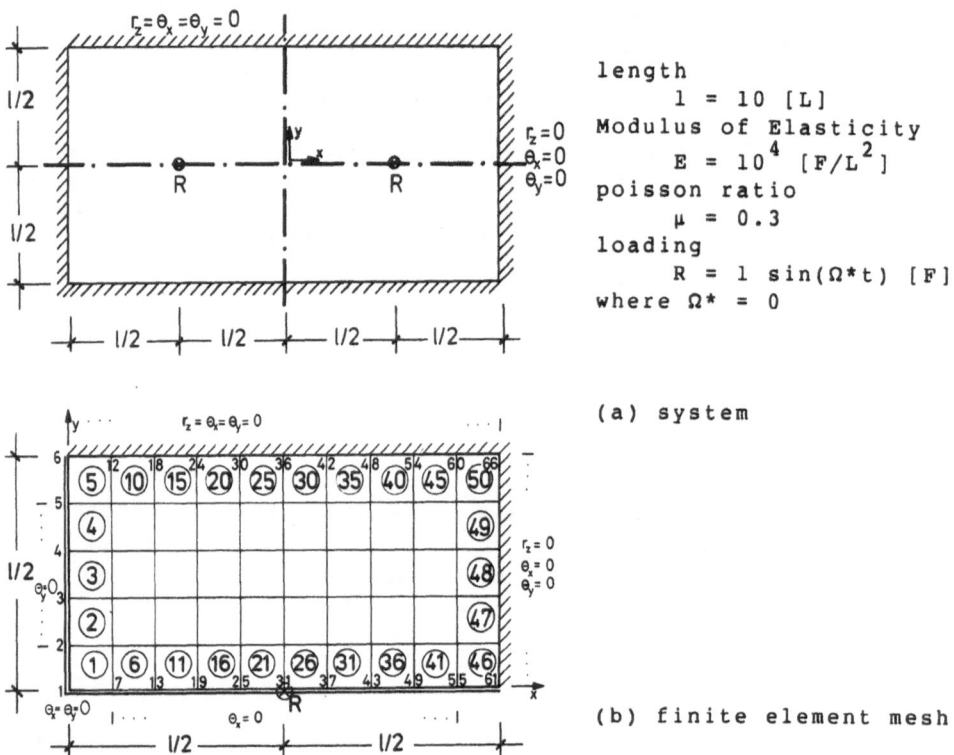

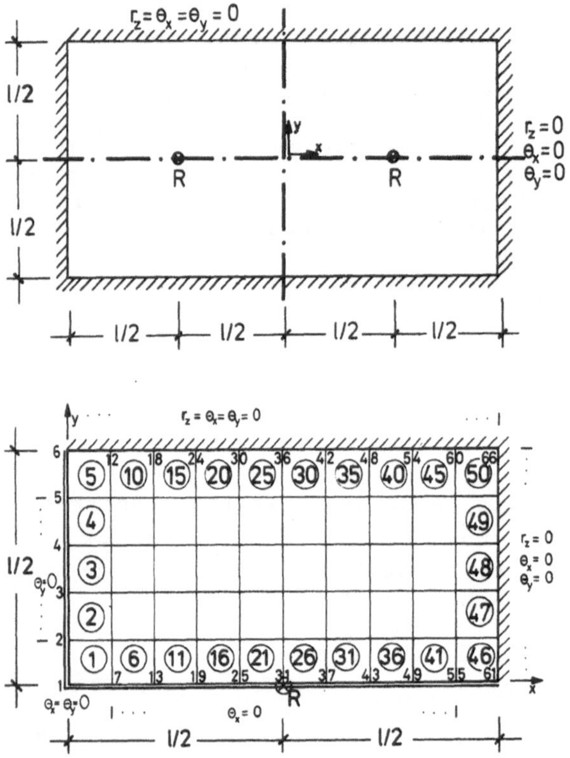

length

$l = 10$ [L]

Modulus of Elasticity

$E = 10^4$ $[F/L^2]$

poisson ratio

$\mu = 0.3$

loading

$R = 1 \sin(\Omega^* t)$ [F]

where $\Omega^* = 0$

(a) system

(b) finite element mesh

Figure 1. Test example

For the frequency analysis a consistent mass matrix is used. For the displacement and stress analysis an identity mass matrix is applicable. Using ten eigenforms in conjunction with a modal analysis yields results within 4% of a static analysis, whereas applying five eigenforms increases the error to 8%. Therefore for the displacement and stress analysis ten eigenforms are subsequently used.

To scale the problem and find pareto optimal solutions a plate with a constant thickness of 0.1 [L] was first analysed.

The lowest natural frequency of this structure was $\omega = 741.4$ $[T^{-1}]$, the maximal displacement at node 31 is $\Delta = 1.44$ [L], and the maximal stress $_{max}\sigma = 358.8$ $[F/L^2]$

The volume of the plate is V = 5.000 [L^3]. These values were taken as boundaries for the optimization. In this way one obtains as explained above different pareto optimal solutions.

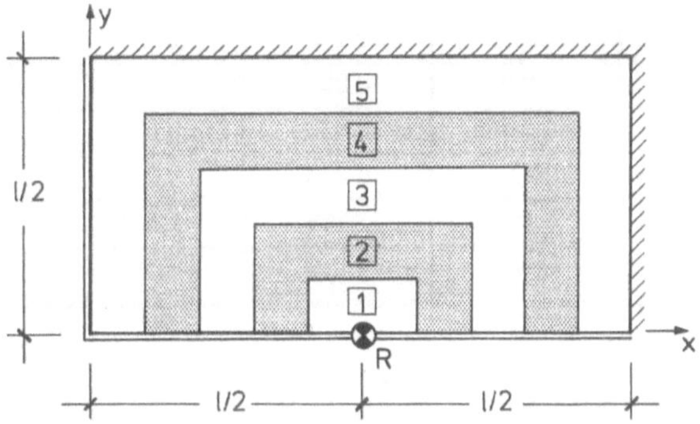

Figure 2. Definition of design variables

In fig. 2 five design groups defining the design variables are chosen. As side constraints we have

$$0.01\ [L] \leq t_i \leq 1.0\ [L].$$

With four different combinations of constraints the optimization was carried out on a SIEMENS 7536 using the finite element programm B&B [5]. By the sequential quadratic programming algorithm of Schittkowski [6] the different nonlinear programming problems were solved.
The combinations were as explained in chapter 3

(a) frequency and side constraints
(b) displacement and side constraints
(c) stress and side constraints and
(d) all four kinds of constraints.

The results are given in table I and fig. 3. The frequency constraint has the strongest influence on the results. Displacement and stress constraints produce similar results.
Since the results depend very much on the choice of the design variables a further computation with 25 design variables has been carried out for the frequency and the displacement constraint alone. The volume is 2.20 [L^3] for the frequency and 1.79 [L^3] for the displacement constraint. The qualitative results are given in fig.4. For further information see [7].

TABLE I

i	t_i^o [L]	(a) t_i [L]	(b) t_i [L]	(c) t_i [L]	(d) t_i [L]
1	0.1	0.101	0.178	0.253	0.145
2	0.1	0.073	0.118	0.127	0.096
3	0.1	0.035	0.059	0.060	0.043
4	0.1	0.065	0.053	0.032	0.071
5	0.1	0.123	0.043	0.031	0.115
Volume	5.0	4.107	3.166	2.871	4.366
CPUs	81	620	2083	1805	2954

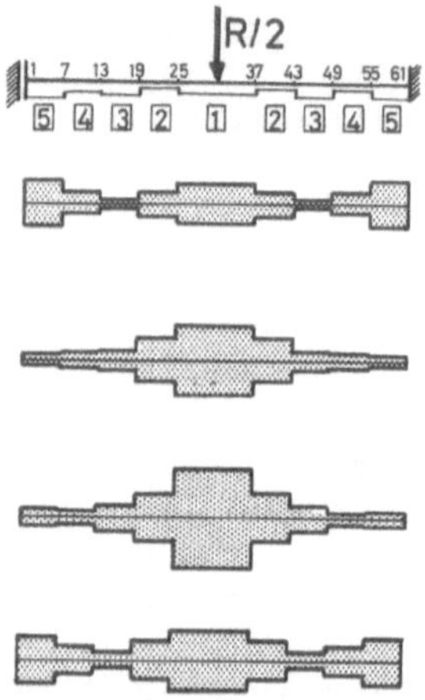

Figure 3. Results with five design variables

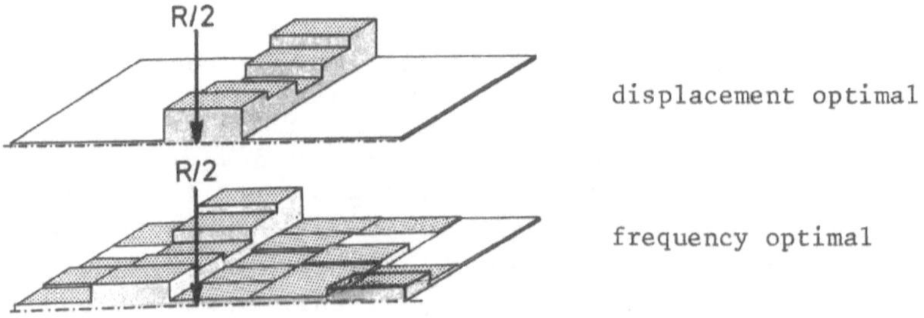

Figure 4. Results with 25 design variables

6. DISCUSSION

A multicriterion plate optimization problem has been formulated and different reduced problems, associated with the constraint method, have been applied to solve Pareto optimal plates. As can be observed from the plate example, the computing cost easily becomes very high even for one Pareto optimal structure. Usually more than ten alternatives must be generated for the decision maker during the design process. Consequently, different kinds of reduced problems should be utilized in generating Pareto optima for the original problem. Generally a module which generates Pareto optimal structures for the designer is an important part of the multicriterion design system.
The example shows that it is possible to get Pareto optimal solutions by using the finite element method, however there is still much to be done before the method can be used in practise. Solving the problem (1) requires a lot of programming to restructure the design and optimization modules of the existing package. To continue with special purpose finite element programs seems therefore to be reasonable.

REFERENCES

[1] Stadler,W.: 'Multicriteria Optimization in Mechanics'. Applied Mechanics Review 37 (1984), 277-286.
[2] Stadler,W.: 'Update to Multicriteria Optimization in mechanics'. Applied Mechanics Review 39 (1986), 417-420.
[3] Chankong,V.; Haimes,Y.Y.: Multiobjective Decision Making, North Holland, New York, 1983.
[4] Klingmüller,O.; Lawo,M.; Thierauf,G.: Matrix Structural Analysis Part II Dynamics (in German). Vieweg, Braunschweig,1983.
[5] Booz,G.; e.a.: B&B a Finite Element Programm for the Analysis and Design of General Structures (in German), Univ. Essen, Research Report 1985.
[6] Schittkowski,K.: 'Software for Mathematical Programming', in Computational Mathematical Programming, Ed. K. Schittkowski, Springer, Berlin, 1985, 383-451.
[7] Lawo,M.: Optimum Structural Design (in German). Vieweg, Braunschweig, 1987.
[8] Eschenauer,H.: Numerical and Experimental Investigations on Structural Optimization of Engineering Desigs, Univ. of Siegen, Bonn+Fries, 1986.
[9] Osyczka, A.: Multicriterion Optimization in Engineering, Ellis Horwood, 1984.

OPTIMIZATION OF SYSTEMS IN BENDING - CONJECTURES, BOUNDS AND ESTIMATES RELATING TO MOMENT VOLUME AND SHAPE

P.G. LOWE
Department of Civil Engineering
University of Auckland
Private Bag
Auckland
New Zealand

ABSTRACT; One of the achievements of optimization theory for plate systems in bending has been the exact solution of a broad range of problems of technical interest. Some of these exact solutions are remarkably simple and could be regarded as moderately practical to actually achieve as built components. But many are complicated, and it is likely that at present unsolved problems will prove to be even more complicated.

In the present study, the moment volume (V_o) is the primary quantity of interest. For some shapes of plates in bending V_o is easily calculated; for others the task is much more difficult. For many problems which have been essentially solved, no V_o values have been calculated, presumably because of the difficulties involved.

Certain conjectures which have been made previously are here applied to obtain V_o estimates, and also to study the role of plate shape as affecting moment volume. Some observations are made about the practicality of some basic solutions, since, it is argued, practicality must be taken into consideration if the potentials of optimization are to be more widely appreciated.

1. Introduction

In this paper only constant depth (thickness), uniformly loaded and edge supported plates in bending will be considered. Primary interest is in optimized plates where the objective function is the moment volume (V_o). This is a measure of the structural material needed to sustain the loading, and the most appropriate physical model is of a tensile weak matrix incorporating suitable (fibre) reinforcement to sustain the tensile stresses generated. A practical application might be to reinforced concrete plates subject to transverse loading.

G. I. N. Rozvany and B. L. Karihaloo (eds.), Structural Optimization, 169–176.

Certain conjectures have been proposed [1]. These are restated below, and are used to obtain bounds for the moment volume required in specific examples, as well as to indicate trends for moment volume requirement as the plate shape is changed.

These conjectures have been applied for the most part to ductile isotropic plates at collapse according to the simplest yield criteria. One feature of these studies has been the paucity of exact solutions against which to compare the conjectures. Since there are very many more known exact solutions for optimized plates, many more relevant comparisons can be made. In turn, if successful, these comparisons may give further confidence about the validity of the conjectures for a range of plate types, including optimized plates.

It is attractive to seek absolute optima, if they are attainable, but in some manner questions of practical realisation of the optimal designs may need to be incorporated early in the problem solution. An instructive example is the obtuse corner in a simply supported plate configuration. If a limit on transverse curvature is not imposed then a solution which is much simpler but not wholly optimal can be obtained. How such a local violation of an optimal requirement affects the moment volume requirements is studied.

2. The Conjectures

Consider a uniform thickness plate, edge supported in a uniform manner and uniformly loaded by a pressure (p) so as to produce a limit state in the plate. In the case of an isotropic plate the limit state pressure to cause collapse in a ductile manner with the formation of sufficient yield lines for a mechanism of collapse to occur. For an optimized plate the limit state pressure will be denoted by p_o, the pressure to cause a uniform tensile reinforcement stress throughout the plate. How the reinforcement is arranged (the layout) will form part of the solution, and implied also is a mechanism of deformation (collapse) which in such cases should be a smooth displaced shape.

In all cases an equilibrium moment field ($M_{\alpha\beta}$, α, β = 1, 2) could or can be found. For the <u>isotropic</u> case the available strength (M_α), the strength assigned in two orthogonal directions (α = 1, 2) at each point of the plate, is the <u>same</u> at <u>all</u> points and in <u>all</u> directions. In the <u>optimized</u> case the available strength (M_α) provided equals the <u>principal values</u> of $M_{\alpha\beta}$ at every point, and this information forms part of the solution.

The mechanical quantity of interest is the <u>Moment Volume</u> (V) which is defined to be

$$V = \int_A (|M_1| + |M_2|)\, dA \,, \tag{1}$$

where the integration extends over the whole (plan) area (A) of the plate.

Then the First Conjecture is that the parameter δ, defined by

$$\delta \equiv \frac{pA^2}{V}, \tag{2}$$

has a lower limiting value [1]. No proof will be offered here; instead various tests will be applied to check the performance of the conjecture.

For a simply supported optimized plate, the lower limit is 4π; if the edge is clamped it is $96\pi/7$. These two values correspond to the known (exact) solutions for the circular shape of plate which attains a minimum (optimum) value of $V_o = \pi/4 \,.\, p_o R^4$ (R = Plate radius) for the simply-supported edge case and uniform limit pressure p_o, and $V_o = 7\pi/96 \times p_o R^4$ for the clamped edge case.

$$\text{Hence} \quad \delta \geq 4\pi \quad \text{or} \quad \frac{96\pi}{7} \tag{3}$$

according to edge support type, and the equality holds only for the circular shape.

As the plate plan-form (shape) changes from circular, the value of δ increases, and continues to increase the more elongated the shape becomes. For isotropic simply supported plates governed by the square yield locus (M_o, yield moment|unit length) the corresponding statement is

$$\delta \equiv \frac{p_c . A^2}{V_I} \geq 1.5\pi \tag{4}$$

Now, $V_I = 4A\,M_o^*$ and hence

$$\frac{p_c . A}{M_o} \geq 6\pi \tag{5}$$

which is the result proved by Schumann [2].

---

* The equivalent expression in [1] reads $V_I = 2A\,M_o$, but this assumes only one face of the plate reinforced rather than both, as assumed by the square yield locus.

3. Isoperimetric Properties

The classical isoperimetric inequality relates the geometrical quantities of area (A) enclosed by the perimeter (B) of a plane closed curve through the inequality

$$\alpha \equiv \frac{B^2}{A} \geq 4\pi , \tag{6}$$

where the equality holds only for the circular shape [3]. As for δ, so too for α, as the shape of A becomes more elongated, less area is enclosed for a given perimeter (B) and hence the value of α rises without limit.

Now it can be seen that the First Conjecture is in the nature of an isoperimetric inequality and is the expression of the physical observation that of all the plates of a given area and support type, the circular shape is the weakest, namely requires the largest moment volume (V) to sustain a given equilibrium (collapse) pressure (p).

Though (3) has usefulness, it is believed that a much more useful parameter is $\underline{\beta}$ defined by

$$\underline{\beta} \equiv \frac{\delta}{\alpha} \tag{7}$$

and leading to the Second Conjecture namely that

$$\underline{\beta} \geq 1 \tag{8}$$

for simply supported optimised plates.

One particular usefulness of $\underline{\beta}$ is that as the shape of a plate becomes more elongated, $\underline{\beta}$ has a finite upper limiting value of 3 ([4], for the isotropic counter part).

Hence, the full second conjecture for edge simply supported optimized plates of arbitrary shape (simply connected)) is

$$1 \leq \underline{\beta} < 3 . \tag{9}$$

The corresponding expression for clamped edge optimized plates is

$$3\frac{3}{7} \leq \underline{\beta} < 8 . \tag{10}$$

The inequalities (9) and (10) are intended to include all possible shapes of edge supported plates. In some circumstances there may be advantage in restricting the shapes in some manner. For example, if shapes are restricted to polygons of n (or fewer) sides then provided that the solution for the <u>regular n sided polygon</u> is available ($\beta = c_n$ say) then

$$c_n \leq \beta < 3 \tag{11}$$

for all n (or fewer) sided polygonal simply supported plates. This is an example of a "little" isoperimetric inequality [3].

For example, suppose n = 4, then the regular shape is the square (side L) and it is known that $V_o = 5/96\ pL^4$ for the simply supported case [5]. But $\alpha = (4L)^2/L^2 = 16$ and hence

$$c_4 = \beta = \frac{p_o (L^2)^2}{\frac{5}{96}\ pL^4} \times \frac{1}{16} = 1.2$$

or $1.2 \leq \beta < 3$ for all four (and three) sided shapes. (12)

A safe (i.e. conservative) conclusion to be drawn is that

$$\beta = 1.2 \tag{13}$$

for all simply supported polygonal shapes with four or three sides.

Hence for a regular triangular shape of plate $\alpha = 20.78$, and an estimate for V_o is

$$V = \frac{pA_o^2}{1.2\alpha} = \frac{5}{6} \cdot \frac{p_o \cdot \left(\frac{\sqrt{3}}{4}\right)^2 L}{20.78} = 0.00752\ p_o L \quad .$$

This compares with the known exact solution of $V_o = 0.00675\ p_o L^4$. Any non-regular triangle can be treated likewise.

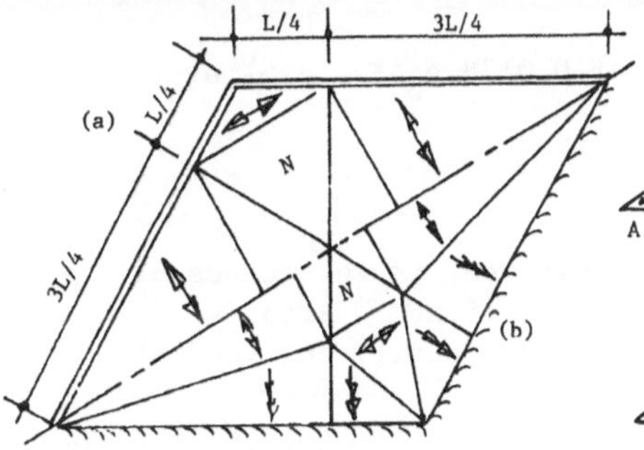

Fig. 1 : Rhombic Plates
(a) Simply Supported (b) Clamped

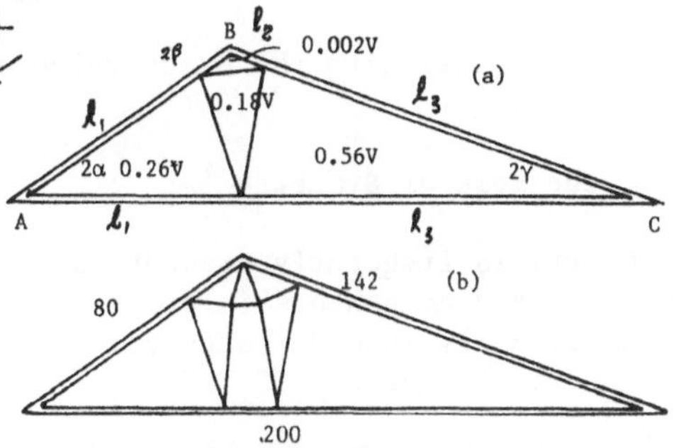

Fig. 2 : Triangular Plate
(a) Restricted (b) Absolute Optimum

Consider the rhombic shape (Fig. 1) and apply (13) where now $A = 2RL = \sqrt{3}L^2/2$, $B = 4L$ ($R = \sqrt{3}L/4$, the radius of the in circle), $\alpha = 18.48$ and hence

$$V = \frac{pA^3}{1.2\ B^2} = \frac{p(\sqrt{3}L^2/2)^3}{1.2(16L)^2} = 0.0338\ pL^4 = 0.0451\ pA^2 \quad .$$

These values compare with the "exact" values of $V_o = 0.0321\ pL^4 = 0.0429\ pA^2$ - showing the appropriate upper bound values to be about 5% conservative. Note, too, that this particular comparison needs further investigation, since the "exact" value is based on a non-optimal obtuse corner zone - see later.

If instead of comparisons for simply supported boundaries, the clamped edge is considered, then (10), when restricted to polygons of four or fewer edges gives for the square

$$\underset{\sim}{\beta} \geq \frac{\delta}{\alpha} = \frac{p_oA^2}{V_o} \cdot \frac{1}{\alpha} = \frac{p_oA^2\ 768}{13\ p_oA^2} \cdot \frac{1}{16} = 3.69$$

or $3.69 \leq \underset{\sim}{\beta} < 8$. (14)

A conservative estimate for V_o for the regular triangular clamped-edge shape would be

$$\underset{\sim}{\beta} = 3.69 \ , \quad \text{or} \quad V = \frac{p_oA^2}{3.69\ \alpha} = \frac{p_oA^2}{3.69} \cdot \frac{1}{20.78} = 0.0130\ p_oA^2 \ .$$

This compares with an exact calculation of $V_o = 0.0129\ p_oA^2$.

Similarly, for the rhombic shape,(Fig. 1), now the estimate for V will be $V = \dfrac{p_oA^2}{3.69\alpha} = \dfrac{p_oA^2}{3.69 \times 18.48} = 0.0147\ p_oA^2$

which compares with the exact value of $0.0138\ p_oA^2$ - a 6% over estimate.

4. The Role of Symmetry

Symmetry is frequently present, or is desired, in the shapes of members and components in practical applications. Also, symmetrically shaped plates are usually more easily analysed than non-symmetrical shapes.

Consider two plates of equal area (A) and similar edge support type, then a consequence of the conjectures is that the Moment Volume requirement for the <u>more</u> symmetrical shape is <u>greater</u> than for the shape with less symmetry. This observation can be tested by numerical examples although a proof is not being offered here.

As an example, consider a series of triangular shapes as specified in Table I. All the shapes have a common area of $\sqrt{3L^2/4}$, and it can be seen that as the shape moves further away from symmetry, so the optimum moment volume reduces.

This trend will therefore endow symmetrical structural components with a certain degree of <u>stability</u> - in the sense that accidental lack of symmetry will reduce demand for moment volume.

ℓ_1/L	ℓ_2/L	ℓ_3/L	2α	2β	2γ	V/pA^2
0.251	1.065	0.407	90.00	26.57	63.43	0.0264
0.273	0.658	0.658	90.00	45.00	45.00	0.0316
0.500	0.500	0.500	60.00	60.00	60.00	0.0361

Table 1 : Shape v Moment Volume

5. Aspects of Practicality

Some optimal layouts are relatively simple, and may be realised in practical situations, whereas others are excessively complicated. Where additional simplicity of layout can be achieved by departing to some degree from the absolute optimum demands, this may well be justified on practicality grounds.

A good example is provided by the obtuse corner in simply supported optimal plates. As is well known, the optimality requirement of $|\kappa_2| \leq k$ = a prescribed constant, where κ_2 is the principal curvature transverse to the spanning direction, is not met by the simple cross corner spanning layout [6] in obtuse corners. But the important question is, what are the consequences for moment volume requirements and how is simplicity of layout affected?

Rozvany [7] et al have proposed a corner layout for such simply supported obtuse corners which does meet the $|\kappa_2| \leq k$ requirement, but there appear to have been no calculations of moment volume made. Such calculations are probably at best likely to be numerical integrations, compared with elementary closed form calculations for the simpler non-optimal alternative.

But indications are that the differences are very small, and the zones affected by the non-optimal obtuse corners, when used as a simpler alternative, are also small and shrink in size as the non-optimality becomes more acute. An example is considered in Fig. 2, where it is seen that the non-optimal corner region is likely to have an insignificant effect on the global moment volume, while at the same time greatly simplifying the spanning regime in the corner region as compared with the absolute optimal alternative. But the comparison is incomplete, since moment volume values for the rigorous case are not available. Note, however, that the conjectures are not violated by the restricted optimal results.

6. Conclusions

Optimization studies provide avenues into understanding more than just the immediate goals of the optimization. Many problems can be solved in detail and in closed form.

But practicality considerations need to be given more prominence. The conjectures studied here perhaps offer some scope to study trends in a new way, and provide a link to non-optimized plate bending studies which fits with practicality needs.

References

1. P.G. Lowe, 'Isoperimetric inequalities in structural mechanics' 9th Australasian Conference on Mechanics of Structures and Materials, 147, Sydney (1984)

2. W. Schumann, 'On isoperimetric inequalities in plasticity', Quart. App. Maths 16, 309 (1988)

3. G. Polya and G. Szego, Isoperimetric Inequalities in Mathematical Physics, Princeton Univ. Press (1951)

4. M.J. Beamish, University of Auckland, Bachelor of Engineering Project Report, Unpublished (1982)

5. C.T. Morley, 'The minimum reinforcement of concrete slabs', Int. J. Mech. Sc. 8, 305 (1966)

6. P.G. Lowe and R.E. Melchers, 'On the theory of optimal constant thickness fibre-reinforced plates II', Int. J. Mech. Sc. 15, 157 (1973)

7. G.I.N. Rozvany, Optimal Design of Flexural Systems, Pergamon Press (1976)

ON THE OPTIMAL DESIGN OF COLUMNS SUBJECTED TO CIRCULATORY LOADS

O. Mahrenholtz

Technical University
of Hamburg- Harburg

R. Bogacz

I.F.T.R.
Polish Academy of Sciences

ABSTRACT

Present paper is an extended review of results previously obtained by the authors, supplemented by some new results dealing with the Beck - Reut column optimization. The possibility of determination of a shape of column with the critical load higher than considered so far as "optimal" is presented.

1. INTRODUCTION AND REVIEW OF PREVIOUS INVESTIGATIONS

The study of the behaviour of structures subjected to circulatory loading is an important problem in engineering practice. In the past, several research workers have investigated this problem from the point of view of stability. Another aspect, which is equally important, is the optimal design of columns. This also has been studied by several authors. Niordson [1] was probably the first who considered the optimal design of vibrating beams. Prager and Taylor [2] have developed a unified theory of optimal design under stability and frequency constraints. The optimal shape of cantilevers under circulatory forces has been investigated by Zyczkowski and Gajewski [3,4] , Vepa [5], Claudon [6], Plaut [7] and others. It may be mentioned here that the optimal forms, which lead to a continuous variation of cross section, are not of much practical utility. On the other hand, stepped columns and beams have always found wider application. The optimal segmentation of a column for a maximum value of the force or optimal design under a given value of follower force for minimum costs has been previously studied by the authors in [8,9] .

The shape of segmented columns obtained by minimization of the volume for the case of bending stiffness proportional to $(mA)^2$ similar as in [5 - 9] and [11,12] is shown in Fig. 1 (Beck's and Reut's columns). The optimization results, obtained for a continuous mass distribution by using approximate methods, were in many respects worse than those obtained in [8], using exact closed form solutions. Application of approximate methods can lead to significant differences in determination of the critical load or optimal shape. E.g., the result of force maximization for a column made of two segments gives a 6% higher critical force than in the case of transfer matrix technique for the same shape [8, 9]; also the critical forces obtained in [6] and [11] using the same optimality criteria differ from each other about 20%, as shown in Fig. 2 (b).

G. I. N. Rozvany and B. L. Karihaloo (eds.), Structural Optimization, 177–184.

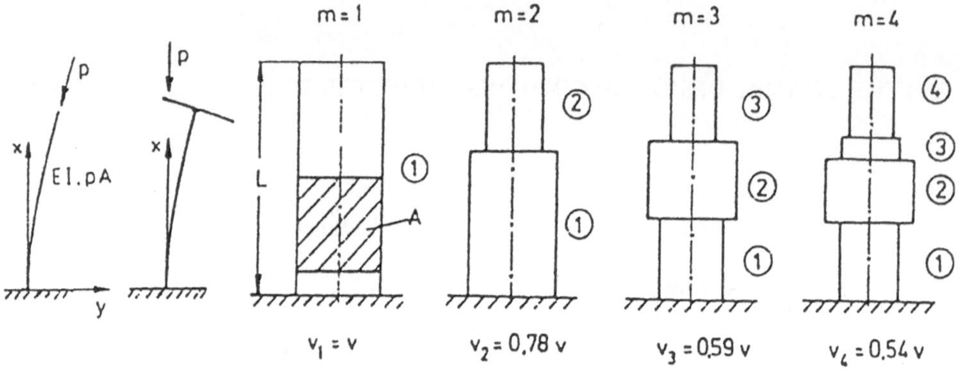

Fig. 1. Beck's and Reut's columns. Results of optimal segmentation for minimum of volume

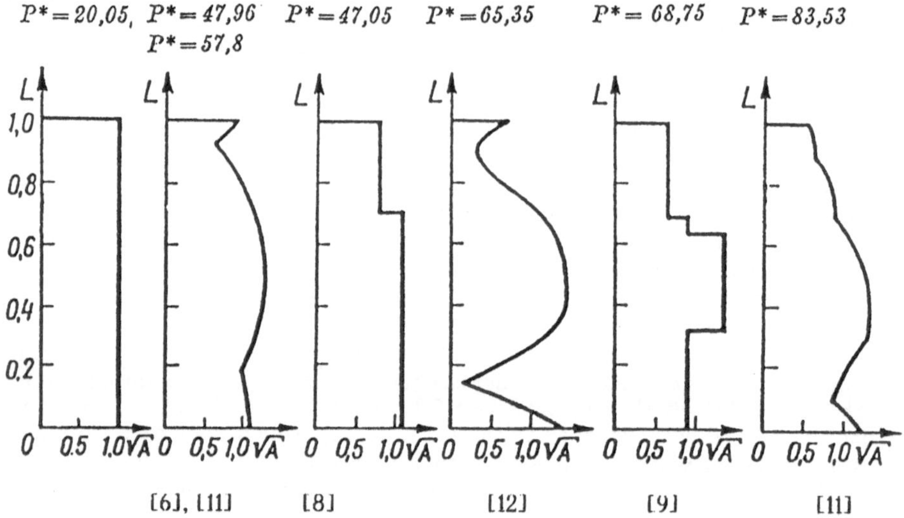

Fig. 2 . Shape of columns subjected to concentrated circulatory load obtained as result of force optimization

Thus, results for the stepped columns consisting of two segments only and, subsequently, for three and four-segment columns were mostly better than the former (Fig. 2). Fig. 2 f shows the latest and best results of optimization obtained by Hanaoka and Washizu [11] for a continuous mass distribution of the column by using a finite element method and optimality condition similar to the one proposed by the authors in [8]. Critical load in this case takes the value of P^*= 83.78 and it is the highest value in the case of continuous variation of the cross-section of column (but without any possibility of error estimation). The resulting optimal shape of the column in the case of maximization of load for a given value of volume (cost) leads to a different result as for the inverse problem- optimal design for a given value of critical load with the constraint $A_j < A$.

This fact is easily observed in Fig. 3, where the result of calculation for Hauger's column are presented. Optimization procedure was similar as in the case of Beck's column but the functional was obtained by use of Leipholz' generalization of the adjointness principle.

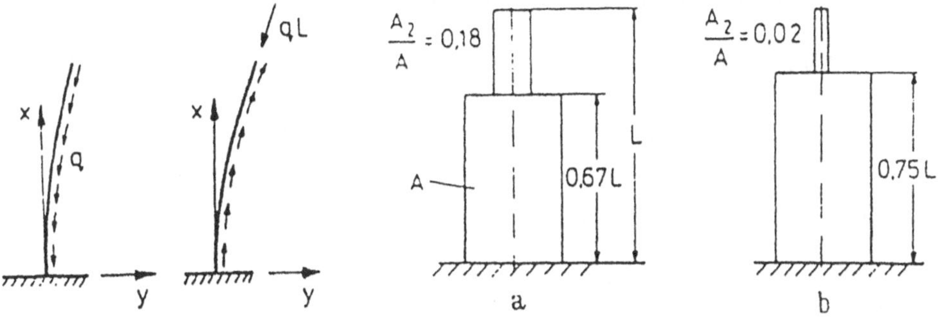

Fig. 3. Columns with distributed circulatory load; optimal shape of Hauger's column composed of two segments for the case of the minimum of volume (a) and the maximum of load (b).

It has been shown in [9], [13, 14] that the stiffness and location of one support affects greatly the critical load, while the stiffness of an elastic foundation has no essential influence on the critical load [15]. Also the dampers used as structural members acting at several distinct points on the column can affect greatly the critical load - much more than the distributed damping [7], [16, 17]. The general case of a structure consisting of elastic segments and viscoelastic or elasto-plastic suports was considered in [14]. The present paper is an extended rewiew of results previously obtained by the authors, supplemented by some new results dealing with the column optimization. Because the variational approach to optimization of columns is much better known than the transfer matrix technique, we confine the considerations in this paper to the latter accounting for a local loss of rigidity in the direction of shear force action as well as the local loss of bending stiffness. We illustrate those problems by the analysis of the critical loading for the Beck - Reut - problem showing that the above - mentioned phenomena may lead to a considerable increment of critical loading.

2. FORMULATION OF BECK'S PROBLEM

In what follows, the structure, shown schematically in Fig. 4, shall be considered. It consists of segments connected by elastic hinge joints located at positions $x_1, x_2, \dots, x_n$ and characterized by the stiffness parameters $k_1, k_2, \dots, k_n$, respectively.

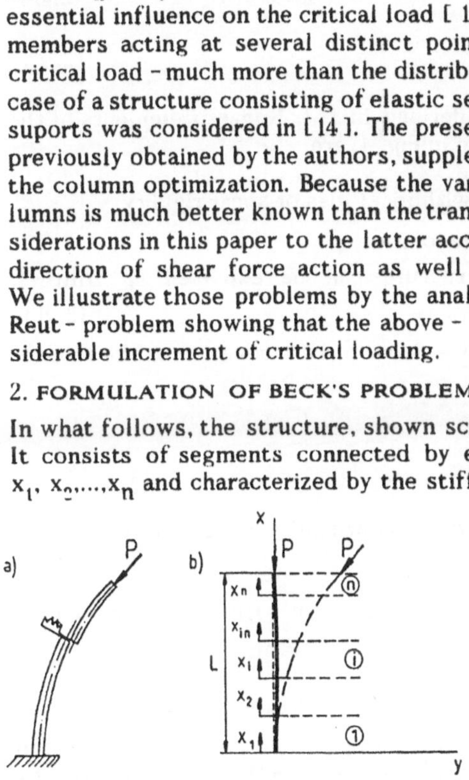

Fig. 4. Segmentation of column

The simplest form of the equation of motion for a uniform segment reads:

$$EI \frac{\partial^4 y}{\partial x^4} + P \frac{\partial^2 y}{\partial x^2} + mA \frac{\partial^2 y}{\partial t^2} = 0, \quad (2.1)$$

where

I - bending stiffness
P - longitudinal force
m - density
A - cross-sectional area.

The boundary conditions for the case of a clamped end are

$$y = 0, \frac{\partial y}{\partial x} = 0. \tag{2.2}$$

For the case of a free end with tangential force we have

$$\frac{\partial^2 y}{\partial x^2} = 0, \quad \frac{\partial}{\partial x}\left(EI\frac{\partial^2 y}{\partial x^2}\right) = 0. \tag{2.3}$$

The exact solution for this segment of constant mass and stiffness distribution has the form:

$$y(x,t) = e^{i\omega t}\left(A_1 \operatorname{sh}\lambda_1 x + A_2 \operatorname{ch}\lambda_1 x + A_3 \sin\lambda_2 x + A_4 \cos\lambda_2 x\right). \tag{2.4}$$

where

$$\lambda_{1/2} = \left(\frac{\pm P}{2EI} + \sqrt{\frac{P^2}{(2EI)^2} + \frac{\rho A\omega^2}{EI}}\right)^{1/2}. \tag{2.5}$$

Since all dependent variables y, φ, M, Q have a similar constitutive form (2. 4) , the state vector S and the partial transfer matrix T_i can be expressed as follows:

$$\mathbf{S} = [\, y, \varphi, M, Q \,]^T = [\, y, y', -EIy'', -EIy''' \,]^T. \tag{2.6}$$

$$\mathbf{S}^o_{i+1} = \mathbf{T}_i \mathbf{S}^o_i \; ; \quad \mathbf{S}^o_j = \mathbf{S}_j(x_j = 0). \tag{2.7}$$

The transfer matrix for the segment is defined in [8]. Nonzero elements of the transfer matrix for such a joint, as shown in Fig. 4, are:

$t_{ii} = 1$, $t_{14} = \varkappa_S$ for the case of joint with local loss of shear rigidity,

and $t_{23} = \varkappa_R$ for the case of a hinge joint.

The transfer matrix for the whole structure can be expresed as follows:

$$\mathbf{T} = \mathbf{T}_n \mathbf{T}_{n-1} \cdots \mathbf{T}_2 \mathbf{T}_1. \tag{2.8}$$

Satisfying the boundary conditions, we get a characteristic equation as the relation between force and frequency,

$$\begin{vmatrix} t_{33} & t_{34} \\ t_{43} & t_{44} \end{vmatrix} = \Phi(P, \omega) = 0. \tag{2.9}$$

Then making use of the gradient method and optimality criteria given in [20] an optimal shape of column is determinated.

3. RESULTS OF THE NUMERICAL CALCULATION FOR THE CASE OF LOCAL LOSS OF STIFFNESS

3. 1 Local discontinuity of displacements

Let us consider the behaviour of a column consisting of two segments connected by an elastic joint with stiffness in the shear force direction, characterized by a parameter $\varkappa_s$,

$$\Delta y\,(x_j) = \frac{1}{\varkappa_{sj}}\,Q\,(x_j) = \gamma_j\,Q\,(x_j), \tag{3.1}$$

where $\varkappa_{sj} = k_{sj}L^3/EI$ and k_{sj} is the stiffness of the j-th joint. The relations illustrating the critical force versus joint location for $\varkappa_{1s} = \varkappa_s \to 0$ are shown in Fig. 5 a and 5 b.

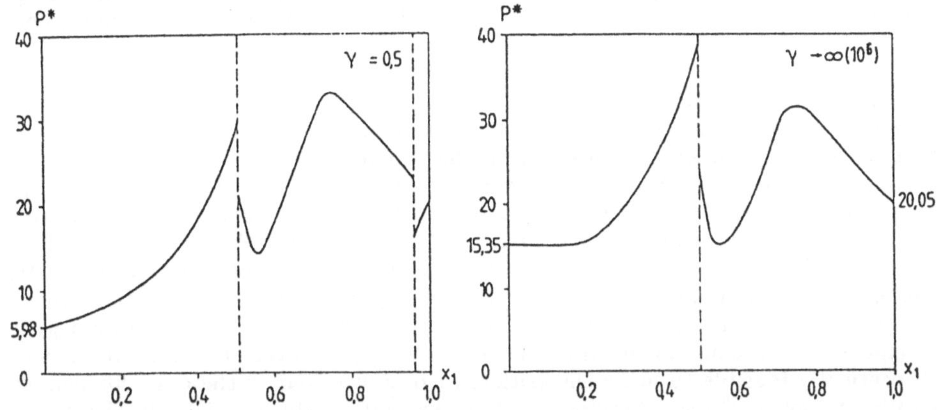

Fig. 5. Critical load versus joint location for $\varkappa_s \to 0$ (10^{-6})

It can be seen that for $x_1 \to 0.5$ the critical value of load increases rapidly for the case of small stiffness of joint and at the value $x_s = 0.5$ a jump of critical force appears. The maximum of the critical load is about two times greater than for the case without joints ($\varkappa_s \to \infty$). The configuration of the characteristic curves in the P^*,ω - plane enables the explanation of the jump phenomenon (Fig. 6).

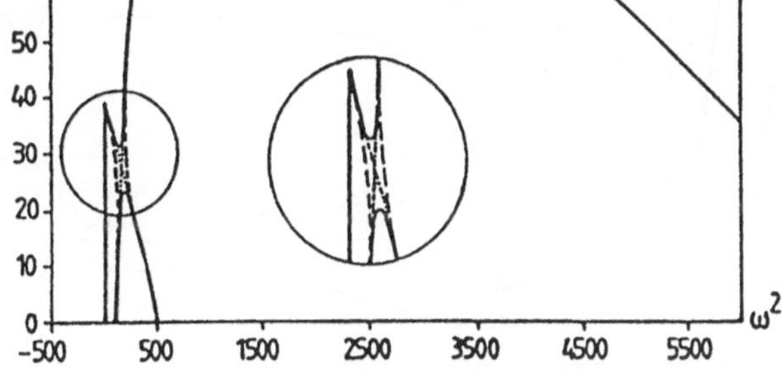

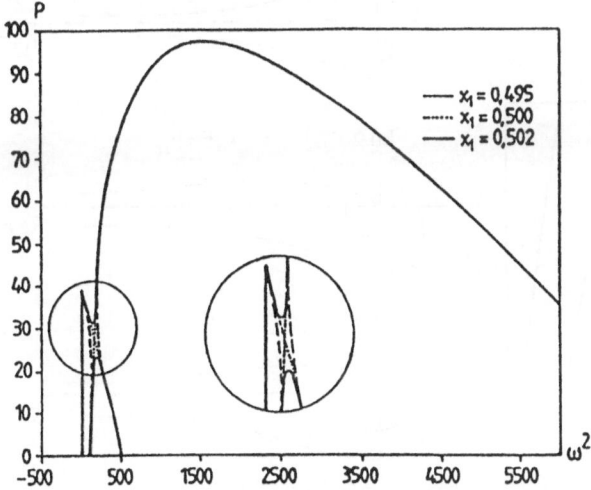

Fig. 6. Switch - over of characteristic curves

The broken line corresponds to the shape of characteristic curves for the case of joint location somewhat lower than that of the jump. The dotted line shape shows a switch-over case which corresponds to the jump of the critical value. Subsequently, for the higher joint location the characteristic curves assume the form shown by the continuous line. In this case, instability occurs with the second and the third natural form, while in the former case the column loses its stability oscillating with the first and the second mode. All details of the jump phenomenon are presented in [18] and , for the case of the elastic hinge-joint, in [19].

3.2 Case of Viscoelastic Hinge-joint

Using the above presented method , a case of elastic hinge-joint was considered by the authors [19]. As it follows from the considerations of such a hinge-joint with a stiffness smaller than the stiffness of uniform column, the value of critical load can be greater than that in the case of the uniform column. The configuraton of characteristic curves for various joint locations x and for the stiffness $\varkappa_s = 10^{-3}$ is presented in [20]. In this case, for $x \to 0.5$, we obtain a critical load $P^* \to 81$, while, after the jump, $P^* = 71$ with an associated jump of the frequency from about $\omega \simeq 4.0$ to $\omega \simeq 2.0$. The above analysis of an elastic behaviour of Beck's column with a hinge-joint location in central position (x= 0.5) yields the conclusion, that it is theoretically feasible to make the critical load higher than in the case considered so far as "optimal" and obtained by Hanaoka and Washizu in [11]. It should be noted that in the case of a dissipative structure, the characteristic equation (2.9) is of the complex form:

$$\Phi(P,\omega) = \Phi_R(P,\omega) + i\,\Phi_I(P,\omega). \tag{3.2}$$

In order to get critical values of P, one may use the generalized Mikhajlov stability criterion. A typical configuration in the force-frequency plane corresponding to (3.2) is shown in Fig. 7.

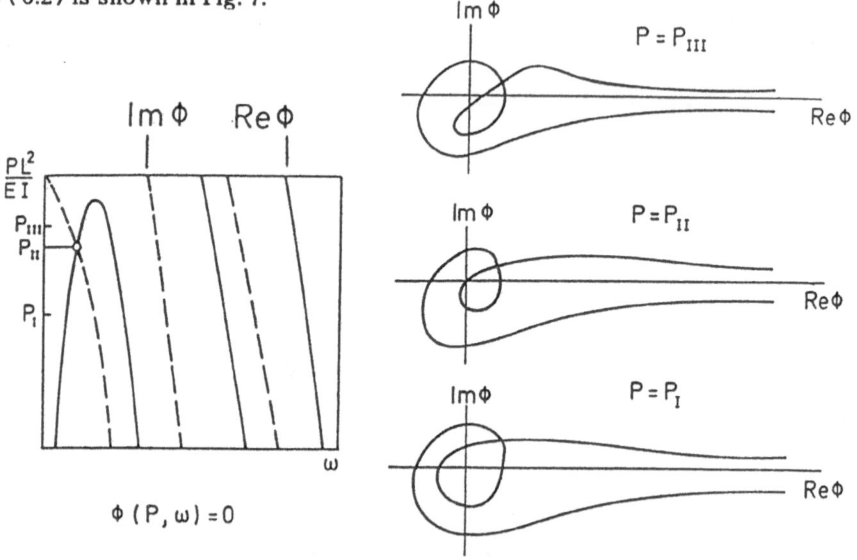

Fig. 7. Force -frequency plane and shapes of Mikhajlov curves for the stable (P_I), critical ($P^* = P_{II}$) and unstable case of the column (P_{III})

The full and dashed lines represent the real and imaginary parts of the characteristic equation, respectively. The point of intersection determines the critical value $P = P_{II}$ as illustrated on the right - hand side of Fig. 7. For $P = P_I$, the column is stable and the shape of Re Φ - Im Φ for $P = P_{III}$ is typical for a supercritical (unstable) case. The critical load versus hinge - joint location for elastic and viscoelastic case is shown in Fig. 8.

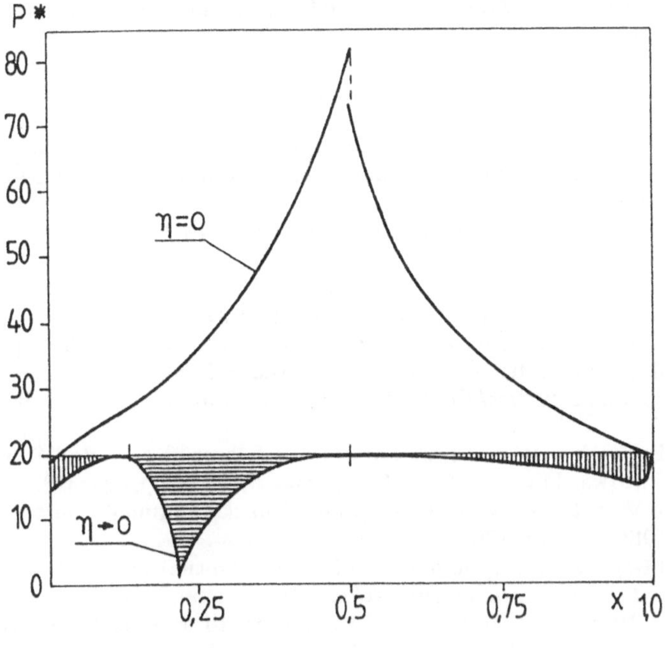

Fig. 8. Critical load versus joint location for elastic ($\eta=0$) and viscoelastic properties of hinge-joint

The computational results have shown that in the case of a single viscoelastic joint the critical force is independent of the damping coefficient, but depends on the joint location. The results for $\eta \to 0$ and $\eta = 0$ are different.

4. CONCLUSION

As the conclusion of the above study it can be stated that stepped structures, treated by transfer matrix technique, may be used as a verification of approximate methods for critical load determination. It is theoretically feasible to obtain a shape of the Beck- Reut column with the critical load higher than considered so far as "optimal". However, such a result would be of academic interest only because the destabilizing influence of damping makes it impossible to realize it under real conditions.

REFERENCES

[1] Niordson F.I.,'The Optimal Design of Vibrating Beam'. *Quarterly of App. Maths.* **23**, 1965, pp.47-53.

[2] Prager W. and Taylor J.E.,'Problems of Optimal Structural Design'. *J.App.Mech.* **35**, 1968, pp.102-106.

[3] Zyczkowski M. and Gajewski A., 'Optimal Structural Design' in: *Non-Conservative Problems of Elastic Stability.* Proc. IUTAM Symp. Herrenalb/Karlsruhe 1969, Springer Verlag 1971, pp. 295-301.

[4] Gajewski A. and Zyczkowski M., 'Optimal Design of Elastic Columns Subjected to General Conservative Behaviour of Loading'. *ZAMP* **21**. 5, 1970, pp.187-198.

[5] Vepa K., *Optimally Stable Structural Forms.* University of Waterloo, Febr. 1972

[6] Claudon J. L., 'Characteristic Curves and Optimum Design of Two Structures Subjected to Circulatory Loads'. *J. de Mechanique* **14**. 3, 1975, pp. 531-543.

[7] Plaut R., 'Optimal Design for Stability under Dissipative, Gyroscopic and Circulatory Loads'. *Optimization in Structural Design,* Warsaw 1973, Springer-Verlag 1975, pp. 168-180.

[8] Bogacz R., Irretier H. and Mahrenholtz O.,'Optimal Design of Structures under Non-Conservative Forces with Stability Constraints'. *Bracketing of Eigenfrequencies of Continous Structures,* Proc. Euromech 112, Hungary 1979, pp.43-65.

[9] Bogacz R. and Mahrenholtz O., 'Optimally Stable Structures Subjected to Follower Forces'. In: *Structural Control,* Ed. H.H.E. Leipholz, North-Holland P.C. 1980, pp.139-157.

[10] Leipholz H.H.E.,'On Variational Principle for the Clamped-Free Rod Subjected to Tangential Follower Forces'. *Mech. Res.Comm.* **5**, 1978, pp. 335-359.

[11] Hanaoka M. and Washizu K., 'Optimum Design of Beck's Column'. *Computer and Structures,* **11**, 1980, pp.473-480.

[12] Błachut J. and Gajewski A.,'Unimodal and Bimodal Optimization of Vibrating Rods and Arcs'. *Proc. Conf. on Optimization,* Ossolineum, Wrocław 1983.

[13] Kounadis A.N., 'Divergence and Flutter Instability of Elastically Restrained Structures under Follower Forces'. *Int. J. Engng. Sci.,* **19**, 1981, pp.553-562.

[14] Bogacz R. and Mahrenholtz O., 'Modal Analysis in Application to Design of Inelastic Structures Subjected to Circulatory Loading'. *Inelastic Structures under Variable Load.* Ed. C. Polizzotto and A. Sawczuk, COGRAS-Palermo 1984, pp.377-386.

[15] Lottati I. and Kornecki A.,'The Effect of an Elastic Foundation and of Dissipative Forces on Stability of Fluid-Conveying Pipes'. *J. of Sound and Vibr.* **109**. 2, 1986, pp. 327-338.

[16] Bogacz R. and Mahrenholtz O.,'On the Optimal Design of Viscoelastic Structures Subjected to Circulatory Loading'. In: *Optimization Methods in Structural Design.* Ed. H. Eschenauer, Wissenschaftsverlag 1983, pp.281-288.

[17] Bogacz R. and Niespodziana A.,'On Stability of Continuous Column with Localized Loss of Stiffness' (in Polish). *IFTR Reports,* **27**, 1987, pp. 3-23.

[18] Bogacz R. and Imiełowski S.,'Stability of Column with Localized Discontinuity of Displacements Subjected to Circulatory Load (in Polish). *IFTR Reports,* 1988.

[19] Bogacz R. and Mahrenholtz O.,'On Stabilityof a Column under Circulatory Load'. *Arch. of Mechanics.* **38**. 3, 1986. pp.281-287.

[20] Mahrenholtz O. and Bogacz R., 'On Configuration of Characteristic Curves for Optimal Structures under Non-Conservative Loads'. *Ing.-Archiv,* 50, 1981, pp.141-148.

STRUCTURAL OPTIMIZATION IN A NON-DETERMINISTIC SETTING

R.E. MELCHERS
Department of Civil Engineering and Surveying
The University of Newcastle
N.S.W. 2308 Australia

ABSTRACT. For probabilistically described absolute structural optimization layout problems it is shown that using a simplified probabilistic framework (socalled First Order Second Moment), some classical solutions for optimal layout remain valid. Some remarks about the more general problem and the difficulty of its solution are made.

1. Introduction

The exact nature of the loading which will be applied to a real structure is no known, in general. The resistance of real structures, even after construction, is also subject to uncertainty, as are matters of modelling and decision. For these reasons a more realistic description of structures is in probabilistic terms (Melchers, 1987a).

The simplest structural reliability problem has one random variable R to model structural resistance. The probability density function $f_R(\ )$ of R is assumed known with mean μ_R and standard deviation σ_R. The loading is correspondingly modelled by just one variable, Q, with known p.d.f. $f_Q(\ )$, mean μ_Q and standard deviation σ_Q (Melchers, 1987a).

The probability that the structure will fail is the probability that its resistance is less than the effect S of the applied load:

$$p_f = P(R < S) = \int_{D:R<S} f_{RS}(\underline{x}) \; d\underline{x} \tag{1}$$

where S is the "load effect" obtained by structural analysis from Q. D describes the domain R<S and $f_{RS}(\ )$ is the joint p.d.f. of Y = (R, S).

Equation (1) cannot be solved in closed form except in some special cases, for example, when R and S are each described by a Gaussian (Normal) distribution;

$$p_f = P(R < S) = \Phi(-\beta) \tag{2}$$

G. I. N. Rozvany and B. L. Karihaloo (eds.), Structural Optimization, 185–191.

with $$\beta = (\mu_R - \mu_S)/(\sigma_R^2 + \sigma_S^2)^{1/2} \tag{3}$$

where $\Phi(\ )$ is the standard Normal N(0, 1) cumulative distribution function and β is also known as the "safety index".

Consider now the absolute optimal design of sandwich beams, grillages and reinforced concrete slabs, for which the unidirectional load carrying material of resistance $R(\underline{x})$ at some generic location $\underline{x}$ is provided at the extremes of the matrix material (of constant, predefined depth).

It is commonly assumed for this class of structure that R is proportional to weight per unit length or area, so that the minimum of the total weight W is

$$\text{min:}\quad W = \int \cdots \int_{D_s} R(\underline{x})\ d\underline{x} \tag{4}$$

(where D_s defines the domain of the structure), subject to the requirements of equilibrium, of the constitutive relations for the material and on a predefined level p_f^* of failure probability. Note that R may have components in each of the direction components of $\underline{x}$. In the classical deterministic analysis, $R = c|M|$, sothat the structural weight is directly proportional to the well-known "Moment-Volume".

2. A Criterion for Absolute Optimal Design

An optimality criterion for the simple type of probabilistically described structures outlined above follows from the discussions of Heyman (1959) and Morley (1966) (see also Prager and Shield, 1967). As in these classical papers, let the structure be subject to bending only. It is now well-known that for given loading, the optimal structure consists of regions of constant curvature, $\kappa_i = dw/dx_i$ = constant (where w is the deflection, x_i is the ith direction), joining smoothly (but not always - see Strang and Kohn, 1983). In the optimal structure the curvature field ("associated structure") corresponds with a statically admissable stress (moment) field. For certain cases this result is also independent of applied load magnitude (but not direction). Such a layout will be termed "kinematically prescribed" herein.

For probabilistically described structures, it is important to note that all possible realizations of loading Q need to be considered. Let the statically admissible moment fields for these be denoted collectively $M(\underline{x})$. These may be related to the resistance $R(\underline{x})$ (which depends on p_f, f_R, f_Q) through

$$R(\underline{x}) = A(\underline{x}).M(\underline{x}) + B(\underline{x}) \tag{5}$$

where $A(\underline{x}) > 0$, $B(\underline{x})$ are location dependent "design" functions with components corresponding to those of R and M.

Using (5), it may be shown (Melchers, 1987b) that the classical optimal layout results remain valid if the requirement on the "associated" structure deformation field:

$$\frac{|\kappa|}{A(\underline{x})} = \text{constant} \tag{6}$$

can be met for all points $\underline{x}$ in the structure and for all realizations of loading Q.

It is important to note that (6) must apply for all loading realizations. This is immediately possible if the optimal structure is "kinematically prescribed". In other cases it means that the design $R(\underline{x})$ and the loading $Q(\underline{x})$ must be linearly related. In view of (5), this means that $M(\underline{x})$ and $Q(\underline{x})$ must be linearly related.

Criterion 6 reduces to the classical result if $A(\underline{x})$ = constant. In general, however, criterion (6) requires the use of iterative techniques, since the "design" functions $A(\underline{x})$ and $B(\underline{x})$ are precisely the objects of the optimization.

3. First Order Second Moment Reliability

For linear structures, subject to one random variable load Q, the bending moment distribution M, is represented by one random variable such that M = aQ. It follows that:

$$\mu_M = a\,\mu_Q \tag{7a}$$

$$\sigma_M = a\,\sigma_Q \tag{7b}$$

and from (3), noting that the load effect S is now M;

$$\mu_R = \mu_M + \beta\sqrt{(\sigma_R^2 + \sigma_M^2)} \tag{7c}$$

Expression (7c) shows that with σ_R and σ_M predefined, μ_R, the mean resistance to be provided in the optimal design, is directly proportional to μ_M, the mean bending moment, as expected. Using (7a, 7b), it is found that

$$\mu_R = a\,\mu_Q + \beta \sqrt{(\sigma_R^2 + a^2\sigma_Q^2)} \tag{7d}$$

which is not linear in the transformation a. However (Ravindra et al, 1969), it may be shown that for $0.3 < \frac{\sigma_R}{\sigma_M} < 3.0$,

$$\mu_R = \mu_M + \beta\alpha\sigma_R + \beta\alpha\sigma_M \tag{8}$$

with $\alpha = 0.7$. Noting that $\sigma_Q = \mu_Q V_Q$ where V_Q is the coefficient of variation of the load Q, a parameter which is often prescribed (Melchers, 1987a), expression (8) for the optimal design becomes

$$\mu_R = a\,\mu_Q\,(1 + \beta\alpha V_Q) + \beta\alpha\sigma_R \tag{9}$$

4. Example

For the clamped beam (Figure 1) it is well known in conventional theory (e.g., Heyman, 1959) that the "associated" strain field for the optimal beam has inflexion points at L/2 and 3L/2 with κ = constant and with slope continuity (see Figure 1b). Clearly the location of the inflexion points will be different in the general reliability based problem because A(x) will not be constant.

In the case of the linearized reliability requirement (9) it follows that R(x) is directly proportional to $|M(x)|$ and hence $|\kappa|/A(x)$ is constant. Thus, the inflexion points in Figure 1 correspond to those of the classical results. The central portion $0 < x < L/2$ in Figure 1 has a maximum value of bending moment $M = QL/4$, with statistical properties μ_M, σ_M, given, for Normal Q, by $\mu_M = \frac{L}{4}\mu_Q$, $\sigma_M = \frac{L}{4}\sigma_M$. The bending moment at other locations along the beam is in direct (linear) proportion. (It will be assumed that $Q \geq 0$, i.e., it does not change direction from that shown in Figure 1a. This strictly conflicts with the assumed Normal statistical property of Q).

To maintain a safety level $p_f = \Phi(-\beta)$ given by (2, 3) and with resistance R (≥ 0) assumed Normal (μ_R, σ_R) at the beam centre (and perfectly correlated elsewhere), the required mean resistance at $x = L/2$ is:

$$\begin{aligned}\mu_R &= \mu_M + \beta \sqrt{(\sigma_R^2 + \sigma_M^2)} \\ &= \frac{L}{4}\mu_Q + \beta \sqrt{(\sigma_R^2 + \frac{L^2}{4^2}\sigma_Q^2)}\end{aligned} \tag{10}$$

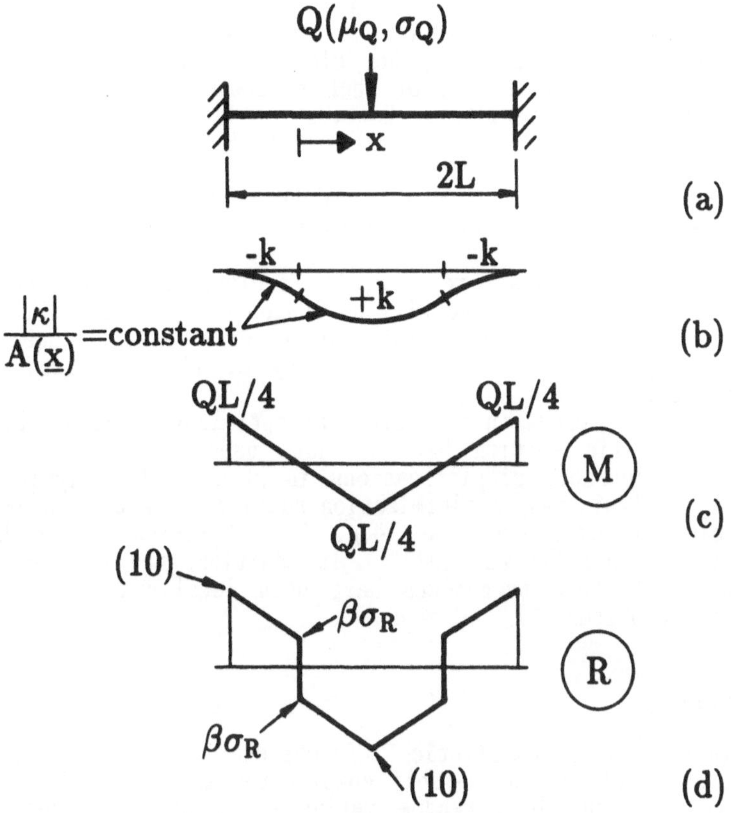

Fig. 1 Clamped Beam Problem, Constant Curvature Layout, Bending Moment Distribution and Optimal Design

Elsewhere, at x, the required resistance is:

$$\mu_R(x) \geq \frac{x}{2}\mu_Q + \beta\sqrt{(\sigma_R^2 + \frac{x^2}{4}\sigma_Q^2)} \tag{11}$$

in order to maintain the required level of reliability. This, and hence the relationship between R(x) and M(x), is nonlinear.

5. Comments

Because A(x) could be approximated as a constant, the criterion for the deformation of the associated strain field could be made to correspond to the classical result $|\kappa|$ = constant. This allows results for the Example to be directly extended to one random variable, but (arbitrary) distributed loading, and to the clamped square and rectangular grillages, with the resistance of such structures represented by just one independent random variable. However, it is important to note that these results apply only to problems for which the "associated" strain field is invariant with respect to load realizations. This is the case if the problem is "kinematically prescribed". It is also, as noted, the case for the special type of linearization between design and loading used herein.

If these assumptions are removed, it is clear that the fortuitously simple results described above will no longer be applicable in general. In addition, more than one limit state will usually need to be considered and this makes the possibility of linearization, as used above, unlikely.

It is for this reason that attempts at optimization of structures in a non-deterministic setting have to date used numerical techniques. Even then considerable simplifications in structural analysis, and/or reliability analysis and/or minimization routines have been necessary to render solutions feasible (see, e.g. Thoft-Christensen and Morutsu, 1986; Rackwitz and Cuntze, 1987). In addition, the problems so far addressed using these techniques have been local rather than global optimization problems.

6. Conclusion

The inclusion of probabilistic information in absolute structural optimization problems considerably complicates their solution. In the special case of one load random variable and one resistance random variable, each described by Normal distributions it was shown that some classical optimal layout solutions remain valid, provided a simplified reliability analysis procedure is accepted.

7. References

Heyman, J., (1959), 'On the Absolute Minimum Weight Design of Framed Structures, *Q.J. Mech. Applied Mech*, *12*, 3, pp.314-324.

Melchers, R.E., (1987a), *Structural Reliability Analysis and Prediction*, Ellis Horwood.

Melchers, R.E., (1987b), 'On Probabilistic Absolute Optimum Design' (submitted).

Morley, C.T., (1966), 'The Minimum Reinforcement of Concrete Slabs', *Int. J. Mech. Sci.*, 8, pp.305-319.

Prager, W. and Shield, R.J. (1967), 'A General Theory of Optimal Plastic Design', *J. Applied Mech.*, 34, 1, pp.184-186.

Rackwitz, R. and Cuntze, R., (1987), 'Formulations of Reliability-Oriented Optimization', *Engineering Optimization*, *11*, 1 and 2, pp.69-72.

Ravindra, M.K., Heany, A. C. and Lind, N.C., (1969), 'Probabilistic Evaluation of Safety Factors', Final Report, *Symp. on Concepts of Safety of Structures and Methods of Design*', London, IABSE, pp.43-66.

Strang, G. and Kohn, R.V., (1983), 'Henky-Prandtl Nets and Constrained Michell Trusses', *Comp. Meth. Applied Mech. Engg.* 36, 2, 207-222.

Thoft-Christenson, P. and Murotsu, Y., (1986), *Application of Structural Systems Reliability Theory*, Springer Verlag.

ON THE SHAPE OPTIMIZATION OF TRUSS STRUCTURE BASED ON RELIABILITY CONCEPT

Y. Murotsu*, M. Kishi** and M. Yonezawa***
* Professor of Aeronautical Engineering, University of Osaka Prefecture, Sakai, Osaka 591, JAPAN
** Lecturer of Naval Architecture, University of Osaka Prefecture, Sakai, Osaka 591, JAPAN
*** Associate Professor of Industrial Engineering, Kinki University, Higashi-Osaka, Osaka 577, JAPAN

ABSTRACT. This paper is concerned with the shape optimization problems of truss structures based on the element and system reliabilities. The algorithmic procedure is proposed for determining the optimum configuration and size of members which minimize the structural weight under the constraints on the failure probabilities of the members. Numerical examples are provided to demonstrate the basic properties of the reliability based optimum design problem and the validity of the proposed method.

1. INTRODUCTION

Many studies have been made on the optimum design of truss structures [1-18]. Some are concerned with the design problem to determine the optimum dimensions of the members so as to minimize the structural weight or cost where the configuration of the structure is assumed to be given [1,4,9,14]. However, it has been shown that the minimum weight in such a design problem is considerably affected by the configuration of the structure. Hence, much effort has been devoted to shape optimization problems, especially based on deterministic failure criteria, where the configuration variables are taken as design variables in addition to the dimensions of the members [2,3,5-8,10-14]. On the contrary, few studies have investigated shape optimization problems based on probabilistic concept [16-18].

This paper deals with a shape optimization problem which minimizes the structural weight or volume under the constraints on failure probabilities. The design variables are the cross-sectional dimensions of the members and the configuration parameters to represent the location of the members. An algorithmic procedure for solving the optimum design problem is proposed which attains the optimum configuration by sequentially deleting the unnecessary members after the optimization of cross-sectional dimensions, starting from the initial configuration. Numerical examples are provided to illustrate the

G. I. N. Rozvany and B. L. Karihaloo (eds.), Structural Optimization, 193–200.

proposed design procedure and to verify its applicability.

2. RELIABILITY-BASED SHAPE OPTIMIZATION OF TRUSS STRUCTURES

2.1. Limit State Functions and Failure Probabilities

Consider a truss structure which consists of n members and ℓ loads applied to its nodes. The members are uniform and homogeneous. It is assumed that there are m failure modes and the limit state functions M_i are expressed in the form:

$$M_i = \sum_{j=1}^{n} a_{ij}(\mathbf{X};\mathbf{Y})R_j - \sum_{k=1}^{\ell} b_{ik}(\mathbf{X};\mathbf{Y})L_k \qquad (i=1,2,\ldots,m) \tag{1}$$

where R_j is the strength of member j, L_k the load acting on the structure, a_{ij} the strength influence coefficient, b_{ik} the load influence coefficient, $\mathbf{X}=(X_1,X_2,\ldots,X_n)^T$ the vector of cross-sectional dimensions of the members, such as cross-sectional areas, thicknesses, etc., and $\mathbf{Y}=(Y_1,Y_2,\ldots,Y_n)^T$ the vector of the configuration parameters to represent the location of the members (Y_j=1 or 0). The coefficients a_{ij} and b_{ik} are determined by the cross-sectional dimensions $\mathbf{X}$ and by the configuration parameters $\mathbf{Y}$, and they are derived by applying a Matrix Method [17]. The strengths R_j of the structural members are determined by the strength parameters C_j (e.g. yielding stresses) of the materials and by the cross-sectional dimensions X_j, and they are given by

$$R_j = R_j(X_j,C_j;\mathbf{Y}) \tag{2}$$

Basic random variables related to the limit state functions are the loads L_k and strength parameters C_j of the materials to be used, and their statistical properties are assumed to be given. The other quantities are treated as deterministic. The limit state function M_i corresponds to the criterion of the i-th member failure for a statically determinate truss, while it does to that of the formation of the i-th mechanism for a statically indeterminate truss. Failure of the i-th mode is assumed to occur if the limit state function is not positive, i.e. $M_i \leq 0$. Therefore, the probability of failure P_{fi} of the i-th failure mode is given by

$$P_{fi} = \mathrm{Prob}[\, M_i \leq 0 \,] \triangleq P_{fi}(\mathbf{X};\mathbf{Y}) \tag{3}$$

Since the failure of the structural system occurs when any one of the limit state functions is not positive, the structural failure probability P_f is given by

$$P_f = \mathrm{Prob}[\, \bigcup_{i=1}^{m} (\, M_i \leq 0 \,)] \triangleq P_f(\mathbf{X};\mathbf{Y}) \tag{4}$$

An approximation of the probability of failure P_{fi} is obtained by using FOSM (First-Order-Second-Moment method). In general it is difficult to calculate the structural failure probability P_f exactly, and thus the upper and lower bounds of the structural failure probability are evaluated by using the stochastically dominant failure modes [17].

2.2. Optimum Design Problems

There are various optimum design problems based on reliability concept. Introduced in this study are the following two shape optimization problems. Both vectors of the cross-sectional dimensions **X** and of the configuration parameters **Y** are taken as design variables.

The first design problem to be considered is:

Problem 1

Find **X** and **Y**

such that $H_c(\mathbf{X};\mathbf{Y}) \rightarrow$ minimize

subject to $P^c_{fi}(\mathbf{X};\mathbf{Y}) \leq P_{fai}$ $(i=1,2,\ldots,m)$ (5)

where H_c is the structural weight or volume, and P_{fai} the allowable failure probability of the i-th failure mode. The objective function H_c is determined as a function of the design variables when the materials to be used are specified. Other equality constraints may be adopted to represent the relations among the design variables, considering a symmetric structure. It will be better to introduce the constraints on reliability indices $\beta_i = \mu_{Mi}/\sigma_{Mi}$, where μ_{Mi} and σ_{Mi} are the mean value and standard deviation of Mi, instead of the constraints on failure probabilities. The reason is that the optimum design problem under the reliability index constraints, i.e. $\mu_{Mi}-\beta^*_i\sigma_{Mi} \geq 0$ (β^*_i: target reliability index), is a convex programming problem when the strength parameters and loads are Gaussian random variables [15].

The other optimum design problem is formulated as follows:

Problem 2

Find **X** and **Y**

such that $H_c(\mathbf{X};\mathbf{Y}) \rightarrow$ minimize

subject to $P^c_f(\mathbf{X};\mathbf{Y}) \leq P_{fa}$ (6)

where P_{fa} is the allowable failure probability of the structure. For a large-scale structure, it takes much time to evaluate the structural failure probability P_f. Moreover, the evaluation is repeated many times in the optimization processes, and the total computation time will become enormous to be practical. Further, it should be noted that the dominant failure modes are changed as the design variables vary. In such cases, the redundant truss structure is approximated as a weakest link system with n members [17], i.e., structural failure occurs caused by failure in any one of the members.

3. PROCEDURE FOR SOLVING THE OPTIMUM DESIGN PROBLEM

The solution to Problem 1 is presented here under the following assumptions:

a) Nodes are specified where the loads are applied and the structure is supported, as well as the nodes connecting the candidate members.
b) A static single loading condition is applied, where the magnitudes of the loads are probabilistic.
c) The initial configuration is specified which is formed by connecting the specified nodes with candidate members.
d) Weakest link approximation is applied to the redundant truss structures.

An algorithmic procedure is given as follows:

Step 1:(Initialization): Specify the initial truss structure (i.e. cross sectional dimensions $\mathbf{X}_0$ and configuration $\mathbf{Y}_0$) and the probability distributions of the loads and strengths together with the allowable failure probabilities P_{faj}.

Step 2:(Optimization of cross sectional areas): By generating the limit state functions M_j of the members through a Matrix Method and calculating the failure probabilities, optimize the cross sectional dimensions $\mathbf{X}$ of the members. This optimum design problem is a nonlinear programming problem, and it is effectively solved by using SLP (Sequential Linear Programming) [17].

Step 3:(Deletion of unnecessary members): Delete the members which satisfy the following conditions: i) The failure probability P_{fj} is very small compared with the allowable value, i.e. $P_{fj} \leq \varepsilon_1 \times P_{faj}$ where ε_1 is a given constant (<<1), or ii) The cross-sectional area $A_j(X_j)$ is very small compared with those of other members, e.g. $A_j \leq \varepsilon_2 \Sigma A_j/n$ where ε_2 is a given constant (<<1), or iii) The volume of the member $V_j = A_j(X_j)\ell_j$ (ℓ_j: length of the member) is large when there are some members with very small failure probability or cross-sectional area.

Step 4:(Iteration and termination): Change the initial configuration to that from which the unnecessary members are deleted in Step 3, and go to Step 2. When there are no members to delete in Step 3 or the structure is a statically determinate, stop the calculation and select the minimum weight structure among the optimized structures.

A solution to Problem 2 will be obtained by sequentially solving Problem 2 and adjusting the allowable failure probabilities of the members so as to satisfy the constraint of the system reliability, as suggested in [17].

4. NUMERICAL EXAMPLES

Consider an initial configuration of a truss structure as shown in Fig. 1 (a). The numerical data are given in Table 1, and all the basic variables are assumed to be Gaussian. The optimization processes are illustrated in Table 2, where the cross-sectional areas, the failure probabilities and the volumes are shown. Based on the optimization at the first stage, member 1 is determined to be deleted because its failure probability is very small. Members 5 and 2 are sequentially deleted in the third and fourth stages, finally resulting in the statically determinate truss structure. The three member redundant truss gives the minimum volume, i.e., the statically determinate truss is not the optimum configuration in this case. The structural failure probabilities are evaluated as $P_{f3}=2.055\times10^{-5}$ and $P_{f2}=2.000\times10^{-4}$ for the three and two member trusses, respectively [17]. From this it is concluded that the reliability of the three member truss is one order of magnitude higher than that of the two member truss. This means that the three member redundant truss can be designed much lighter than the two member truss under the same system reliability level. It may suggest the importance of the optimum shape design based on the structural failure probability. Further, the effect of the correlation coefficient of the loads on the optimum design is shown in Table 3, which illustrates that the optimum configuration is sensitive to the correlation coefficient of the loads. The two member statically determinate truss becomes the optimum configuration when the correlation coefficient of the loads is high, i.e. $\rho_{L1L2}=0.7$.

Next consider another numerical example where the initial configuration is illustrated in Fig. 2 (a). The

Table 1 Numerical data

$\bar{\sigma}_{yj}$ (kN/cm^2)	$CV_{\sigma yj}$	$\bar{L}_1$ (kN)	$\bar{L}_2$ (kN)	CV_{Lk}	ρ_{L1L2}	P_{faj}
18.0	0.05	50.0	259.8	0.1	-0.7	10^{-4}

((j=1,2,...,5), (k=1,2))

Table 2 Optimization processes

Optimization stage	Cross-sectional area & failure probability					Volume
	A_1 (cm^2) (P_{f1})	A_2 (cm^2) (P_{f2})	A_3 (cm^2) (P_{f3})	A_4 (cm^2) (P_{f4})	A_5 (cm^2) (P_{f5})	H_c (cm^3)
1 (a)*	0.6538×10^{0} (0.1247×10^{-7})	0.1183×10^{1} (0.2739×10^{-5})	0.1404×10^{2} (0.9999×10^{-4})	0.8229×10^{1} (0.1000×10^{-3})	0.9796×10^{-1} (0.1000×10^{-3})	0.3844×10^{4}
2 (b)*		0.2123×10^{1} (0.2727×10^{-5})	0.1341×10^{2} (0.1000×10^{-3})	0.8459×10^{1} (0.1000×10^{-3})	0.1608×10^{0} (0.9962×10^{-4})	0.3821×10^{4}
3 (c)*		0.2026×10^{1} (0.4200×10^{-5})	0.1346×10^{2} (0.1000×10^{-3})	0.8679×10^{1} (0.1000×10^{-3})		0.3814×10^{4}
4 (d)*			0.1649×10^{2} (0.1000×10^{-3})	0.9072×10^{1} (0.1000×10^{-3})		0.3996×10^{4}

* : correspond to the configurations of Fig. 1

Table 3 Effect of correlation coefficient of loads on the optimum configuration

	Optimization stage	Correlation coefficient of loads ρ_{L1L2}		
		-0.7	0.0	0.7
H_c (cm^3)	1 (a)*	0.3844×10^{4}	0.3761×10^{4}	0.3599×10^{4}
	2 (b)*	0.3821×10^{4}	0.3761×10^{4}	0.3598×10^{4}
	3 (c)*	0.3814×10^{4}	0.3759×10^{4}	0.3594×10^{4}
	4 (d)*	0.3996×10^{4}	0.3815×10^{4}	0.3590×10^{4}

* : correspond to the configurations of Fig. 1

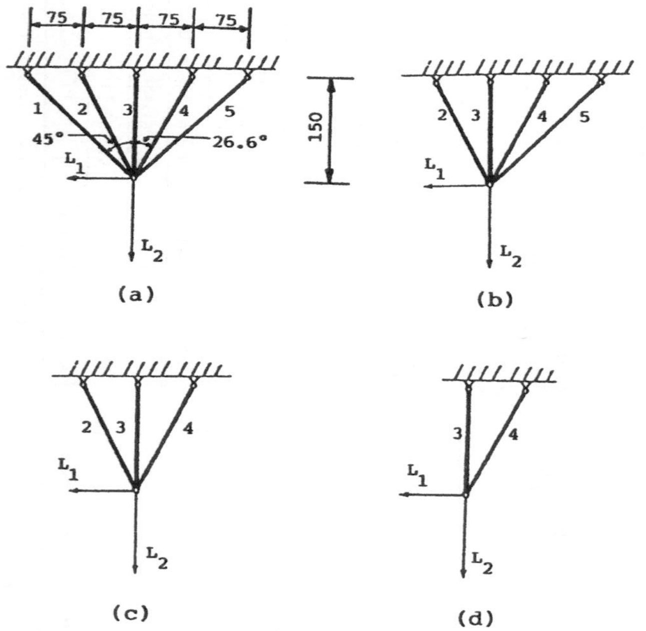

Fig. 1 Initial and transitional configurations (example 1)

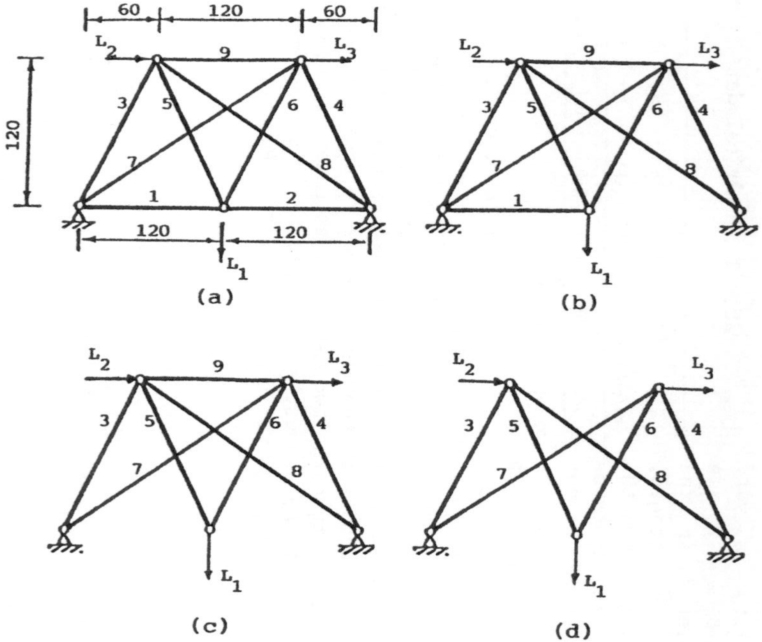

Fig. 2 Initial and transitional configurations (example 2)

transition of the configurations is also shown in the figure and the numerical values are listed in Table 4. It is noted that the initial configuration is optimum in this example although the volumes of each configuration are not so much different.

Table 4 Transition of the configurations

Optimization stage		Cross-sectional area (cm^2) & failure probability					Volume
		A_1 (P_{f1}) A_2 (P_{f2})	A_3 (P_{f3}) A_4 (P_{f4})	A_5 (P_{f5}) A_6 (P_{f6})	A_7 (P_{f7}) A_8 (P_{f8})	A_9 (P_{f9})	H_c (cm^3)
1	9 members	0.5704×10^1 (0.9944×10^{-4})	0.1533×10^2 (0.1000×10^{-3})	0.1041×10^2 (0.1000×10^{-3})	0.1140×10^2 (0.1000×10^{-3})	0.8745×10^1 (0.9988×10^{-4})	0.2036×10^5
	(a)*	0.2507×10^1 (0.9944×10^{-4})	0.2771×10^2 (0.1000×10^{-3})	0.2455×10^2 (0.1000×10^{-3})	0.2494×10^2 (0.1000×10^{-3})		
2	8 members	0.8208×10^1 (0.1000×10^{-3})	0.1533×10^2 (0.1000×10^{-3})	0.1041×10^2 (0.1000×10^{-3})	0.1141×10^2 (0.1000×10^{-3})	0.8743×10^1 (0.1000×10^{-3})	0.2036×10^5
	(b)*		0.2771×10^2 (0.1000×10^{-3})	0.2455×10^2 (0.1000×10^{-3})	0.2495×10^2 (0.9999×10^{-4})		
3	7 members		0.1176×10^2 (0.1000×10^{-3})	0.1693×10^2 (0.1000×10^{-3})	0.1836×10^2 (0.1000×10^{-3})	0.5169×10^1 (0.1000×10^{-3})	0.2107×10^5
	(c)*		0.2232×10^2 (0.1000×10^{-3})	0.1693×10^2 (0.1000×10^{-3})	0.3402×10^2 (0.1000×10^{-3})		
4	6 members		0.1276×10^2 (0.1000×10^{-3})	0.1693×10^2 (0.1000×10^{-3})	0.2058×10^2 (0.1000×10^{-3})		0.2172×10^5
	(d)*		0.2281×10^2 (0.9999×10^{-4})	0.1693×10^2 (0.1000×10^{-3})	0.3678×10^2 (0.9999×10^{-4})		

* : correspond to the configurations of Fig. 2

($\bar{\sigma}_{cyj}=12.0$ kN/cm^2, $\bar{L}_k=200$ kN, $CV_{\sigma yj}=0.05$, $CV_{Lk}=0.2$, $P_{faj}=10^{-4}$) ((j=1,2,....,9), (k=1,2,3))

5. CONCLUDING REMARKS

The shape optimization problems based on the element and system reliabilities are formulated for truss structures and the algorithmic procedure has been proposed for the first type of the design problem, i.e., optimum determination of the configuration and the dimensions of the members under the constraints of the member failure probabilities. The numerical examples are provided to illustrate the basic properties of the optimization problem and the validity of the proposed method. Although this paper is concerned with the simple case where the redundant truss structures are approximated as a weakest link system, consideration of redundancy is essential as briefly commented in the previous section. Buckling failure mode and non-normality of the basic variables are also important factors to be taken into account in the shape optimization of truss structures.

ACKNOWLEDGEMENTS. The authors give their sincere thanks to Messrs. H. Matsui and S. Moriuchi for their help in the numerical calculations.

REFERENCES

1. J. Drymael, 'The Design of Trusses and Its Influence on Weight and Stiffness,' J. Royal Aeronautical Society, **46** (1942), pp. 297-308.
2. W. C. Dorn, R. E. Gormory and H. J. Greenberg, 'Automatic Design of Optimal Structures,' J. Mec., 3, 1 (1964), pp. 25-52.
3. M. W. Dobbs and L. P. Felton, 'Optimization of Truss Geometry,' J. Struct. Div., ASCE, **95, ST10** (1969), pp. 2105-2118.
4. J. M. Chern and W. Prager, 'Minimum-Weight Design of Statically Determinate Trusses Subject to Multiple Constraints,' Int. J. Solids Structures, 7 (1971), pp. 931-940.
5. P. Pedersen, 'On the Optimal Layout of Multi-Purpose Trusses,' Comput. Struct., 2 (1972), pp. 695-712.
6. G. N. Vanderplaats and F. Moses, 'Automated Design of Trusses for Optimum Geometry,' J. Struct. Div., ASCE, **98, ST3** (1972), pp. 671-690.
7. D. J. Sheppard and A. C. Palmer, 'Optimal Design of Transmission Towers by Dynamic Programming,' Comput. Struct., **2** (1972), pp. 455-468.
8. C. Y. Sheu and L. A. Schmit, 'Minimum Weight Design of Elastic Redundant Trusses under Multiple Static Loading Conditions,' AIAA J., **10** (1972), pp. 155-162.
9. W. S. Hemp, Optimum Structures, Oxford University Press (1973).
10. W. R. Spillers, 'Iterative Design for Optimal Geometry,' J. Struct. Div., ASCE, **101, ST7** (1975), pp. 1435-1442.
11. M. P. Saka, 'Shape Optimization of Trusses,' J. Struct. Div., ASCE, **106, ST5** (1980), pp. 1155-1173.
12. Y. Seguchi, Y. Tomita and M. Iwasaki, 'Optimum Design of Frame Structures (in Japanese),' Trans. of JSME, **43, 374** (1977), pp. 3769-3772.
13. K. Yamazaki and J. Oda, 'Optimum Layout of Truss Structures by Finite Element Method (in Japanese),' Trans. of JSME, **A, 46, 411** (1980), pp. 1230-1236.
14. U. Kirsch, Optimum Structural Design, McGraw Hill (1981).
15. M. Yonezawa, Y. Murotsu, F. Oba and K. Niwa, 'Optimum Reliability and Structures in Reliability-Based Design,' Archives of Mechanics,' **30, 3** (1978), pp. 227-241.
16. N. Shiraishi and H. Furuta, 'On Geometry of Truss,' Memoirs of the Faculty of Engineering, Kyoto University, **XLI, 4** (1979), pp. 498-517.
17. P. Thoft-Christensen and Y. Murotsu, Application of Structural Systems Reliability Theory, Springer Verlag (1986).
18. P. Thoft-Christensen, 'Application of Optimization Methods in Structural Systems Reliability Theory,' Structural Reliability Theory Paper, **53**, Institute of Building Technology and Structural Engineering, Aalborg University (1987).

ON OPTIMALITY IN STRUCTURAL AND MATERIAL COMPOSITION OF BAMBOO

J. ODA
Department of Mechanical Engineering
Kanazawa University
2-40-20, Kodatsuno, Kanazawa, 920
Japan

ABSTRACT. In this paper the structural and material systems of bamboo are analyzed from the viewpoint of strength morphology. From the results, it appears that the node parts of bamboo not only are an optimum stiffener to prevent buckling fracture, but also an optimum arrester for any cracks initiated in the axial direction. The vascular bundles of bamboo, which correspond to the reinforced fiber of FRP,distribute uniquely in the axial and radial directions.The variation of volume percentage in the radial direction is based on the minimum weight design criterion.

1.Introduction

It is generally well known that an analysis of structural or material systems of plant or animal origin is a useful way of studying design techniques in engineering, because all living things are made as an optimum system for the longest survival period.

In this study the structural and material systems of bamboo are analyzed from the viewpoint of strength morphology. From the results, the following points of interest emerge:

(i) The node parts of bamboo not only are an optimum stiffener to prevent buckling fracture, but also an optimum arrester for any cracks initiated in the axial direction;

(ii) The vascular bundles of bamboo, which correspond to the reinforced fiber of FRP, distribute uniquely in the axial and radial directions. The variation of volume percentage in the radial direction is based on the minimum weight design criterion.

G. I. N. Rozvany and B. L. Karihaloo (eds.), Structural Optimization, 201–208.

2.Kinds of Bamboo and the Dynamical Environment

Throughout the world there are about 200 kinds of bamboo. More than half of these grow in the south-eastern part of Asia. The size range is very large; the arundinaria bamboo is used in this study. The arundinaria bamboo is the largest in Japan with a maximum diameter and height of about 0.13 and 15m respectively, which are adequate to make a test specimen. Figure 1 shows the coordinate system of bamboo and geometrical notation.

Next, consider the load conditions governing the bamboo in nature. It is obvious that wind or snow has the strongest effect on the structural strength of bamboo. The body force of bamboo must also be considered; it reaches the maximum at the root. For many arundinaria bamboos, however, the body force value is only about 300 $\sim$ 500N which is small in comparison with that of wind or snow. [1].

3.Analysis of Geometrical Structure

On a macroscopic level bamboo has the following geometrical characteristics:

(i) It is cylindrical in shape with, many nodes;

(ii) Diameter and wall thickness decrease with increasing height.

These geometrical characteristics presumably correspond to the load conditions mentioned above. For example, the cylindrical shape would show that bamboo is isotropic in r–θ plane and that the cross sectional modulus is larger than that of solid cylinders, whereas the nodes act as stiffeners to prevent buckling fracture by the bending moment due to wind or snow. Variation of diameter and wall thickness with height makes the stresses uniform throughout the length of bamboo, and moreover decreases the critical buckling moment. Figure 2(b) shows the distribution of the critical buckling moment M_{cr} for two bamboos used in the experiment. M_{cr} is calculated by using the following formula [2] for a cylindrical shell under a uniform bending moment

$$M_{cr} = CRt^2 \qquad (1)$$

where $C=2\sqrt{2}\pi E/9\sqrt{1-\nu^2}$, $R=a+t/2$, E is Young's modulus of bamboo and ν is Poisson's ratio. Furthermore, in Fig.2(b) the distribution of Z as section modulus, given by $\pi(b^4-a^4)/4b$, is shown. These distributions are similar to the bending moment diagram obtained approximately under the assumption of a uniform wind force in Fig.2(a). Thus, it may be said that the dimensions, R and t in all parts of bamboo correspond to the load conditions in nature. This is also obvious from the distribution of $\Delta l/2b$ in Fig.2(b), which shows the number of nodes per unit length.

Moreover, it is interesting to note that the node functions to arrest any cracks initiated in the z-direction. The node's function is due to its unique shape and material composition. Figure 3 shows the cross section of the node in r-z plane. The shape is similar to the crack arrester model shown in Fig.4. This model is used for plates subjected to tensile stresses in shipbuilding.

4.Analysis of Material Composition

Figure 5 shows the material composition in r-θ plane of the cylindrical part. From Figs.3 and 5, it is obvious that bamboo is composed of fibers distributed uniformly in z-direction except at the nodes. Therefore, it splits easily in the z direction. On the other hand, at the nodes the fibers distribute uniquely in z and θ directions. This corresponds to the function of the node as a crack arrester. By utilizing these characteristics of bamboo, many traditional daily necessities have been made in Japan. Figure 6 shows the example of a tea-mixer called a 'chasen' and which is usually used to mix tea powder in boiled water in the Japanese tea ceremony. In this bamboo mixer, one must pay attention to the position of node. It is obvious that the crack arrester function is due to its characteristic shape and material composition.

Next, let us consider the relation between the distribution of fibers in the r-direction and the tensile strength. In the r-direction the fibers are distributed more closely at the parts near the outer surface. These fibers are called the vascular bundle and they are stronger than that of the matrix. From this, it may be assumed that the distribution corresponds to the stress distribution induced by the external load. Figure 7 shows the distribution in the r-direction with respect to the values of the tensile strength σ_B in the z-direction. Figure 8 shows the relation between $(r-a)/(b-a)$ and the percentage (V_f) of the vascular bundles at varying height from the base. This results is very similar to the distribution of σ_B shown in Fig.7. On the other hand, by using the results of Fig.8 the relation between V_f and Young's modulus E_z in the z-direction is obtained as shown in Fig.9. It is seen that E_z changes almost linearly with V_f. This relation is very similar to the law of mixtures for composite materials such as GFRP and CFRP. Therefore, if this law is applied directly to bamboo under the criterion of minimum weight, one can obtain the optimum fiber distribution by using the model of a multi-laminate beam as shown in figure 10[3]. That is, the following optimum design problem is formulated:

$$\text{minimize} \qquad W=\sum_{j=1}^{n} \{\gamma_f V_{fj} + \gamma_m (1-V_{fj})\} A_j \tag{2}$$

$$\text{subject to} \qquad \sigma_i \leqq \sigma_{fi} \tag{3}$$

where W is weight of bamboo per unit length. V_{fi} and A_i are the fiber volume percent and the cross-sectional area of lamina i, respectively. γ_f and γ_m are the specific gravities of fiber and matrix of bamboo respectively. σ_i is the axial stress in lamina i

$$\sigma_i = E_i M r_i / \sum_{j=1}^{n} E_j I_j \tag{4}$$

where M is the bending moment. E_i and I_i are Young's modulus and moment of inertia of lamina i, respectively. The law of mixtures of composite materials states that

$$E_i = E_f V_{fi} + E_m (1-V_{fi}) \tag{5}$$

where E_f and E_m are Young's moduli of fiber and matrix, respectively. On the other hand, σ_{fi} in equation (3) which depends on V_{fi}, can be expressed using the law of mixtures on material strength:

$$\sigma_{fi} = \sigma_{ft} V_{fi} + \sigma_{mt} (1-V_{fi}) \tag{6}$$

where σ_{ft} and σ_{mt} are the tensile and compressive strengths of fiber and matrix

From the above formulation, it is obvious that the objective function W is linear in the design variable V_{fj} but that the constraint condition is nonlinear. If the mechanical properties of the matrix and fiber composing the bamboo correspond to the values of GFRP composing glass fiber and epoxy resin, the following material constants may be used in the calculations:

$$\left.\begin{array}{ll} E_f = 72.5\text{GPa}, & E_m = 3.43\text{GPa} \\ \sigma_{ft} = 1960\text{GPa}, & \sigma_{mt} = 78.4\text{MPa} \\ \gamma_f = 2.60 & \gamma_m = 1.15 \end{array}\right\} \tag{7}$$

Figure 11 shows the optimum distribution of V_{fi} in the direction of thickness for the bamboo model of $a=0.032$m and $b=0.04$m under $M=9.81\times10^3$N·m. For the same case, the distributions of σ_i and σ_{fi} are shown in Fig.12. From these figures, it is obvious that all the laminae contain fiber and that the distributions of V_{fi} and σ_{fi} are nonlinear in the r direction of thickness. The results are very similar qualitatively to the ones of Figs.7 and 8.

5. Conclusions

In this paper, as example of analysis of a natural structure the arundinaria bamboo was used. The structural shape and material composition were analyzed from the viewpoint of strength morphology leading to the following points of interest:

(i) The nodes of bamboos not only are an optimum stiffner to prevent buckling fracture, but also are an optimum arrester for any cracks initiated in the axial direction.

(ii) The vascular bundles of bamboo, which correspond to the reinforced fiber of FRP, distribute uniquely in the axial and radial directions. The variation of volume percentage in the radial direction is based on the minimum weight design criterion.

From the above, it is concluded that structural shape and material composition of bamboo have many functions satisfying simultaneously the design object of bamboo.

References

1. Oda, J., 'Morphological Analysis of bamboo from the viewpoints of Engineering', Trans.JSME, 46, Sep. 1980, pp. 997-1006.
2. Hayashi, T. (ed.), Handbook of Structural Stability, Koronasha, Japan, 1971.
3. Oda, J., 'Minimum Weight Design Problems of Fiber-Reinforced Beam Subjected to Uniform Bending' Trans. ASME, Ser. R., 107, No. 1, March 1985, pp. 88-93.

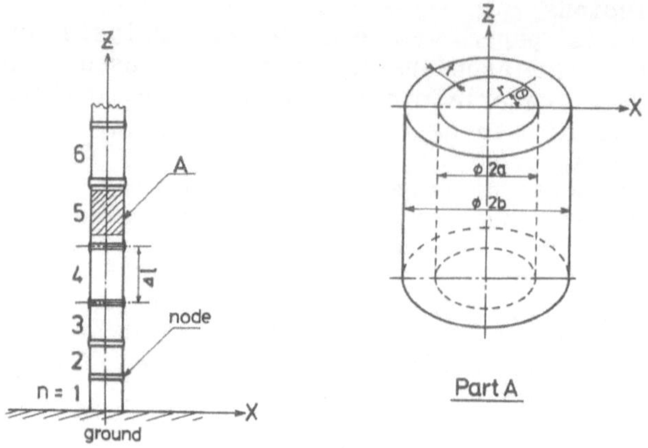

Figure 1. Coordinate system of bamboo

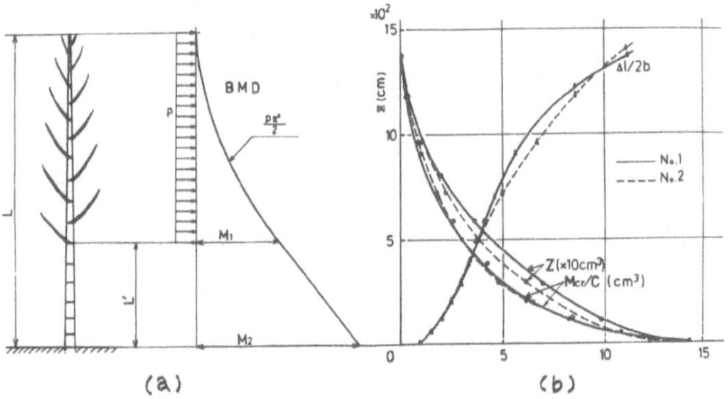

Figure 2. Bending and critical buckling moments of bamboo

(a) Bending moment due to wind force

(b) Critical buckling moment and others

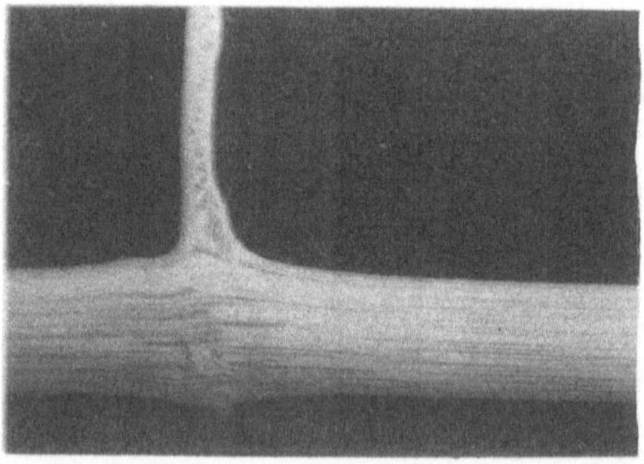

Figure 3. Material composition in r-z plane of bamboo

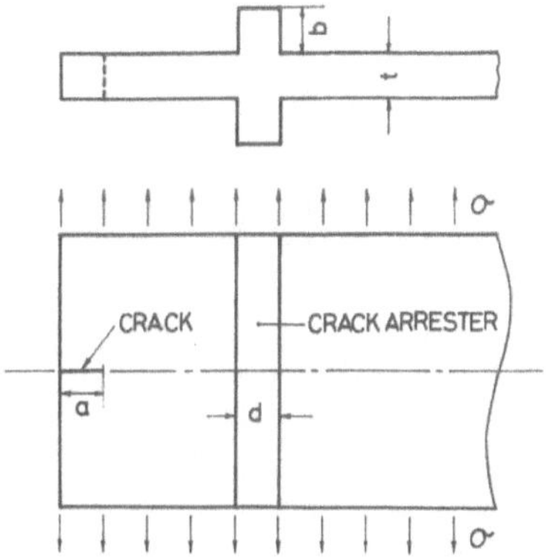

Figure 4. Crack arrester model

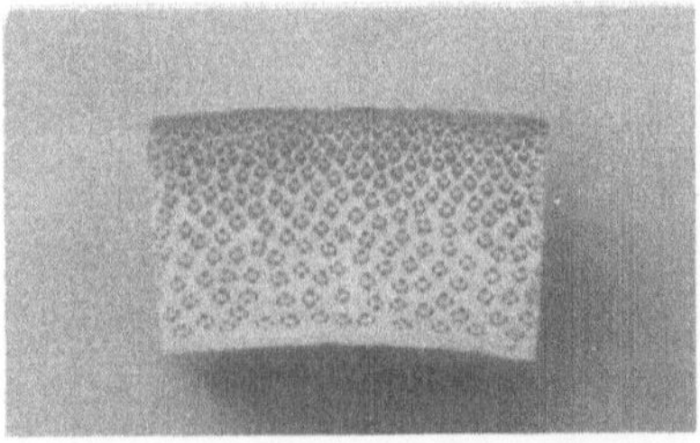

Figure 5. Material composition in r–θ plane of bamboo

Figure 6. Tea-mixer made of bamboo

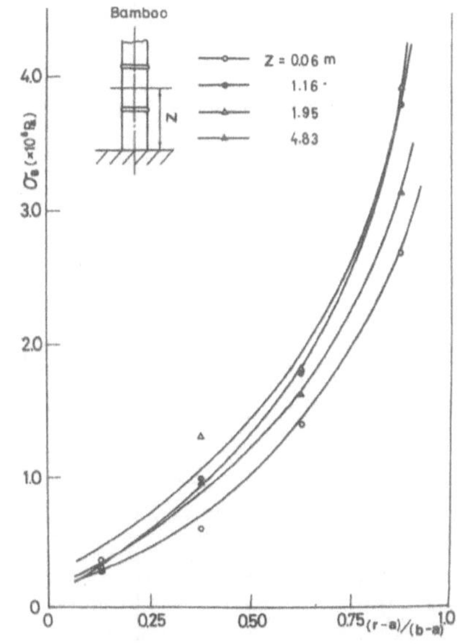

Figure 7. Distribution of tensile strength of bamboo

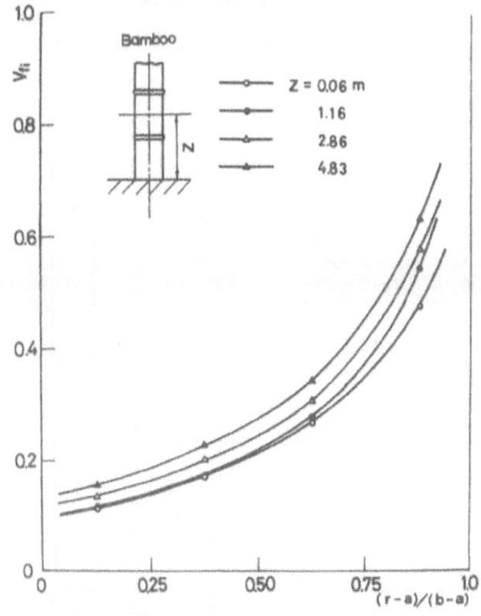

Figure 8. Volume percent of vascular bundle in bamboo

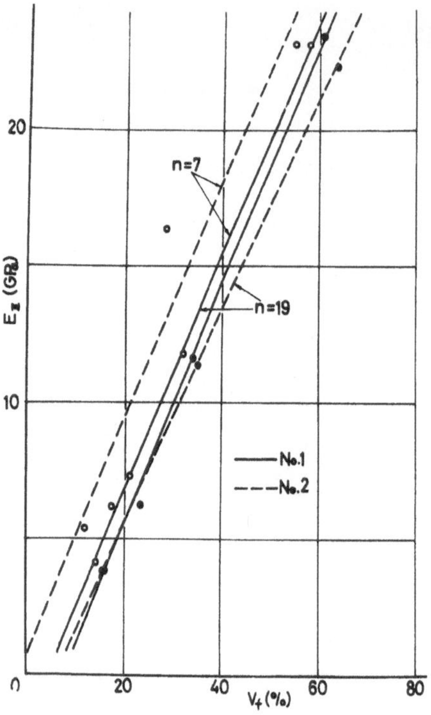

Figure 9. Relation between V_f and E_Z in bamboo

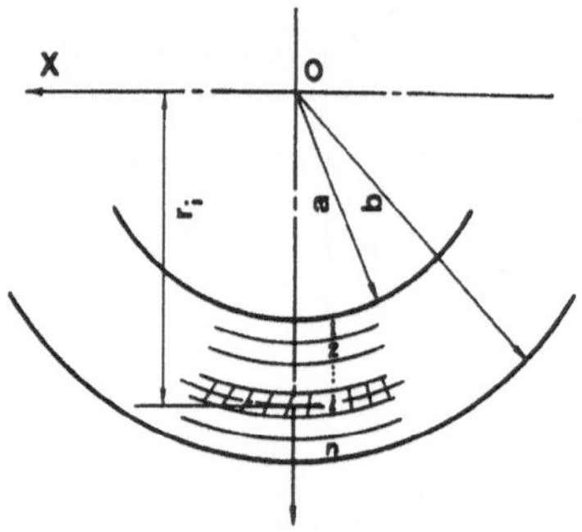

Figure 10. Multi-laminate beam model simulating bamboo

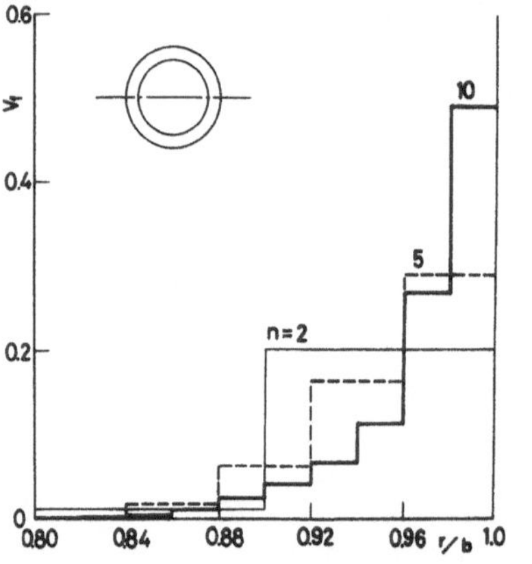

Figure 11. Optimum distribution of V_f in simulate model

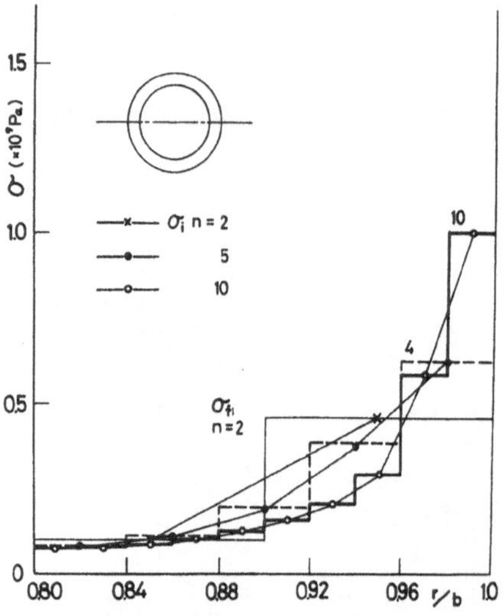

Figure 12. Distributions of σ_i and σ_{fi} in optimum cylindrical beam

OPTIMIZATION OF A HOLLOW BEAM-SHAFT WITH PRESCRIBED INNER CONTOUR

R. D. Parbery
Department of Mechanical Engineering
University of Newcastle
New South Wales 2308
Australia

ABSTRACT. This paper considers the problem of minimizing the cross-sectional area of a prismatic bar with a hollow cross-section required to withstand either a twisting moment or a bending moment in a prescribed plane. In earlier work a first perturbation approximation was found. In this paper the second approximation is presented giving a significant improvement in some cases.

1. INTRODUCTION

Previous authors [1],[2],[3] considered certain problems in the optimization of prismatic bars subjected to torsion or torsion and bending. Banichuck [1] applied perturbation techniques to find an approximate solution for the cross-sectional shape of a hollow bar with maximum torsional stiffness. Parbery and Karihaloo [3] considered the minimization of cross-sectional area for a hollow bar of given bending and torsional stiffness. A first approximation to the solution was found using perturbation methods for the case when the bar contained a circular hole.

This paper is an extension to the work reported in [3]. The second order perturbation solution is reported and a practical design procedure without iteration has been found.

2. MATHEMATICAL FORMULATION

Consider the problem of minimizing the area S of the solid region D of the cross-section of a prismatic hollow bar. D is bounded by an inner contour Γ_i and an outer contour Γ_0. The bar is required to have minimum values J^0, I^0 of torsional and bending stiffness respectively.

In [3] the optimality condition for the problem was shown to be

$$\phi_x^2 + \phi_y^2 = \mu_1 + \mu_2 y^2 \qquad (x,y) \in \Gamma_0 \qquad (1)$$

Subject to the design constraints

G. I. N. Rozvany and B. L. Karihaloo (eds.), Structural Optimization, 209–216.

$$J = 2\left[\iint_D \phi \, dxdy + C\,\Omega\right] \geq J^0 \tag{2}$$

$$I = \iint_D y^2 \, dxdy \geq I^0 \tag{3}$$

Here ϕ is the Prandtl stress function for torsion, x,y are Cartesian co-ordinates, Ω is the area contained by Γ_i, C is the (constant) value of ϕ on Γ_i and μ_1, μ_2 are Lagrange parameters.

3. SOLUTION

The above discussion applies to an inner boundary Γ_i of any given shape. In the sequel, attention is restricted to the case where Γ_i is circular with radius R.

As was demonstrated in [3] the equalities apply in equations (2), (3) for $I^0/J^0 \geq \frac{1}{2}$. For $I^0/J^0 < \frac{1}{2}$, the outer boundary Γ_0 is circular and the torsional stiffness is greater than specified.

3.1 Perturbation Formulation

It is convenient to change to an orthogonal curvilinear co-ordinate system (s,t) where s is the distance along Γ_i from the x-axis and t is the distance along the outward normal to Γ_i. Using these co-ordinates and setting $\phi = 0$ on Γ_0 we may re-write the optimality condition (1) as [3]:

$$\phi_t\,(s,h) = -\mu_3 \sqrt{\left[1 + \mu_4 \left(1 + \frac{h}{R}\right)^2 \sin^2 \frac{s}{R}\right]} \tag{4}$$

where h(s) is the thickness in the t direction and μ_3, μ_4 are new parameters given by $\mu_3 = \sqrt{\mu_1}$ and $\mu_3^2\,\mu_4 = \mu_2$.

The cross-section is now assumed to be thin walled so that $\max_s h(s) = H \ll L$, where L is the perimeter of Γ_i, and $\varepsilon = H/L$ is a small parameter. The problem is non-dimensionalised by introducing the following variables

$$s = Ls'; \quad t = Ht'; \quad h = Hh'; \quad \phi = HL\phi'; \quad \Omega = L^2\Omega'; \quad S = HLS';$$
$$R = LR'; \quad J = HL^3J'; \quad I = HL^3I'; \quad \mu_3 = L\mu_3'; \quad C = HLC'. \tag{5}$$

Primes are omitted in the sequel when there is no confusion between dimentioned and non-dimensional variables. In terms of the non-dimensional variables, the equations defining ϕ, the optimality condition (4), the design restrictions (2), (3), and the cross-sectional area become:

$$T\phi_{tt} + \frac{\varepsilon^2}{T}\phi_{ss} + \frac{\varepsilon}{R}\phi_t = -2\varepsilon T; \qquad T = 1 + \varepsilon\frac{t}{R} \tag{6}$$

$$\phi\,(s,h) = 0 \tag{7}$$

$$\phi\,(s,0) \quad = \quad C \tag{8}$$

$$\int_0^1 \phi_t\,(s,0)\,ds \quad = \quad -2\Omega \tag{9}$$

$$\phi_t\,(s,h) \quad = \quad -\mu_3 \sqrt{[1 + \mu_4\,(1 + \varepsilon\,\frac{h}{R})^2\,\sin^2\frac{s}{R}]} \tag{10}$$

$$J \quad = \quad 2\,(\varepsilon \int_0^1\int_0^h T\phi\,dt\,ds + C\Omega) \quad \geq \quad J^0 \tag{11}$$

$$I \quad = \quad R^2 \int_0^1\int_0^h T^3\,\sin^2\,(\frac{s}{R})\,dt\,ds \quad \geq \quad I^0 \tag{12}$$

$$S \quad = \quad \int_0^1\int_0^h T\,dt\,ds \quad \rightarrow \quad \min \tag{13}$$

3.2 Perturbation Solution

Because of space limitions, only a brief outline of the perturbation solution can be given, details of the first approximation are given in [3]. The variables ϕ, C, h, μ_3, μ_4 and S are expanded in powers of ε in the form

$$v \quad = \quad v_0 + \varepsilon\,v_1 + \varepsilon^2\,v_2 + \ldots \quad , \tag{14}$$

and substituted into the governing equations (6) - (13). The equations governing the first and second approximations are then found by equating the powers of ε^0 and ε^1 respectively. The solution for the first approximation is given by [3]:

$$\mu_{3_0} \quad = \quad \frac{1}{4\,E\,[\sqrt{(-\mu_{4_0})}]}\,, \qquad C_0 \quad = \quad 2\,\pi\,J^0 \tag{15}$$

$$\phi_0 \quad = \quad 2\,\pi\,J^0\,(1 - \frac{t}{h_0}) \tag{16}$$

$$\frac{h_0}{J^0} \quad = \quad \frac{8\,\pi\,E\,[\sqrt{(-\mu_{4_0})}]}{\sqrt{1 + \mu_{4_0}\,\sin^2\theta}}\,, \qquad \theta \; = \; s/R \tag{17}$$

$$S_0 = 16\, J^0\, E\,[\surd(-\mu_{4_0})]\; F\,[\surd(-\mu_{4_0})] \qquad -1 < \mu_{4_0} \le 0 \quad (18)$$

Parameter μ_{4_0} is determined from:

$$\frac{I^0}{J^0} = \frac{4}{\pi^2}\, \frac{E[\surd(-\mu_{4_0})]}{\mu_{4_0}}\, \{E[\surd(-\mu_{4_0})] - F[\surd(-\mu_{4_0})]\}$$

$$\text{for } -1 < \mu_{4_0} \le 0 \quad (19)$$

where F( ) and E( ) represent complete elliptic integrals of the first and second kind respectively.

The equations required to determine the second approximation are listed below.

$$\phi_{1_{tt}} + \frac{1}{R}\,\phi_{0_t} = -2 \qquad (20)$$

$$\phi_1(s,h_0) + h_1\,\phi_{0_t}(s,h_0) = 0 \qquad (21)$$

$$\phi_1(s,0) = C_1 \qquad (22)$$

$$\phi_{1_t}(s,h_0)\,\phi_{0_t}(s,h_0) = \mu_{3_0}{}^2\,[A + (B + \mu_{4_0}\frac{h_0}{R})\,\sin^2\frac{s}{R}] \qquad (23)$$

where $A = \dfrac{\mu_{3_1}}{\mu_{3_0}}$ and $B = \dfrac{\mu_{4_1}}{2} + \dfrac{\mu_{3_1}}{\mu_{3_0}}\,\mu_{4_0}$

$$\int_0^1 \phi_{1_t}(s,0)\,ds = 0 \qquad (24)$$

$$\int_0^1\int_0^{h_0} \phi_0\,dt\,ds + C_1\Omega = 0 \qquad (25)$$

$$\int_0^1 (h_1 + \frac{3}{2}\,\frac{h_0{}^2}{R})\,\sin^2(\frac{s}{R})\,ds = 0 \qquad (26)$$

Equation (21) differs from the corresponding equation in [3]. The latter is not quite correct as it leads to an error of order $O(\varepsilon)$ in equation (7).

The solution to (20) - (26) is as follows [4]:

$$\frac{C_1}{J^{0^2}} = -64\,\pi^2\, E[\sqrt{(-\mu_{4_0})}]\; F[\sqrt{(-\mu_{4_0})}] \tag{27}$$

$$\mu_{3_1} = \mu_{3_0} A \quad \text{and} \quad \mu_{4_1} = 2(B - \mu_{4_0} A) \tag{28}$$

$$\phi_1(s,t) = -t^2 - 2\,\pi\, C_0\, (t - \frac{t^2}{2h_0}) + t\, f(s) + C_1 \tag{29}$$

where $f(s) = 2h_0 - \mu_{3_0}^2 \frac{2h_0}{C_0} [A + (B + \mu_{4_0}\, 2\,\pi\, h_0) \sin^2 \frac{s}{R}]$.

$$h_1(s) = -\pi\, h_0^2 + \frac{h_0^3}{C_0} [1 - \frac{\mu_{3_0}^2}{C_0} (A + (B+\mu_{4_0}\, 2\pi h_0) \sin^2 \frac{s}{R}] + + C_1 \frac{h_0}{C_0} \tag{30}$$

$$S_1 = \frac{C_1}{C_0} I_A + (1 - \frac{\mu_{3_0}^2 A}{C_0}) \frac{I_H}{C_0} - \frac{\mu_{3_0}^2}{C_0^2} (BI_D + 2\pi\, \mu_{4_0}\, I_I) \tag{31}$$

In (29) to (31) A and B are determined from the simultaneous linear equations

$$A.I_A + B.I_B + \frac{2\pi\, C_0^2}{\mu_{3_0}^2} - \frac{2C_0 I_A}{\mu_{3_0}^2} - 2\pi\, \mu_{4_0}\, I_C = 0 \tag{32}$$

and

$$AI_D + BI_E - \frac{C_1 C_0 I_B}{\mu_{3_0}^2} - \frac{2\pi\, C_0^2}{\mu_{3_0}^2} I_C - \frac{C_0}{\mu_{3_0}^2} I_D - \frac{2\pi\, \mu_{3_0}^2\, \mu_{4_0}\, I_F}{C_0^2} = 0\,, \tag{33}$$

where the integrals I_A, I_B, ..., are evaluated in the Appendix. The optimal thickness distribution and optimal cross-sectional area are found by

$$h \simeq h_0 + \varepsilon\, h_1; \qquad S \simeq S_0 + \varepsilon\, S_1\ . \tag{34}$$

4. DISCUSSION AND CONCLUSION

The second perturbation approximation to the optimal shape has now been found. It will be noted that the perturbation parameter ε contains the maximum thickness H which is also unknown until the solution is found. However it is possible [4] to arrange a design procedure which overcomes this difficulty without the need for iteration. The savings of the optimal shape compared to a circular annulus increase as the ratio I^0/J^0 increases ($I^0/J^0 > 0.5$). The additional savings obtained by using the second approximation compared to the first depends both on the ratio I^0/J^0 and on the actual values of I^0 and J^0.

Consider by way of illustration the following numerical example : $R = 25$ mm, $J^0 = 4 \times 10^4$ mm, $I^0/J^0 = 0.8$. The solution in this case gives:

$$S_0 = 77.825 \text{ mm}^2$$

$$S \approx S_0 + \varepsilon S_1 = 75.677 \text{ mm}^2$$

For comparison, a hollow circular section with the same R and satisfying the stiffness constraints would have an area of 99.9 mm^2. The saving obtained in this case by using the second approximation instead of the first is 4.1%, whereas if the value of J^0 was doubled while I^0/J^0 and R remained the same, it would be 8.4%.

5. REFERENCES

[1] N. V. Banichuk, 'Optimization of Elastic Bars in Torsion', *Int. J. Solids Structures*, 12, 1976, pp.275-286.

[2] N. V. Banichuk and B. L. Karihaloo, 'Minimum-Weight Design of Multi-Purpose Cylindrical Bars', *Int. J. Solids Structures*, 12, 1976, pp.267-273.

[3] R. D. Parbery and B. L. Karihaloo, 'Minimum-Weight Design of Hollow Cylinders for Given Lower Bounds on Torsional and Flexural Rigidities', *Int. J. Solids Structures*, 13, 1977, pp.1271-1280.

[4] R. D. Parbery, 'On Multiconstraint and Multiload Optimization of Mechanical Elements', Ph.D. Thesis, University of Newcastle, N.S.W., 1985.

6. APPENDIX

In this section the integrals needed in finding the perturbation solution are given. F( ) and E( ) are complete elliptic integrals of the first and second kind respectively, and $R = 1/2\pi$.

$$I_A = \int_0^1 h_o \, ds$$

$$I_A = \frac{2RJ^o}{\pi\Omega^2} E[\sqrt{(-\mu_{4_o})}]F[\sqrt{(-\mu_{4_o})}] \; ; \qquad -1 < \mu_{4_o} \leq 0$$

$$= \frac{2RJ^o}{\pi\Omega^2} E[\sqrt{(\frac{\mu_{4_o}}{1+\mu_{4_o}})}]F[\sqrt{(\frac{\mu_{4_o}}{1+\mu_{4_o}})}] \qquad 0 \leq \mu_{4_o}$$

$$I_B = \int_0^1 h_o \sin^2 (s/R) \, ds$$

$$= \frac{4J^o}{\mu_{3_o}\mu_{4_o}} \{ E(\sqrt{(-\mu_{4_o})}] - F[\sqrt{(-\mu_{4_o})}]\} \; ; \quad -1 < \mu_{4_o} \leq 0$$

$$= \frac{4J^o}{\mu_{3_o}\mu_{4_o}} \{ E[\sqrt{(\frac{\mu_{4_o}}{1+\mu_{4_o}})}] - F[\sqrt{(\frac{\mu_{4_o}}{1+\mu_{4_o}})}]\}; \; \mu_{4_o} \geq 0$$

The remaining integrals are evaluated for $-1 < \mu_{4_o} \leq 0$ only.

$$I_C = \int_0^1 h^2 \sin^2 (s/R) \, ds$$

$$= \frac{4RC_o^{\,2}}{\mu_{3_o}^{\,2}\mu_{4_o}} [\frac{\pi}{2} (1 - \frac{1}{\sqrt{(1+\mu_{4_o})}})]$$

$$I_D = \int_0^1 h_o^{\,3} \sin^2 (s/R) \, ds$$

$$= \frac{4RC_o^{\,3}}{\mu_{3_o}^{\,3}\mu_{4_o}} [F(\sqrt{(-\mu_{4_o})}) - \frac{E(\sqrt{(-\mu_{4_o})})}{1+\mu_{4_o}}]$$

$$I_E = \int_0^1 h_o^{\,3} \sin^4 (s/R) \, ds$$

$$= \frac{4RC_o^3}{\mu_{3_o}^3 \mu_{4_o}^2} [E(\sqrt{(-\mu_{4_o})}) (\frac{2 + \mu_{4_o}}{1 + \mu_{4_o}}) - 2 F (\sqrt{(-\mu_{4_o})})]$$

$$I_F = \int_o^1 h_o^4 \sin^4 (s/R) ds = \frac{RC_o^4 \pi}{\mu_{3_o}^4 \mu_{4_o}^4} [2 - \frac{(2 + 3\mu_{4_o})}{(1 + \mu_{4_o})^{3/2}}]$$

$$I_G = \int_o^1 h_o^2 ds = \frac{2 \pi R C_o^2}{\mu_{3_o}^2 \sqrt{(1 + \mu_{4_o})}}$$

$$I_H = \int_o^1 h_o^3 ds = \frac{4RC_o^3}{\mu_{3_o}^3} \frac{E \sqrt{(-\mu_{4_o})}}{1 + \mu_{4_o}}$$

$$I_I = \int_o^1 h_o^4 \sin^2 (s/R) ds = \frac{\pi R C_o^4}{\mu_{3_o}^4 (1 + \mu_{4_o})^{3/2}}$$

DYNAMIC OPTIMIZATION OF MACHINE SYSTEMS CONFIGURATION

Z.A. Parszewski, Dr.Eng.Sc., DIC., FIE., FIMechE,M.ASME
Professor of Mechanical Engineering
T.J. Chalko, Dr.Eng.Sc., MIE
D-X. Li, M.Eng.Sc., Ph.D student
The University of Melbourne, Parkville
Victoria, 3052, Australia

1. INTRODUCTION

Speed, reliability and accuracy require new dynamic influences to be considered in system design, analysis and maintenance. These influences represent some dynamic interaction between the subsystems of a machine or structure, or any dynamic system, or between them and other machines or structures not included in the considered system and representing the environment [1].

The relative positions of the subsystems representing the system's configuration, can be described by a set of parameters. They may correspond to the connecting co-ordinates i.e. co-ordinates of interaction between the subsystems.

Examples are numerous in all areas of engineering, as e.g.:

- positions of the joints of a truss or a frame
- bearings transverse positions of a rotor or a crankshaft mechanism
- microprocessors and digital systems interconnections for a given pattern of concurrency of data transfer
- a multiple input-output plant with a specified range of parameter uncertainty e.g. an aircraft

Depending on the problem, the subdivision can be considered on various levels of the structure. Of major importance and hence of main interest in the approach presented in this paper, is when the subsystems represent assemblies or even whole machines or other structures manufactured separately and joined at the assembly stage, often on site.

The described parameters can hence usually be easily varied on design or are subject to manufacturing or assembly errors (e.g. rods' lengths in the truss; bearing pedestals' dimensions or positions in the

G. I. N. Rozvany and B. L. Karihaloo (eds.), Structural Optimization, 217–224.

rotor case; eccentric sleeves can be introduced in the crankshaft case).

The possible small variations of these parameters introduce hence only slight change to the system configuration without changing the subsystems. They are hence called the configuration parameters [2].

If however their number exceeds the number of the system equilibrium conditions, additional conditions can be imposed on the system that may greatly influence the system dynamics. E.g. small variations of the rods' lengths in a truss will change only slightly the nods' positions but will change, in this case, the force distribution in the rods in a major way.

Still greater may be the influence of bearings small displacements (shifts and tilts) on e.g. the distribution of bearings loads, shaft bending moments, rotor deflection etc. In the cases of trusses, rolling element bearing rotors and other elastic structures, the system dynamic characteristics (stiffness, damping) remains practically unchanged. The system is anisotropic but homogeneous.

However, in the very important cases of hydrodynamic oil bearings the shifts and tilts introduce changes of the eccentricity ratio c in the bearings. As the stiffness and damping matrices of journal bearings depend on the eccentricity ratio c (Fig.2,3), the dynamic characteristics of the system will vary. The system is hence anisotropic and nonhomogeneous. The described small variations of the system configuration may introduce hence important changes of its critical speeds, instability threshold and dynamic response at working speed.

2. SYSTEM DYNAMICS IN TERMS OF THE SYSTEM CONFIGURATION PARAMETERS

For hyperstatic systems the extra constraints parameters, describing the system configuration and allowing for simple variation (for design) or subject to errors (for tolerance analysis) are chosen as the independent configuration parameters. Instead of fixing them, additional conditions are imposed, in this method, on the system allowing for the selection of its dynamics at working speeds.

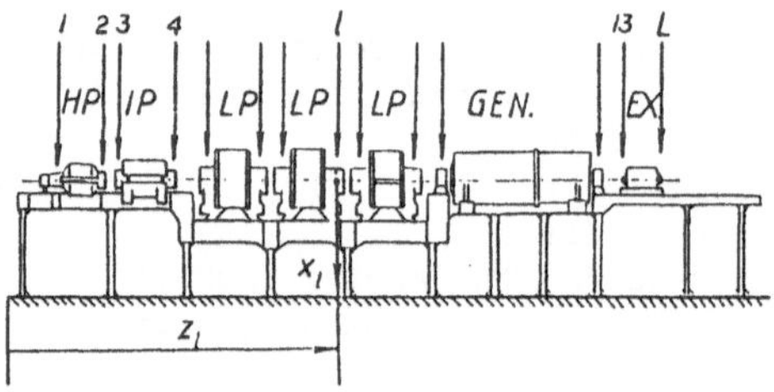

Figure 1.

Figure 1 represents a turbogenerator set with the line of rotors supported on fourteen fluid film bearings. For a multi-bearing rotor-support system like that, several possibilities exist. In general the rotor line and each bearing can be considered separately if displacements of all bearings are considered.

Different subdivisions are used for machines installed on frames [2] or where not all bearings are manipulated.

3. PRINCIPLES OF CONFIGURATION PARAMETERS SELECTION FOR ROTOR SYSTEMS.

Even two bearing single rotor system is of interest (n>m) if tilts as well as shifts of bearings are of importance. If tilts are neglected (short bearing) some configuration parameters selection (optimisation) is possible for more than two bearing rotors and for any coupled systems.

The configuration parameters selection will be then based on the following considerations:

1. steady state equilibrium consideration of the system at working speeds
2. dynamics consideration:
   - 2a dynamic response of the system at working speed
   - 2b stability of the system at working speeds

3.1 Steady State Consideration

(a) The additional conditions at steady state may regard the rotor considerations:
- some relations in bending moments distribution (if rotor strength is a limiting factor)
- deflection distribution (if stiffness of the rotor is the limiting factor).

or

(b) the bearing conditions:
- some relations in bearing reactions distribution.

All the above additional conditions in the rotor as well as in the bearings area, can be expressed by the bearing reactions distribution. Once the reactions are selected they will give the bending moments and deflections for the rotor as well as the corresponding eccentricity ratios for the bearings. Hence corresponding positions of the journal centres and the bearing centres can be found and bearing position fixed. Pedestal, casings, foundation deformations under load and their thermal expansion can be also included.

The selection conditions for steady state should however leave, if possible, sufficient number of configuration parameters for the selection of the system dynamics.

Modern journal bearings are of anisotropic type (multi-lobe or eccentric partial sleeves). Their best characteristics correspond to one load direction (Fig. 2). The best configuration of the system can hence be arranged when the bearings reactions directions are known for example, all parallel. The additional conditions imposed on steady state can hence require plane deflection of the rotor and hence coplanar/parallel bearing reactions at steady state.

The actual eccentricity ratios in the bearings will be given by reactions (values) distribution. This will be defined by the equilibrium conditions and the remaining additional conditions related to the system dynamics.

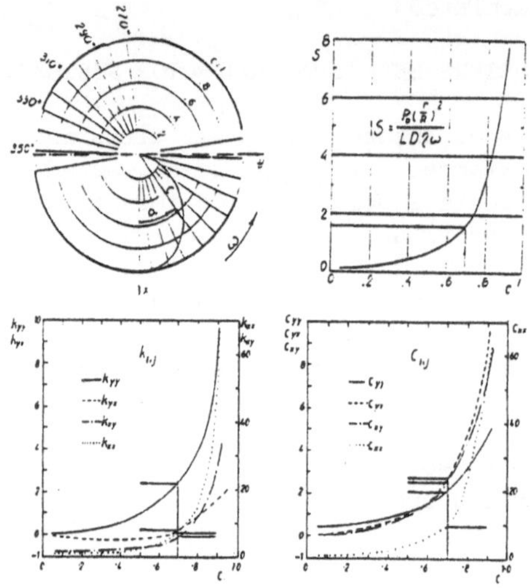

Figure 2

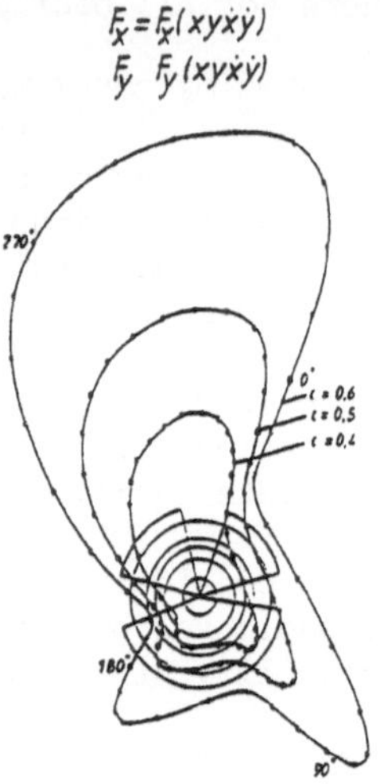

Figure 3

3.2 Dynamic Consideration

The bearing reaction values distribution is found next on the basis of the system selected (optimal) dynamic response and stability check.

Depending on the problem the selection can be achieved

(1) on the basis of the system response at working speed to:
  (a) the machine excitations e.g. disbalance
  (b) environmental excitation transferred to it from other structures not included in the considered system

or (c) a compromise of both.

(2) Stability at working speed provides additional conditions for selection.

4. OPTIMAL LAY-OUT OF A ROTOR-BEARING SYSTEM - Visualizing Example

The following equations provide the program "analysis" that is next coupled with the optimising program e.g. COPES [3], (Vanderplaats, 1979). The independent variables are the bearing reaction ratios and as the goal function the average square of the vibration amplitudes at the selected points is taken. The bearing reaction distribution defines the system configuration uniquely i.e. the rotor's elastic curve and the relative positions of the bearings. A program synthesizing the methods has been developed. For better visualization, however, the results will be first presented here in the intermediate graphic form. This is not necessary in the normal configuration selection (optimisation) process, but may be used for sensitivity analysis and tolerance selection. The sensitivity analysis (to configuration parameters variation) is hence only a part of the general approach presented.

An example is taken for a four bearing rotor system shown in Fig. 4.

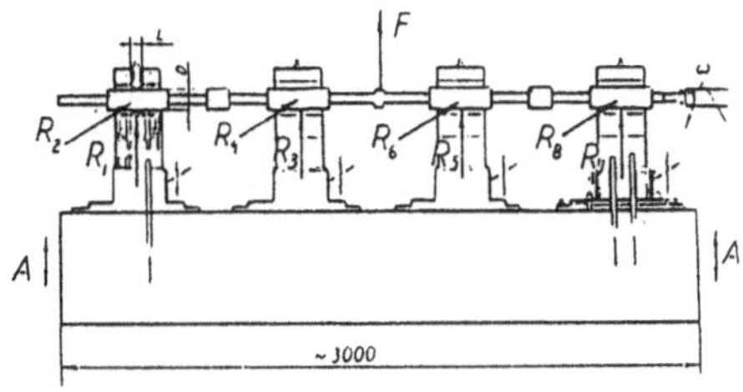

$D = 50\,mm \quad L/D = 0{,}83 \quad c = 0{,}3$

$\eta = 0{,}02\,Ns/m^2 \quad \omega = 300\,rad/s \quad G = 160\,N$

Figure 4.

The equations of vibratory motion of the system for its forced response (to unbalance force F, the response to environmental excitation A as well as for the stability analysis of the system) are as follows in

$$\left[\alpha_s(R_3,R_5)\right]^{-1}\begin{bmatrix}x_1\\x_2\\\vdots\\x_j\\x_{j+1}\\\vdots\\x_i\end{bmatrix} = \begin{bmatrix}\left[\alpha_e\right]^{-1}\begin{bmatrix}y_1\\y_2\\\vdots\\y_j\end{bmatrix}\\0\end{bmatrix} + \begin{bmatrix}F\end{bmatrix}$$

$$\alpha_s^{-1}(R_3,R_5)x = E\cdot F \qquad (1)$$

$$\{\ddot{\eta}\} + diag[\alpha^2]\{\eta\} = [\Phi]^T(f(\eta,\dot{\eta})) + E + F \qquad (2)$$

Figure 5 Figure 6

Figure 5 for linear approach and Figure 6 for non-linear approach and limit cycle, in modal coordinates $\{\eta\} = [\Phi]\{{x \atop y}\}$ as basis for configuration selection, with the transient bearing force F from Figure 3.

Corresponding results in terms of the (independent) bearing reaction ratios R_3 and R_5 in the vertical plane (all horizontal reactions are to disappear) are presented in Figure 7. Figure 7b and 7a show the response (quadratic mean of the amplitude in vertical and horizontal directions) due to unbalance force (Fig. 4) on the coupling and due to environmental excitation on the supporting structure respectively.

The subsystems composition method was used for the computation of the amplitudes of forced vibrations.

The assessment of stability margin r (Fig. 7c the smallest negative value of the real parts of the characteristic roots) was also done.

The three diagrams in Figure 7 form a base for the selection of the two remaining independent reaction ratios R_3 and R_5 and hence for the selection of positions of corresponding bearings. As a compromise, the following ratios were selected

$$R_3^s = 0.3550 \quad ; \quad R_5^s = 0.1875$$

The corresponding amplitudes of vibration and stability margin are marked in Figure 7 by 'A^S' and 'r^S' respectively.

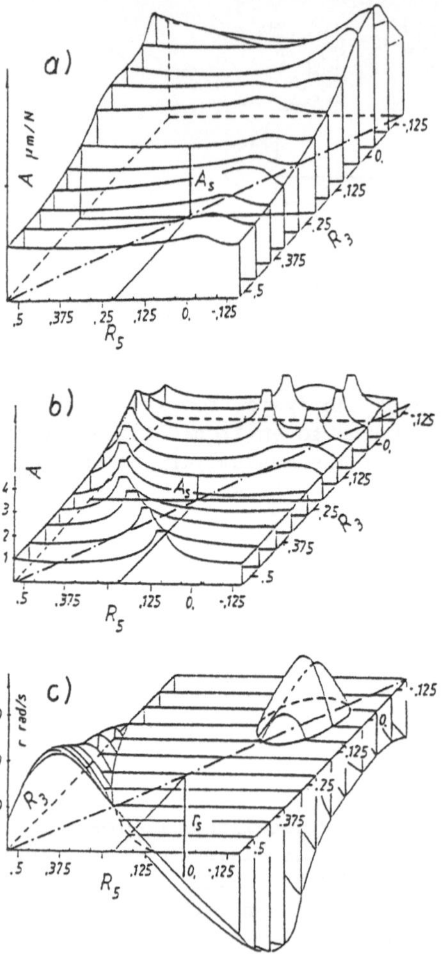

Selected in that way (together with equilibrium equations) reaction vector defines the system configuration uniquely i.e. the rotor's elastic curve and relative positions of the bearings. The last ones are given by the non-dimensional bearing load i.e. the Sommerfield number S and eccentricity ratio C relation for the bearing (e.g. Fig. 3).

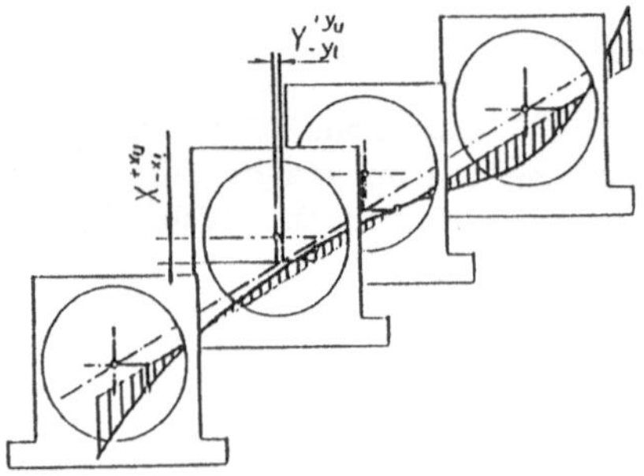

Figure 8.

Figure 8 gives, in perspective, the configuration of the system corresponding to its selected dynamics. Any deformation of the supporting structure under load and thermal delation can be easily included in the bearing positions, compensating hence for all those effects.

Figure 7

3. OPTIMIZATION

3.1 Dynamic Response of the System - Subprogram ANALIZ

Optimization program COPES [3], (A FORTRAN Control Program for Engineering Synthesis, Vanderplaats, 1979) was compiled and applied.

COPES uses constrained function minimization program CONMIN which employs the Fletcher-Reeves Conjugate Direction Method and Zoutendijks Method of Feasible Directions.

The optimization procedure asks users to supply subprogram ANALIZ which initializes design variables (in our case, the described configuration parameters), evaluates the values of the ojective function, Fig. 9.

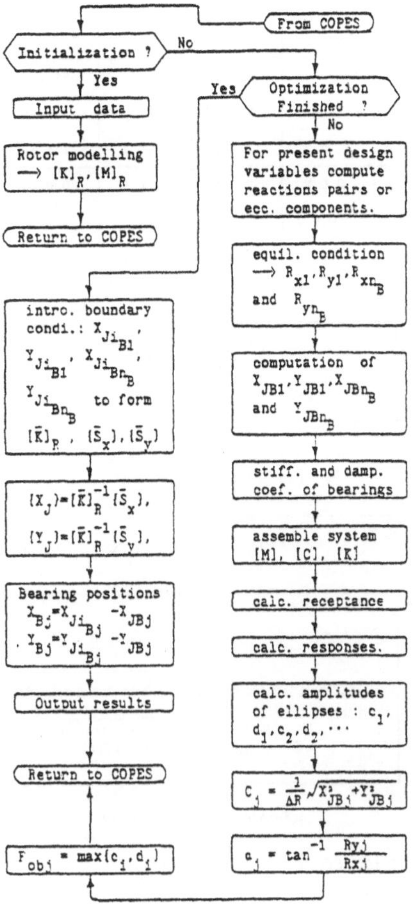

Figure 9.

Averaging function is proper in most cases. To show the approach in a clear way, and as an alternative possibility, maximum amplitude of rotor vibration is considered here:

$$F_{obj} = \{c_1, d_1, c_2, d_2, \ldots\} \quad (2)$$

Here $c_j d_j$ are the axes of journal centre trajectory in the jth bearing.

The optimization procedure minimizes the objective function for whole rotor.

The constraints are some practical limits imposed on the bearing eccentricity ratios.

4. A PRACTICAL APPLICATION EXAMPLE

Application is presented for the same rotor bearing installation shown in Figure 4.

$$\phi_{o1} = 100°$$
$$\phi_{n1} = 150°$$
$$\phi_{o2} = 284.8°$$
$$\phi_{n2} = 65.5°$$
$$\phi_{o3} = 20.8°$$
$$\phi_{n3} = 47°$$
$$D = 50 \text{ mm}$$
$$L/D = 0.83$$
$$\Delta R = 0.3 \text{ mm}$$

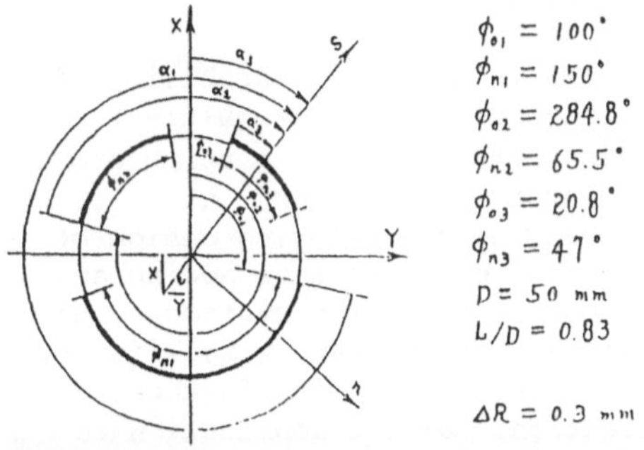

Figure 10.

The bearings are short three-sleeve journal bearings (Fig. 10) modelling bearings used in turbo-generators [4] (Parszewski et al., 1987). An excitation force of 1.0 N, corresponding to the rotor unbalance is applied as shown in Figure 4. The static load is only gravity force of the rotor, and the reaction direction constraints are taken so that all resultant reactions are opposite to the gravity force.

The starting values are given in Table 1, which represent approximately uniform reaction distribution.

The objective function history for successive iterations is shown in Figure 11. The optimal result is given in Table 2, and the corresponding bearing configuration is shown in Figure 12. The result shows 33% improvement with respect to the starting configuration.

Table 1

No.	Brg. 1	Brg. 2	Brg. 3	Brg. 4
X_{JB}	- 78.8µm	- 57.0µm	- 77.2µm	- 78.5µm
Y_{JB}	-101.1µm	- 89.3µm	-100.3µm	-101.0µm
R_X	40.9 N	30.0 N	40.0 N	40.7 N
R_Y	0.0 N	0.0 N	0.0 N	0.0 N
X_B	0.0µm	- 49.4µm	- 24.6µm	0.0µm
Y_B	0.0µm	- 11.8µm	- 0.7µm	0.0µm
Objective Function	46.35 µm			

Table 2

No.	Brg. 1	Brg. 2	Brg. 3	Brg. 4
X_{JB}	- 32.2µm	-100.8µm	- 99.3µm	- 39.0µm
Y_{JB}	- 72.4µm	-111.3µm	-110.6µm	- 77.4µm
R_X	20.0 N	55.2 N	54.0 N	22.5 N
R_Y	0.0 N	0.0 N	0.0 N	0.0 N
X_B	0.0µm	634.3µm	639.1µm	0.0µm
Y_B	0.0µm	37.1µm	35.1µm	0.0µm
Objective Function	31.10 µm			

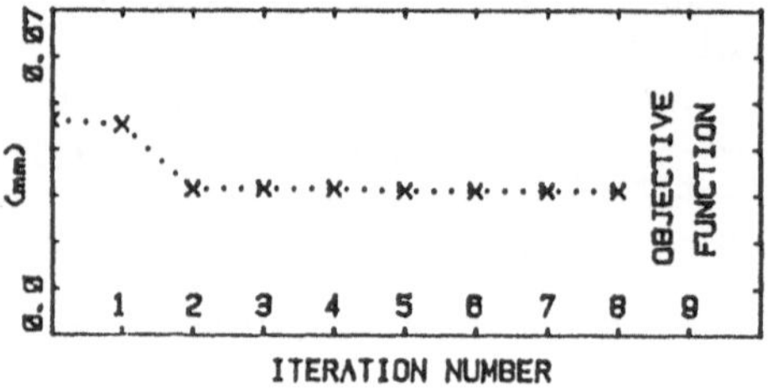

Figure 11

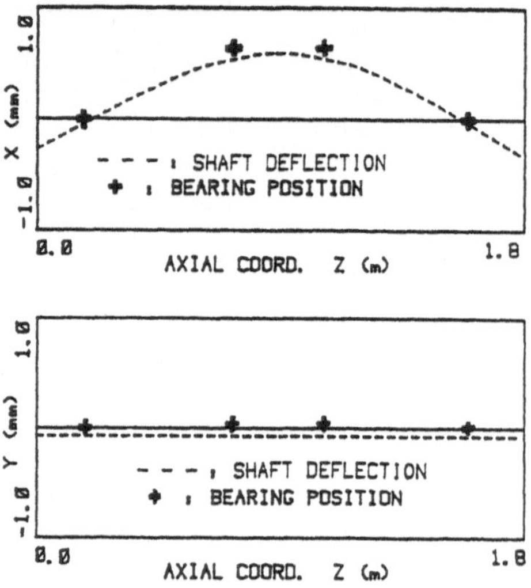

Figure 12

4. CONCLUSIONS

The results show the feasibility of dynamic optimization of multi-bearing rotor systems in terms of configuration parameters, hence without changing any existing design of the system components or adding any new components to the systems. The optimized results can be realized by adjustments of the bearing positions within practical limits in the order of the bearing clearance and rotor deflection. Direct application to other machines and structures has been also explained.

REFERENCES

[1] Parszewski, Z.A., Krodkiewski, J.M., and Skoraczynski, J. 1984. 'Computer-experiment interface in rotor-support dynamics'. Computers in Engineering, Vol.I, ASME, New York.

[2] Parszewski, Z.A. 1987. 'System configuration and its dynamic response'. Proc. ASME Vibration Conference, Boston.

[3] Vanderplaats, G.M. 1979. COPES - A FORTRAN control program for engineering synthesis.

[4] Parszewski, Z.A., Nan, X. and Li, D.X. 1987. 'Steady and unsteady bearing forces with thermal effects'. Proceedings of International Conference on Mechanical Dynamics, Shenyang, China.

DESIGN FOR MINIMUM STRESS CONCENTRATION - SOME PRACTICAL ASPECTS

Pauli Pedersen
Department of Solid Mechanics
The Technical University of Denmark
DK-2800 Lyngby
Denmark

ABSTRACT. The procedure necessary to obtain the shape which returns minimum stress concentration involves many choices, and we shall here sum up common personal experiences from solutions to different problems. Parametrization of the design shape is carried out using global expansion functions, very much like modal expansion in vibrational analysis. Finite element modelling is treated independent of design modelling and too simple finite elements are avoided. Linear stress elements are obtained from constant stress elements and this is utilized in the sensitivity analysis, that to a large extent is performed analytically.

1. INTRODUCTION AND PROBLEM FORMULATION

The problem of shape design for minimum stress concentration is highly nonlinear, and must be solved iteratively. Thus we may immediately convert the problem to a sequence of problems of optimal redesign, i.e. how do we change a given shape to a better "neighbouring" shape? The solution to this involves three steps: finite element stress analysis for the given shape - sensitivity analysis with respect to the parameters that describe the shape - decision of optimal redesign.

Even with such a specified procedure, a number of options are available and the success of our optimization to a large extent depend on our choice of:

- design parameters
- finite element model
- method of sensitivity analysis
- method of optimization

The goal of the present paper is to argue for the choices behind the work reported in [1], [2], [3], and to describe some extensions in present work on anisotropic materials and laminates. Thus the intentions of the paper are in a way rather narrow.

In mathematical terms the objective of our shape design is to

G. I. N. Rozvany and B. L. Karihaloo (eds.), Structural Optimization, 225–232.

$$\underset{\begin{Bmatrix}\text{over feasible}\\ \text{shapes}\end{Bmatrix}}{\text{Minimize}} \left[\underset{\begin{Bmatrix}\text{over structure}\\ \text{and load cases}\end{Bmatrix}}{\text{Maximum}} (F) \right] \tag{1.1}$$

where the stress F may be the von Mises reference stress σ_M, or the maximum stress λ (maximum stress eigenvalue), or some linear combinations of these. This Min-Max-problem is converted to a Min-problem with constraints by treating F_{max} as a further unknown, i.e.,

$$\text{Minimize } F_{max} \tag{1.2}$$

subject to the constraint that the stress level everywhere, for all load cases m, is limited by F_{max} when design changes Δt_i are performed

$$F_m(x) + \sum_i \frac{\partial F_m(x)}{\partial t_i} \Delta t_i - F_{max} \leq 0 \tag{1.3}$$

for $m = 1,2,\ldots,M$ and the total structural space x.

This may seem to constitute many equations, but in actual design there are normally only a few critical points and corresponding load cases. The word "critical" means close (say $> 0.95\ F_{max}$) to the F_{max} of the recent stress analysis. These stresses are localized by the finite element analysis and the actual points may differ during different redesigns. Assembling the critical stresses in the vector $\{F\}$, ineq. (1.3) is rewritten to

$$\{F\} + [\nabla F]\{\Delta T\} - \{1\}F_{max} \leq \{0\} \tag{1.4}$$

The elements of the gradient matrix are $\partial F_j / \partial t_i$, i.e. the partial derivatives of the reference stress at point j (for a given loading case) with respect to design variables i.

The problem of minimizing F_{max} subject to (1.4) constitutes a linear programming problem which can be solved by the Simplex method. We shall not further discuss this choice of optimization method, and thus concentrate on discussions related to: design parameters - finite element model - sensitivity anslysis.

2. DESIGN PARAMETRIZATION

The design description for shape optimization is by no means unique, and the success of our optimization to a large extent depend on our chosen description. In order to classify the possibilities we may at the one extreme have completely local design parameters like the direct space coordinates of the design shape (in practice discretized to a finite, although large number of points). At the other extreme we have completely global design parameters which indirectly give the space coordinates of the design shape. A number of in-between possibilities also exists, where design parameters control a certain part of the design space in a somewhat indirect way (like by master nodes). Naturally, our choice of design description must depend on the actual optimization problem. For minimization of stress concentration we have for several years had good experience with global design variables, that enforce smoothness and desired connections to neighbouring shapes, which are not subjected to design optimization. This approach also protects against the inherent weaknesses of the necessary finite element stress analysis. Therefore, this design description shall be presented here, mathematically in two dimensions by a curve C

$$C = C_0 + \sum_{i=1}^{I} t_i f_i(s) \tag{2.1}$$

obtained from a basic curve C_0 which is modified by the preselected design functions $f_i(s)$ where s is the natural parameter for the curve C_0. The design parameters then are the linear combination factors t_i. The basic curve C_0 may also depend on some design parameters, like for the case of a superelliptic curve C_0 where the lengths of the half-axes and the power of the superellipse are three further design parameters. In fact, the basic curve C_0 may be dealt with in an updated way.

In [1] and [2] a similar approach in cylindrical coordinates with an initial circular design of radius r_0:

$$r(\theta) = r_0 + \sum_{i=1}^{I} t_i f_i(\theta) \tag{2.2}$$

and in [3] in polar coordinates with an initial spherical design, again with radius r_0

$$r(\theta,\phi) = r_0 + \sum_{i=1}^{I} t_i f_i(\theta,\phi) \tag{2.3}$$

where applied. Naturally these descriptions have limitations, that are not actual for the more general approach (2.1), which is illustrated in fig. 2.1 with only two design parameters.

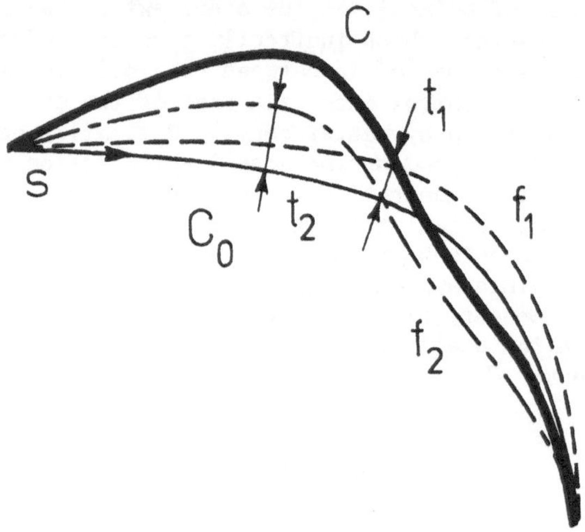

Figure 2.1. Shape C described as a sum of the basic shape C_0 and two modification functions f_1 and f_2.

3. FINITE ELEMENT MODELLING

Also the finite element modelling is by no means unique, and the success of the analysis depends not only on our computational resources, but even more on the way we use them, i.e. what elements and what nodal positions we choose. Again there is no straightforward answers and a good finite element model is strongly related to the actual problem which constitutes the design as well as the load case.

In problems of stress concentration for linear elastic problems the lowest order elements should not be used. Therefore the work in [1], [2] and [3] is based on quadratic displacement assumptions. Even in recent papers on shape optimization are linear displacement elements used, but this creates the problem that the result of the analysis very much depends on the actual modelling. Adaptive finite element modelling may be a repairing attitude towards this, but we must remember that in relation to practical problems with multiple load cases we then have to work with several finite element models for the same design.

In the author's opinion quadratic displacement elements constitute a good compromise between accuracy of results and simplicity of usage. As shown in [4] the results with linear stress elements are far less sensitive to modelling, than those with constant stress elements. Furthermore, as shown in [4] for two-dimensional elements and in [5] for three-dimensional elements, there exists simple linear relations between these elements

$$[S_e^L] = \mathcal{L}[S_e^C] \tag{3.1}$$

which in reality means that results based on a linear stress assumption (quadratic displacement) are obtained with efforts corresponding to stiffness generation at the level of constant stress (linear displacement). In relation to sensitivity analysis the simple relations (3.1) are especially valuable.

Although adaptive finite element modelling should not just compensate for a too simple element and/or model, it should be used to improve the finite element model and to make sure, that as the design shape evolves the accuracy of the finite element is kept at an acceptable level. An example from automatic mesh generation is shown in fig. 3.1.

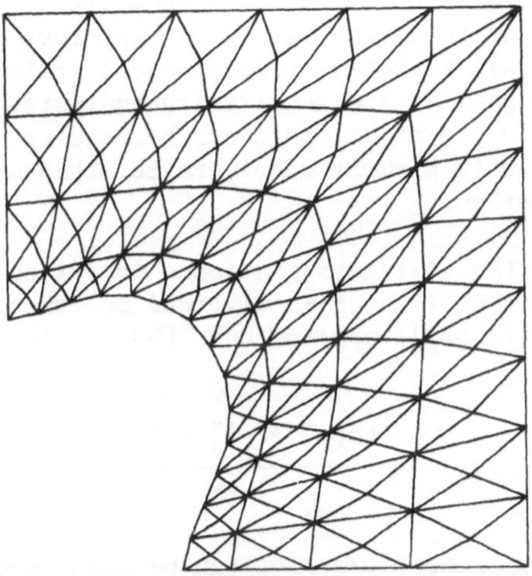

Figure 3.1. Generated triangular mesh with equal nodal spacing on the curved design shape.

4. SENSITIVITY ANALYSIS

The key information for the optimal redesign process is to know how stresses change with the parameters of the shape design. As stated before we concentrate on a number of critical points, located in the ordinary finite element analysis, where the displacements $\{D\}$ are obtained from the known stiffness matrix $[S]$ and the external loads $\{A\}$. The stress matrix $[Q_e]$ of a given element e, together with the element displacements $\{D_e\}$ (contained in $\{D\}$) then gives the stresses $\{\sigma\}$

$$\{\sigma\} = [Q_e]\{D_e\} \quad : \quad [S]\{D\} = \{A\} \tag{4.1}$$

With "," as notation for partial derivatives, the stress sensitivities with respect to design parameter t_i are

$$\{\sigma\}_{,t_i} = [Q_e]_{,t_i}\{D_e\}+[Q_e]\{D_e\}_{,t_i} \quad ; \quad [S]\{D\}_{,t_i} = \{A\}_{,t_i}-[S]_{,t_i}\{D\} \tag{4.2}$$

where only the stiffness gradient matrix may give rise to practical problems. In [2] we have used simple overall difference calculations, but in [1] and [3] the linear relations (3.1) were used to obtain a highly accurate and quick procedure:

$$[S]_{,t_i} = \sum_e [S_e^L]_{,t_i} = \sum_e \mathcal{L}\left[[S_e^C]_{,t_i}\right] \tag{4.3}$$

Furthermore, it should be noted that the summation is only over the elements connected to the actual design shape. In other papers the summation is extended to all elements, because nodal redistribution is treated as an integrated part of the sensitivity analysis. This is not only unnecessary, but it may even be dangerous because we then may maximize the errors of the finite element model, rather than minimize the physical stress concentration.

Not in all cases is the stiffness gradients analytically available like in [1] and [3], and then the semi-analytical approach is a good alternative. The eq. (4.2) is still applied but $[S]_{,t_i}$ is in this method approximated by

$$[S]_{,t_i} \approx \frac{[S(t_i+\Delta t_i)] - [S(t_i)]}{\Delta t_i} \tag{4.4}$$

and black-box finite element programs can be used. Accuracy problems have been reported [6], [7], telling us that the size of Δt_i should be carefully selected.

The von Mises stress σ_M is obtained by

$$(\sigma_M)^2 = \frac{1}{2}\{\sigma\}^T[Z]\{\sigma\} \tag{4.5}$$

where $[Z]$ is a symmetric matrix of integer constants. Having determined the actual stress derivatives $\{\sigma\}_{,t_i}$ the derivative of the von Mises stress $(\sigma_M)_{,t_i}$ is evaluated by

$$(\sigma_M)_{,t_i} = \frac{1}{2\sigma_M} \{\sigma\}^T[Z]\{\sigma\}_{,t_i} \tag{4.6}$$

Dealing with a stress eigenvalue λ and its corresponding directions $\{\Gamma\}$ as solutions to

$$\Big[[\sigma] - \lambda[I]\Big]\{\Gamma\} = \{0\} \tag{4.7}$$

(actual stresses now in two-dimensional notation), then sensitivity analysis for eigenvalues is applied, and we have

$$\lambda_{,t_i} = \frac{\{\Gamma\}^T[\sigma]_{,t_i}\{\Gamma\}}{\{\Gamma\}^T\{\Gamma\}} \tag{4.8}$$

5. CONCLUDING REMARKS

Only few details could be given in this paper, but the remaining ones can be found in the references. Also the results for a number of solved problems can be found in these papers. Many of these solutions point towards superelliptic designs, which are therefore taken as starting designs in a project in progress for anisotropic materials.

A common feature for the designs of minimum stress concentration is a minimum change of curvature at the boundaries, which also indicate, that the traditional fillet design with straight lines connected to circular curves is not a good design.

REFERENCES

[1] Kristensen, E.S. & Madsen, N.F.: 'On the optimum shape of fillets in plates subjected to multiple in-plane loading cases'. Int. J. Numer. Meth. Engng. 10, 1007-1019, 1976.

[2] Pedersen, P. & Laursen, C.L.: 'Design for minimum stress concentration by finite elements and linear programming'. J. Struct. Mech. 10, 243-271, 1982-83.

[3] Dybbro, J.D. & Holm, N.C.: 'On minimization of stress concentration for three-dimensional models'. Comp. & Struct. 4, 637-643, 1986.

[4] Pedersen, P.: 'Some properties of linear stress triangles and optimal finite element models'. Int. J. Numer. Meth. Engng. 7, 415-429, 1973.

[5] Pedersen, P.: 'On computer-aided analytic element analysis and the similarities of tetrahedron elements'. Int. J. Numer. Meth. Engng. 11, 611-622, 1977.

[6] Barthelemy, B. & Haftka, R.T.: 'Accuracy analysis of the semi-analytical method for shape sensitivity calculation'. To appear in Proceedings of AIAA, SDM Conference, Virginia, Apr. 1988.

[7] Pedersen, P., Cheng, G. & Rasmussen, J.: 'On accuracy problems for semi-analytical sensitivity analysis'. DCAMM Report No. 367, 15 p.,Dec. 1987.

OPTIMALITY CONDITIONS FOR MULTIPLE LOADED STRUCTURES - INTEGRATING CONTROL AND FINITE ELEMENT METHOD

Alija Pičuga
University "Džemal Bijedić"
Faculty of Mechanical Engineering
B. Parovića bb
88000 Mostar
Yugoslavia

ABSTRACT. The minimum cost design of structures (beam, truss or frame) is discussed. The structure has unknowns: area of cross section and location of some points (for example: supports). For the same structure multiple load is considered: vibration, buckling and static load - with prescribed frequency, buckling force and compliance, respectively.

Optimality conditions for the unknowns are derived by variational approach. The problem is formulated in the optimal control form, so obtained results are valid for an arbitrary structure, load and boundary conditions. Optimality equations for specific structure can be solved numerically: by combination of the finite element method and the iterative algorithm. An illustrative example is presented and efficiency of the optimal design is judged by comparing it with a uniform cross-sectional area design.

1. INTRODUCTION

We are going to consider the problem of optimal synthesis with the following characteristics: (a) both local geometry and global topology are unknowns, (b) arbitrary multiple loads exist and (c) the formulation is in the optimal control form.

The published papers have been dealing only with parts of this problem. The exception, regarding entirety of the approach are works by Mróz Z., Rozvany G.I.N. (1975) [3], Carmichael D.G, Clyde D.H. (1977) [1] .

The intent of presented paper is to bring a unifying and systematic approach to both the formulation and solution of the problem.

2. PROBLEM FORMULATION

Let the structure (beam, truss or frame) with NM members, while performing its function, assume NS different mechanical states. These are eigenvalue problem and static load.

G. I. N. Rozvany and B. L. Karihaloo (eds.), Structural Optimization, 233–239.

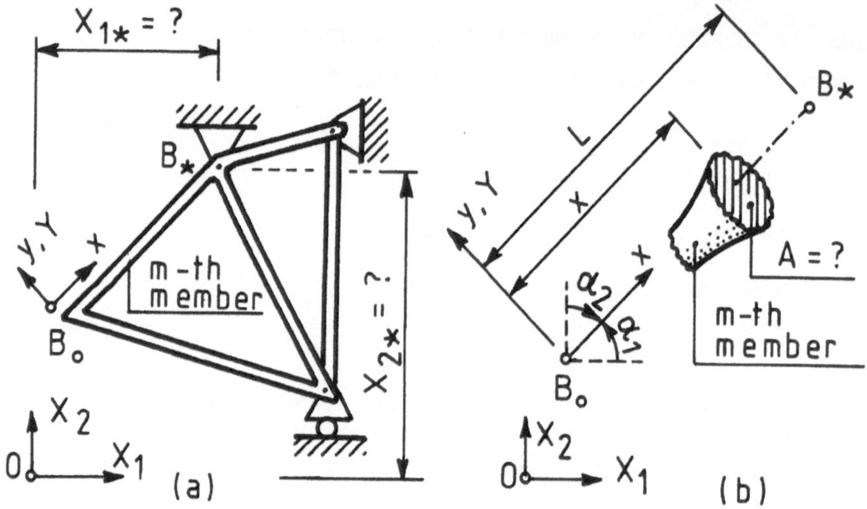

Fig. 1. An example of the plane structure (a) and details of one member (b).

We formulate the problem of optimal synthesis:

among all possible unknowns: constants X_{j*},
continuous function A, y, Y, $\dot{y}$, $\dot{Y}$ on $0 \leq x \leq L$ (1a)

find such, which satisfy:

- state equations: $\dot{y} - f = 0, \quad \dot{Y} - F = 0,$ (1b)
  (with boundary conditions at x = 0 and x = L),
- global restriction[1] to the compliance K, $(K = \sum_m \int Q^T y \, dx)$:

$$\sum_m \int Q^T y \, dx - K_p = 0, \tag{1c}$$

and unknown A gives minimum total cost C of the structure:

$$\min C, \quad (C = \sum_m c \int A \, dx). \tag{1d}$$

Here are: vector-functions: state variables y, Y and f, F, Q:

$y = y^{m,s}(x) = \{y_1 \; y_2 \; \dots \; y_{NI}\}^T$ - generalized displacements,

$Y = Y^{m,s}(x) = \{Y_1 \; Y_2 \; \dots \; Y_{NI}\}^T$ - generalized internal forces,

---

1 The eigenvalues: frequency ω and buckling force P exist in Eqs. (1b). So, we shall insert restriction for eigenvalue directly into Eqs. (1b): $\omega = \omega_p$ and $P = P_p$.

Also, restriction to minimum and maximum of control A, X_{j*} bring to "switching" equations (see [6] pp. 376). So, we can put this restriction directly in numerical solution (into move-limits).

$f = f^{m,s}(y,\ Y,\ A) = \{f_1\ f_2\ \dots\ f_{NI}\}^T$ and similar for F,
$Q = Q^{m,s}(x,\ A) = \{Q_1\ Q_2\ \dots\ Q_{NI}\}^T$ - distributed load,
$A = A^m(x)$ - area of the cross-section of m-th member,
$c = c^m$ - cost per unit volume of m-th member,
$x = x^m$ - local coordinate axis of m-th member oriented to point B_*,
system x - (y, Y) is clockwise from x to (y, Y), $(\dot{\ }) = d(\ )/dx$,
X_j, (j = 1, 2, ...) - global coordinate system; X_{j*} - unknown coordinate of the point B_*,
$(\ )^{m,s}$ - value ( ) related to member m and load s; m = 1, 2, ..., NM; s = 1, 2, ..., NS,
$(\ )_p$ - prescribed value of ( ), superscript T means transpose,
K, ω, P, K_p, ω_p, P_p have specific value for every s,
$\sum_m (\ )$ - sum of all values ( ) computed for every m, $\int (\ )\,dx = \int_0^L (\ )\,dx$; $L = L^m$ - length of m-th member,

Assumption 1. Elements in y, Y and Q are ordered in a specific way:

if, for specific values m, s i the value of y_i is:	then for the same m, s, i: Y_i and Q_i act in the same positive direction as y_i and Y_i is:	Q_i is:
linear displacement	force	distributed force
angle	moment	distributed moment

Note. Positive Y_i is mesured on the positive cross-section.

3. GOVERNING EQUATIONS

We are introducing additional unknowns - Lagrange multipliers: vector-function

$$\lambda = \lambda^{m,s}(x) = \{\lambda_1\ \dots\ \lambda_{NI}\}^T, \qquad \Lambda = \Lambda^{m,s}(x) = \{\Lambda_1 \cdots \Lambda_{NI}\}^T$$

constants B, ($B = B^s$) and forming a new, auxiliary functional C_a :

$$C_a = C + \sum_s \Big(\sum_m \int \big(\lambda^T(\dot{y} - f) + \Lambda^T(\dot{Y} - F)\big)\,dx + B^2(K - K_p)\Big). \quad (2)$$

In that way, we replace finding of conditional extremum (1) by finding of the extremum of the functional (2) without restrictions. To derive the necessary optimality conditions, setting the first variation ΔC_a to zero:

$$\Delta C_a = -\sum_s \sum_m \Big(\int \big((\dot{\lambda} + H_{,y})^T\,\delta y + (\dot{\Lambda} + H_{,Y})^T\,\delta Y\big)\,dx +$$

$$+ (\lambda^T \Delta y + \Lambda^T \Delta Y)_* - (\lambda^T \delta y + \Lambda^T \delta Y)_o) -$$

$$- \sum_m \int H_{,A} \, dx \, \delta A - \sum_j \sum_{m_j} H_* \cos \alpha_j \, \delta X_{j*} = 0. \tag{3}$$

Here are:

$$H = H^m(y, Y, A) = - cA + \sum_s (\lambda^T f + \Lambda^T F - B^2 Q^T y), \tag{4}$$

δ, Δ - variation and total variation, $(\ )_{,A} = \partial(\ )/\partial A$,
$(\ )_{,p} = \left\{ \partial(\ )/\partial p_1 \quad \partial(\ )/\partial p_2 \ \ldots \right\}^T$ if p is an vector,
$(\ )_o$ or $(\ )_*$ - value ( ) computed at the point B_o or B_*,
$\sum_j (\ )$ - sum of all values ( ) computed for every j, m_j - set of members which contain the definite point B_* with unknown X_{j*},
$\alpha_j = \alpha_j^m$ - angle between axis X_j and x.

We shall find optimality conditions for unknowns λ, Λ, A and X_{j*} - from Eq. (3). This Eq. contains groups of members with arbitrary and mutually independent variations. So, in order to satisfy Eq. (3) every group must be zero.

3.1. General Solution for λ, Λ

Setting members with δy and δY in Eq. (3) to zero and using Eq. (4), we obtain:

$$\dot{\lambda} + f^T_{,y} \lambda + F^T_{,y} \Lambda - B^2 Q = 0, \qquad \dot{\Lambda} + f^T_{,Y} \lambda + F^T_{,Y} \Lambda = 0. \tag{5}$$

Eqs. (5), for fixed m, s - are independent from others, with different m or s. But, what is the solution for λ, Λ ?

Assumption 2. Functions f, F can be expressed in linear form:

$$f = Uy + VY, \qquad F = Zy - U^T Y - Q, \tag{6}$$

where U, V, Z are quadratic matrices of order NI x NI and V, Z are diagonal (or null) matrices.

Theorem. If Assumption 2. is satisfied, then the general solution for λ, Λ in Eqs. (5) is:

$$\lambda = - B^2 Y, \qquad \Lambda = B^2 y, \qquad \text{(for every m and s).} \tag{7}$$

Proof. The derivatives of f, F are (from Eqs. (6)):

$$f^T_{,y} = U^T, \qquad F^T_{,y} = Z^T = Z,$$
$$f^T_{,Y} = V^T = V, \qquad F^T_{,Y} = (-U^T)^T = -U. \tag{8}$$

Inserting Eqs. (7-8) in (5), we obtain equations:

$$B^2(\dot{y} - (Uy + VY)) = 0, \qquad B^2(\dot{Y} - (Zy - U^TY - Q)) = 0,$$

which are identically zero, according to Eqs. (6) and (1b).

Note. The members in Eqs. (3) with variations of y, Y in points B_o and B_* are identically zero. The proof is possible by Eqs. (7) and taking the general form of boundary conditions for y, Y in these points.

3.2. Final Form of the Function H. Optimality Conditions for A, X_{j*}

With solution (7), function H in Eq. (4) takes the form:

$$H = - cA + \sum_s B^2(-Y^Tf + y^T(F - Q)). \tag{9}$$

Setting members with variations δA and δX_{j*} in Eq. (3) to zero, we are getting the next form of optimality conditions for A and X_{j*} :

$$- cL + \int_o^L \sum_s B^2(-Y^Tf_{,A} + y^T(F_{,A} - Q_{,A})) \, dx = 0, \tag{10}$$

(for every member m),

$$\sum_{m_j} H_* \cos \alpha_j = 0, \quad \text{(for every unknown coordinate } X_{j*}). \tag{11}$$

4. ITERATIVE GENERATION OF THE OPTIMAL DESIGN

We cannot find the solution of the problem in a closed form, so we use the next II-phase numerical process.

I. It is possible to solve the state variables y, Y from Eqs. (1b) if the control variables A, X_{j*} are known. Here we use the finite element method (FEM). After that it is possible to compute constants B.

II. For known y, Y, B we can find new, "better" control variables from Eqs. (10-11). Because we use FEM, L in Eq. (10) is now the length of the finite element.

Now, we return to phase I. and repeat the complete procedure (using also move-limits, [8] pp. 138, [7]) until convergency will be satisfied.

We cannot find constants B and coordinates X_{j*} directly. Let us denote these unknowns shortly as ( ). At first, we compute an additional control U related to every unknown ( ) separately. These values U are:

$$(K - K_p), \qquad (\omega - \omega_p), \qquad (P - P_p)$$

for constants B and the left side of Eqs. (11) for X_{j*}. Now, we compute every unknown ( ) on k-th iteration by the following secant procedure:

$$(\ )^k = (\ )^{k-1} - U^{k-1} \frac{(\ )^{k-1} - (\ )^{k-2}}{U^{k-1} - U^{k-2}} . \tag{12}$$

5. AN EXAMPLE

<u>Mechanical state</u>: first Fig. (3a), second (3b). The structure is in Fig. (2).

<u>General data.</u>

$I = aA^3$, $a = 23.148\ m^{-2}$, eight finite elements per member 1 and 2, one element for member 3. For all member are the same: c and E (Young's modulus).

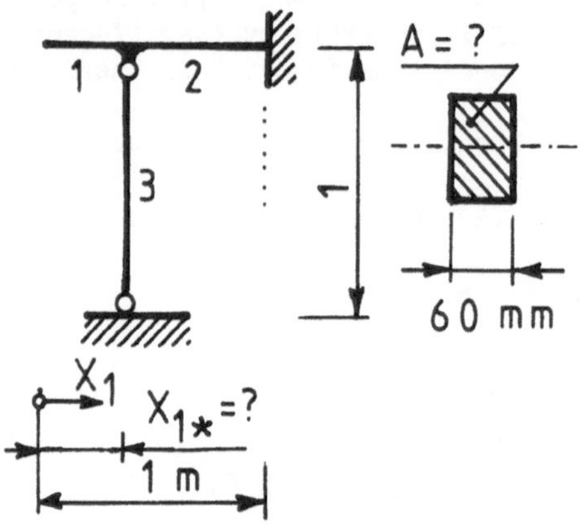

Fig. 2. Example; unknowns are A, X_{1*} .

		minimum allowed value	maximum allowed value
A m^2	for member 1, 2:	$12 \cdot 10^{-4}$	$60 \cdot 10^{-4}$
	for member 3:	$0.54 \cdot 10^{-4}$	$5 \cdot 10^{-4}$
X_{1*} m		0.2	0.6

<u>Comparative design</u>: all dimensions on the mean level (min + max)/2, cost is C_{uni} , compliance under first load is K^1_{uni} and under second load K^2_{uni} .

<u>Problem is:</u> compute A and X_{1*} for the minimum cost design; its cost is C_{opt}, compliance under first load is K^1_{opt} and under second load is K^2_{opt}. It must also be:

$K^1_{opt} = K^1_{uni}$, $K^2_{opt} = K^2_{uni}$.

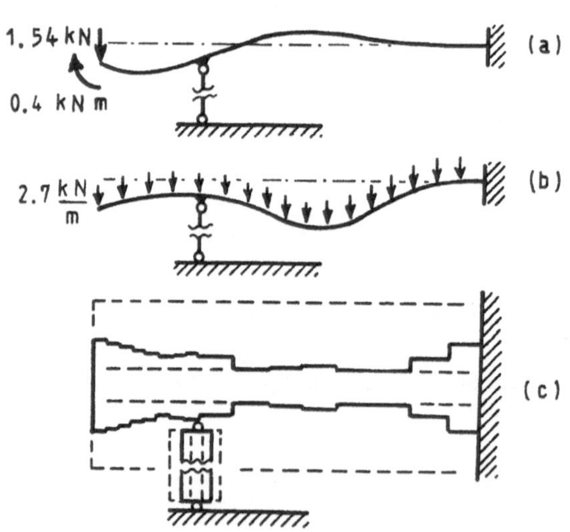

Fig. 3. Solution. First (a) and second (b) mechanical state with deflections for the optimal design. Optimal solution for A, X_{1*} (c). $C_{uni}/C_{opt} = 1.76$, [4].

Number of iterations 12-15.

6. CONCLUSION

Formulation of the problem of minimum cost synthesis of multi-purpose structure with global restriction only is presented. Unknowns are: area of the cross-section and location of some points.

The problem is formulated in the optimal control form and in that way we are getting a unifying and systematic approach to both the formulation and solution of the problem. So, the form of derived optimality conditions (7) and (10-11) enables their application for arbitrary mechanical states and boundary conditions.

The optimality equations, for specific structure, can be solved numerically: by combination of finite element method, secant (Eqs. (12)) and move-limits procedure.

An illustrative example is presented and efficiency of the optimal design is judged by comparing it with a uniform structure. For this example cost of the uniform design is 1.76 times greater than optimal.

REFERENCES

[1] Carmichael D.G., Clyde D.H.: 'Observations on the theory of optimal load transmission by flexure', 6-th Austral. Conf. Mech. of Struct. and Mater., Univ. of Canterbury: (383-390) 1977.

[2] Karihaloo B.L.: 'Optimal design of multi-purpose tie column of solid construction', Int. J. Solids Structures, Vol. **15** : (103-109) 1979.

[3] Mróz Z., Rozvany G.I.N.: 'Optimal design of structures with variable support conditions', J. of Optimisat. Theory and Applic., Vol. **15** No. 1: (85-101) 1975.

[4] Pičuga A.: 'Minimum-cost synthesis of multi-purpose beams', XVI-th IUTAM Congress, Lyngby, Denmark, 1984.

[5] Pičuga A.: 'An introduction to finite element method' (in Serbo-Croatian), Svjetlost, Sarajevo, (pp. 116) 1985.

[6] Olhoff N.: 'Optimal design against structural vibration and instability', The Techn. Univ. of Denmark, Lyngby, 1978.

[7] Keng-Tung Cheng: 'Optimal design of solid elastic plates', The Techn. Univ. of Denmark, Lyngby, (DCAMM), Rep. S17: (pp. 97) 1980.

[8] Grinev V.B., Filipov A.N.: 'Optimizacya elementov konstrukcij po mekhanicheskim kharakteristikam' (in Russian), Naukova Dumka, Kiev, (pp. 292) 1975.

A VARIATIONAL PRINCIPLE USEFUL IN OPTIMIZING RECTANGULAR-BASE SHALLOW SHELLS

Raymond H. Plaut
Department of Civil Engineering
Virginia Polytechnic Institute and State University
Blacksburg, Virginia 24061, U.S.A.

and

David T. Young
Department of Civil Engineering
University of North Carolina at Charlotte
Charlotte, North Carolina 28223, U.S.A.

ABSTRACT. Thin, elastic, shallow shells with constant thickness and rectangular base are considered. The edges are clamped or simply supported, with no in-plane displacements. Buckling under a uniformly distributed load is analyzed. A variational principle is presented and then utilized to optimize the forms of shells with given surface areas. In some cases, the optimal forms possess significantly higher buckling loads than standard forms.

1. INTRODUCTION

There has been extensive research on the optimal design of shells. Reference [1] presents an excellent survey of this field. Recent work includes references [2-7]. The present paper treats shallow shells having a rectangular base. A variational principle is formulated, and some optimal forms for maximum buckling load are obtained.

The shells are assumed to be thin and elastic, with constant thickness and specified surface area. They are supported along a rectangular base, and the four edges are either all simply supported or all clamped. A vertical load is applied uniformly over the shell, and its buckling value is maximized.

2. FORMULATION

The base of the shell lies in the X,Y plane, with $0 \le X \le a$, $0 \le Y \le b$, and the height of the middle surface is $Z(X,Y)$, with $Z=0$ along the

G. I. N. Rozvany and B. L. Karihaloo (eds.), Structural Optimization, 241–248.

edges X=0,a and Y=0,b. The shell has thickness h, density ρ, Young's modulus E, and Poisson's ratio ν. A downward, uniformly distributed load q is applied. The given area of the middle surface is S, and for shallow shells one can write [7]

$$ab + \frac{1}{2}\int_0^a \int_0^b (Z_X^2 + Z_Y^2)\, dX\, dY = S \tag{1}$$

where subscripts represent partial derivatives.

Let W(X,Y) and ϕ(X,Y) denote the downward deflection and a stress function, respectively. It is convenient to define the quantities

$$D = Eh^3/[12(1-\nu^2)], \quad x = X/a, \quad y = Y/a, \quad c = b/a,$$

$$z = Z\sqrt{12(1-\nu^2)}/h, \quad w = W\sqrt{12(1-\nu^2)}/h, \quad \psi = \phi h/D,$$

$$p = qa^4\sqrt{12(1-\nu^2)}/(Dh), \quad u = w - z, \tag{2}$$

$$\beta^2 = 24(1-\nu^2)(S-ab)/(h^2).$$

Condition (1) then becomes

$$\frac{1}{2}\int_0^1 \int_0^c (z_x^2 + z_y^2)\, dx\, dy = \beta^2. \tag{3}$$

Using Marguerre's shallow shell equations [8], the equation of equilibrium is

$$A \equiv \nabla^4 w - R(u,\psi) - p = 0 \tag{4}$$

and the compatibility equation is

$$B \equiv \nabla^4\psi - R(w,z) + \frac{1}{2} R(w,w) = 0 \tag{5}$$

where

$$R(F,G) = F_{xx}G_{yy} + F_{yy}G_{xx} - 2F_{xy}G_{xy},$$

$$\nabla^4 = \nabla^2\nabla^2, \quad \nabla^2 = \frac{\partial^2}{\partial x^2} + \frac{\partial^2}{\partial y^2}. \tag{6}$$

The edges of the shell are either clamped or simply supported, with no in-plane displacements. This leads to the following boundary conditions at x=0 and x=1:

$$w = 0; \quad w_x = 0 \text{ or } w_{xx} = 0; \quad \psi_{xx} - \nu\psi_{yy} = 0;$$
$$\psi_{xxx} + (2+\nu)\psi_{xyy} = 0 \tag{7}$$

(see Appendix A). Interchanging x and y furnishes the conditions at y=0 and y=c.

In order to obtain the buckling equations, replace w by w+g (and, hence, u by u+g), where g is the buckling mode, and replace ψ by ψ+f, where f is the corresponding stress function, in Eqs. 4 and 5. Terms not involving g and f are then eliminated with the use of Eqs. 4 and 5. The resulting equations are linearized in g and f, yielding the buckling equations

$$F \equiv \nabla^4 g - R(g,\psi) - R(u,f) = 0, \tag{8}$$

$$H \equiv \nabla^4 f + R(g,u) = 0. \tag{9}$$

The functions g and f satisfy the same boundary conditions as w and ψ, respectively.

3. VARIATIONAL PRINCIPLE

Consider the functional

$$\begin{aligned} V = \int_0^1 \int_0^c \{ & (\nabla^2 g)^2 - (\nabla^2 f)^2 + g_x^2\psi_{yy} + g_y^2\psi_{xx} - 2g_x g_y \psi_{xy} + \\ & + 2g_x f_x u_{yy} + 2g_y f_y u_{xx} - 2(g_x f_y + g_y f_x) u_{xy} \} dx\, dy + \\ & + \int_0^c [f_x f_{xx} - f f_{xxx} + f_x f_{yy} - f f_{xyy}]_{x=0}^{x=1} dy + \\ & + \int_0^1 [f_y f_{yy} - f f_{yyy} + f_y f_{xx} - f f_{xxy}]_{y=0}^{y=c} dx \end{aligned} \tag{10}$$

where u is defined in Eq. 2. The functions w, ψ, g, and f satisfy their respective boundary conditions, and z=0 on the edges of the shell. At the corners of the shell, the condition f=0 is imposed.

With the use of integration by parts, Eqs. 8 and 9, and the boundary conditions, one can show that V=0. Also, one can show that V is stationary with respect to variations in g and f satisfying the boundary and corner conditions (see Appendix B). A similar variational principle for shallow shells with a circular boundary was given in reference [9].

4. OPTIMIZATION

The buckling value of the load p is to be maximized. If the optimal solution is unimodal, one can construct an appropriate augmented functional J by appending the following to p: functions A and B from Eqs. 4 and 5; the functional V from Eq. 10; the constant-area constraint from Eq. 3; and the normalization condition

$$\int_0^1 \int_0^c g^2 \, dx \, dy = 1. \tag{11}$$

This leads to

$$\begin{aligned} J = p - \int_0^1 \int_0^c \lambda_1 B \, dx \, dy - \int_0^1 \int_0^c \lambda_2 A \, dx \, dy + \lambda_3 V - \\ - \lambda_4 \{\tfrac{1}{2} \int_0^1 \int_0^c (z_x^2 + z_y^2) \, dx \, dy - \beta^2\} - \\ - \lambda_5 \{\int_0^1 \int_0^c g^2 \, dx \, dy - 1\} \end{aligned} \tag{12}$$

where $\lambda_1(x,y)$, $\lambda_2(x,y)$, λ_3, λ_4, and λ_5 are Lagrange multipliers.

Stationarity of J with respect to w and ψ furnishes the adjoint equations

$$\nabla^4 \lambda_1 - R(\lambda_2, u) + \lambda_3 R(g,g) = 0, \tag{13}$$

$$\nabla^4 \lambda_2 + R(\lambda_1, u) - R(\lambda_2, \psi) + 2\lambda_3 R(f,g) = 0, \tag{14}$$

where the boundary conditions on λ_1 and λ_2 are the same as those on ψ and w, respectively, on the edges of the shell, and $\lambda_1=0$, ψ_x=constant, and ψ_y=constant at the corners. Stationarity of J with respect to the design function z(x,y) leads to the optimality condition

$$\lambda_4 \nabla^2 z = R(\lambda_2, \psi) - R(\lambda_1, w) - 2\lambda_3 R(f,g). \tag{15}$$

Numerical results were obtained in reference [10] for shells with all edges simply supported or clamped, for aspect ratios c=1 and c=1.5, and for surface area parameter values of β^2 = 50, 100, and 150. Typical optimal forms are depicted in Figures 1 and 2 for simply supported and clamped edges, respectively. In the latter case, the optimal form is horizontal at the edges. When compared to a reference shell with $z(x) = \Lambda\sin(\pi x)\sin(\pi y/c)$ and the same value of β^2, the optimal critical load is 4%-50% higher in the cases with simply supported edges and 26%-68% higher in the cases with clamped edges.

Similar results were also computed in reference [10] for buckling due to a load which is uniformly distributed over a central rectangular region of the shell.

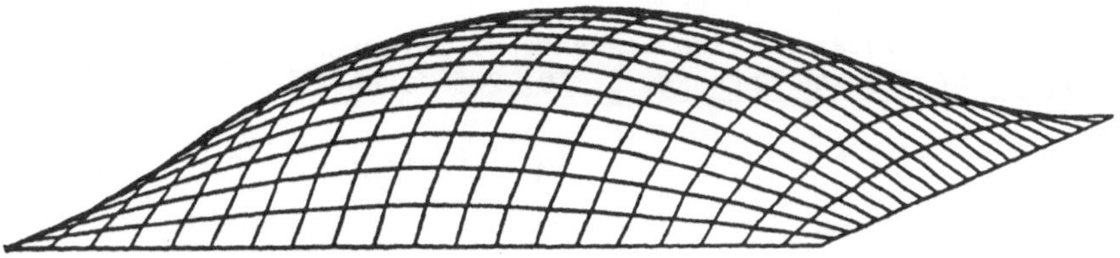

Figure 1. Optimal shell with simply supported edges ($\beta^2 = 100$).

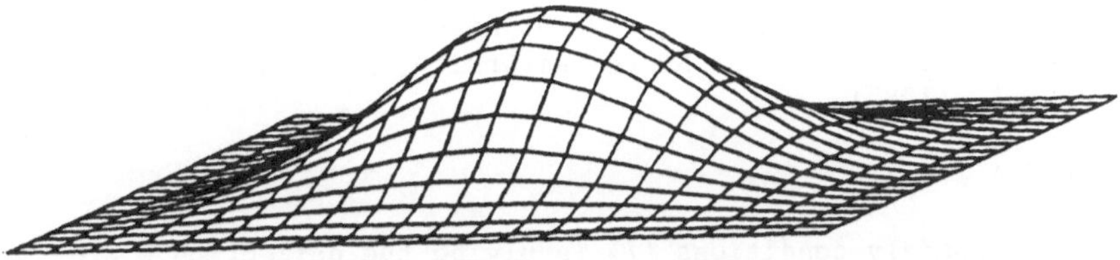

Figure 2. Optimal shell with clamped edges ($\beta^2 = 100$).

Acknowledgment. This material is based upon work supported by the U.S. National Science Foundation under Grant CEE-8210222.

5. REFERENCES

1. Kruzelecki, J., and Życzkowski, M., 'Optimal Structural Design of Shells - A Survey,' SM Archives, 10, 101-170 (1985).

2. Reitman, M. I., 'Optimum Design of Concrete Shells,' Annals of the New York Academy of Sciences, 410, 263-269 (1983).

3. Jakubowska, B., and Lesniak, Z. K., 'Shape Optimization of Double-Curved Shells,' Optimization Methods in Structural Design, Proceedings of Euromech Colloquium 164, Siegen, West Germany, 1982 (Editors H. Eschenaur, N. Olhoff), Mannheim: Bibliographisches Institut, 1983, pp. 210-215.

4. Stadler, W., 'Natural Structural Shapes in Shell Theory: An Application of Multicriteria Optimization,' Presented at Euromech Colloquium 165, Flexible Shells, Munich, West Germany, 1983.

5. Plaut, R. H., Johnson, L. W., and Parbery, R., 'Optimal Forms of Shallow Shells with Circular Boundary,' Parts 1-3, Journal of Applied Mechanics, 51, 526-539 (1984).

6. Plaut, R. H., Johnson, L. W., and Olhoff, N., 'Optimal Forms of Shallow Cylindrical Panels with Respect to Vibration and Stability,' Journal of Applied Mechanics, 53, 135-140 (1986).

7. Banichuk, N. V., and Larichev, A. D., 'Optimal Design Problems for Curvilinear Shallow Elements of Structures,' Optimal Control Applications and Methods, 5, 197-205 (1984).

8. Chia, C.-Y., Nonlinear Analysis of Plates, New York: McGraw-Hill, 1980.

9. Huang, N. C., 'Unsymmetrical Buckling of Thin Shallow Spherical Shells,' Journal of Applied Mechanics, 31, 447-457 (1964).

10. Young, D. T., 'Optimal Forms of Rectangular-Base, Shallow Shells with Respect to Buckling,' Ph.D. Dissertation, Virginia Polytechnic Institute and State University, Blacksburg, Virginia, U.S.A. (1985).

APPENDIX A

The boundary conditions (7) involving the deflection w are standard, and only the boundary conditions on ψ need to be discussed.

The stress-strain relations are

$$\varepsilon_x = (\sigma_x - \nu\sigma_y)/E, \quad \varepsilon_y = (\sigma_y - \nu\sigma_x)/E, \quad \gamma_{xy} = \tau_{xy}/G, \tag{A1}$$

where $G = E/(2+2\nu)$. (In this Appendix, subscripts do not denote partial derivatives.) In terms of the stress function ϕ,

$$\sigma_x = \frac{\partial^2\phi}{\partial Y^2}, \quad \sigma_y = \frac{\partial^2\phi}{\partial X^2}, \quad \tau_{xy} = -\frac{\partial^2\phi}{\partial X\partial Y}. \tag{A2}$$

Since in-plane displacements are assumed to be zero along the edges,

$$\varepsilon_y = 0 \quad \text{at} \quad x = 0,1. \tag{A3}$$

The third condition in Eq. 7 then follows from Eqs. A1-A3 and the definition of ψ in Eq. 2.

Let U and V denote in-plane displacements in the X and Y directions, respectively. The strain-displacement relations are [8]

$$\begin{aligned} \varepsilon_x &= \frac{\partial U}{\partial X} + \frac{1}{2}\left(\frac{\partial W}{\partial X}\right)^2 - \left(\frac{\partial W}{\partial X}\right)\left(\frac{\partial Z}{\partial X}\right), \\ \varepsilon_y &= \frac{\partial V}{\partial Y} + \frac{1}{2}\left(\frac{\partial W}{\partial Y}\right)^2 - \left(\frac{\partial W}{\partial Y}\right)\left(\frac{\partial Z}{\partial Y}\right), \\ \gamma_{xy} &= \frac{\partial U}{\partial Y} + \frac{\partial V}{\partial X} + \left(\frac{\partial W}{\partial X}\right)\left(\frac{\partial W}{\partial Y}\right) - \left(\frac{\partial W}{\partial X}\right)\left(\frac{\partial Z}{\partial Y}\right) - \left(\frac{\partial W}{\partial Y}\right)\left(\frac{\partial Z}{\partial X}\right). \end{aligned} \tag{A4}$$

At the edges X = 0 and X = a, due to the vanishing of Z and the displacements, one can write conditions such as

$$Z = \frac{\partial Z}{\partial Y} = \frac{\partial^2 Z}{\partial Y^2} = U = \frac{\partial U}{\partial Y} = \frac{\partial^2 U}{\partial Y^2} = W = \frac{\partial W}{\partial Y} = \frac{\partial^2 W}{\partial Y^2} = 0. \tag{A5}$$

With the use of Eqs. A4 and A5, one can show that

$$\frac{\partial \gamma_{xy}}{\partial Y} - \frac{\partial \varepsilon_y}{\partial X} = 0 \quad \text{at} \quad X = 0,a. \tag{A6}$$

Then, with the use of Eqs. A1, A2, and the definition of ψ in Eq. 2, Eq. A6 leads to the last boundary condition in Eq. 7.

APPENDIX B

In order to prove the variational principle, replace g by $g+\varepsilon\eta$ and f by $f+\varepsilon\alpha$ in Eq. 10, where ε is a scalar parameter. Then take the first partial derivative of V with respect to ε, and set $\varepsilon = 0$. This yields δV. After integrating by parts to eliminate all derivatives on the functions η and α inside the double integrals, and combining terms, one obtains

$$\begin{aligned}
\delta V = {} & 2 \int_0^1 \int_0^c (\eta F - \alpha H)\, dx\, dy \; + \\
& + 2 \int_0^c \left[\eta_x \nabla^2 g - (\nabla^2 g)_x \eta\right]_{x=0}^{x=1} dy \; + \\
& + 2 \int_0^1 \left[\eta_y \nabla^2 g - (\nabla^2 g)_y \eta\right]_{y=0}^{y=c} dx \; + \\
& + 2 \int_0^c \left[(g_x \psi_{yy} - g_y \psi_{xy})\eta\right]_{x=0}^{x=1} dy \; + \\
& + 2 \int_0^c \left[(g_y \psi_{xx} - g_x \psi_{xy})\eta\right]_{y=0}^{y=c} dx \; + \\
& + 2 \int_0^c \left[(g_x u_{yy} - g_y u_{xy})\alpha + (f_x u_{yy} - f_y u_{xy})\eta\right]_{x=0}^{x=1} dy \; + \\
& + 2 \int_0^1 \left[(g_y u_{xx} - g_x u_{xy})\alpha + (f_y u_{xx} - f_x u_{xy})\eta\right]_{y=0}^{y=c} dx + K + L
\end{aligned} \tag{B1}$$

where

$$K = \int_0^c \left[f_x \nabla^2\alpha - (\nabla^2\alpha)_x f - \alpha_x \nabla^2 f + (\nabla^2 f)_x \alpha\right]_{x=0}^{x=1} dy, \tag{B2}$$

$$L = \int_0^1 \left[f_y \nabla^2\alpha - (\nabla^2\alpha)_y f - \alpha_y \nabla^2 f + (\nabla^2 f)_y \alpha\right]_{y=0}^{y=c} dx, \tag{B3}$$

and F and H are defined in Eqs. 8 and 9, respectively. From the boundary conditions, it follows that g, g_y, g_{yy}, η, η_y, η_{yy}, u, u_y, u_{yy}, g_x or g_{xx}, and η_x or η_{xx} are zero at x=0 and x=1. Interchanging x and y gives conditions at y=0 and y=c. Using these boundary conditions and Eqs. 8 and 9, all terms on the right side of Eq. B1 vanish except K and L.

At x=0 and x=1, Eq. 7 leads to the boundary conditions

$$f_{xx} = \nu f_{yy}, \quad \alpha_{xx} = \nu\alpha_{yy}, \; f_{xxx} = -(2+\nu) f_{xyy}, \quad \alpha_{xxx} = -(2+\nu)\alpha_{xyy}. \tag{B4}$$

Using Eq. B4, one can write

$$K = (1+\nu) \int_0^c \left[f_x \alpha_{yy} + f\alpha_{xyy} - f_{yy}\alpha_x - f_{xyy}\alpha\right]_0^1 dy. \tag{B5}$$

Integrating by parts and combining terms, one can show that

$$\begin{aligned} \frac{K}{(1+\nu)} = {} & f_{xy}(0,c)\eta(0,c) - f_{xy}(0,0)\eta(0,0) + f_{xy}(1,0)\eta(1,0) - \\ & - f_{xy}(1,c)\eta(1,c) - f(0,c)\eta_{xy}(0,c) + f(0,0)\eta_{xy}(0,0) - \\ & - f(1,0)\eta_{xy}(1,0) + f(1,c)\eta_{xy}(1,c). \end{aligned} \tag{B6}$$

From the corner condition f=0, it follows that η=0 and then K=0. Similarly, one can show that L=0. Hence, δV=0, and V is stationary with respect to variations in g and f satisfying the boundary and corner conditions.

MINIMAX ALGORITHMS FOR STRUCTURAL OPTIMIZATION

Prof. E. Polak
Department of Electrical Engineering
and Computer Sciences
University of California
Berkeley, California 94720
U.S.A.

ABSTRACT. In this paper we highlight the salient features of our recently developed theory for the construction of broad classes of nondifferentiable optimization algorithms. These algorithms can be used for the solution of a wide variety of unconstrained and constrained minimax problems, such as those occurring in the design of structures subjected to dynamic loads, floor planning and layout problems, control system and electronic circuit design.

1. INTRODUCTION

To motivate our discussion of minimax algorithms as a tool for the solution of structural design problems, let us consider an idealized base isolation problem. Thus, suppose that we are required to design a passive base isolation system for a structure that must be built on top of a metropolitan underground train station. Since the motions will be very small, the structure can be modeled as a linear second order differential equation of the form (see [5])

$$M\ddot{y}(t) + C(x)\dot{y}(t) + K(x)y(t) = BF(t) \tag{1.1}$$

where $y(t) \in \mathbb{R}^{3q}$ is a vector of floor displacements, with three components per floor (q is the number of floors), two for horizontal motion and one for vertical motion. Next, M, is a mass matrix, while $C(x)$ and $K(x)$ are damping and spring action matrices, which we assume to be continuously differentiable in $x \in \mathbb{R}^n$, the design parameter vector of the base isolation device. The three dimensional ground acceleration forces are described by the time dependent function $F(t) \in \mathbb{R}^3$, and B is the coupling matrix.

Now suppose that the passing trains leave several very specific and repeatable signatures, i.e., they produce a set of excitation functions $\{F_j(t)\}_{j=1}^{p}$, defined on the interval $[0,T]$, where T is the maximum duration of the disturbance caused by the trains. Assuming that the components of the designable parameter x must satisfy a constraint of the form $x \in \mathbf{X} \triangleq \{x \in \mathbb{R}^n \mid \underline{x}^i \leq x^i \leq \overline{x}^i,\ i = 1,\dots,n\}$, the base isolation design problem can be expressed in the form

$$\min_{x \in \mathbf{X}} \max_{j \in \mathbf{p}} \max_{t \in [0,T]} \| y(t;x,F_j) \| \,, \tag{1.2}$$

where $\mathbf{p} \triangleq \{1,2,\dots,p\}$, and $y(t;x,F_j)$ denotes the solution of (1.1) corresponding to the given value of the base isolation design vector x and ground motion $F_j(t)$.

Alternatively, we may have determined that the frequency spectrum of the disturbances is contained in an interval $[\omega',\omega'']$ and that its magnitude is bounded by a function $b(\omega) > 0$ on

G. I. N. Rozvany and B. L. Karihaloo (eds.), Structural Optimization, 249–256.

that interval. Then we can attempt to design the base isolation device using frequency domain techniques by solving the problem

$$\min_{x \in X} \max_{\omega \in [\omega', \omega'']} w(\omega) \| H(x, j\omega) \| , \tag{1.3}$$

where $w(\omega) = 1/b(\omega)$ and $H(x,j\omega)$ is the $3q \times 3$ complex valued transfer function matrix from the ground acceleration $F(t)$ to the floor displacement vector $y(t)$. Note that $\|H\|^2$ is the largest eigenvalue of the positive semidefinite matrix $H^* H$.

For more extensive treatments of modeling design problems as minimax optimization problems and numerical results, see [1], [3], and [11].

2. UNCONSTRAINED MINIMAX ALGORITHMS

We will develop a family of minimax algorithms by extension of the method of steepest descent which solves problems of the form

$$\min_{\mathbb{R}^n} f(x) \tag{2.1}$$

where $f : \mathbb{R}^n \to \mathbb{R}$ is a continuously differentiable function. We begin by recalling the method of steepest descent and its convergence properties [9].

STEEPEST DESCENT ALGORITHM 2.1 :

Step 0 : Select an $x_0 \in \mathbb{R}^n$ and set $i = 0$.

Step 1 : Compute the *search direction*

$$h_i = -\nabla f(x_i) = \arg \min_{h \in \mathbb{R}^n} \{ f(x_i) + \langle \nabla f(x_i, h \rangle + \tfrac{1}{2} \| h \|^2 \} . \tag{2.2}$$

Step 2 : If $\nabla f(x_i) = 0$, stop. Else compute the *step size* $\lambda_i \in \lambda(x_i) \triangleq \arg\min_{\lambda \geq 0} f(x_i + \lambda h_i)$.

Step 3 : Set $x_{i+1} = x_i + \lambda_i h_i$, replace i by $i + 1$, and go to Step 1. ■

In practice one uses the Armijo step size rule [9] which is much more efficient, but somewhat harder to analyze than the one dimensional minimization rule, used in Step 2, above.

THEOREM 2.1 : If $\{x_i\}_{i=0}^{\infty}$ is an infinite sequence constructed by Algorithm 2.1, then every accumulation point $\hat{x}$ of $\{x_i\}_{i=0}^{\infty}$ satisfies $\nabla f(\hat{x}) = 0$.

PROOF : Suppose that $x_i \xrightarrow{K} \hat{x}$ as $i \to \infty$ and that $\nabla f(\hat{x}) \neq 0$. Then the directional derivative

$$df(\hat{x}; h(\hat{x})) = -\| \nabla f(\hat{x}) \|^2 < 0 . \tag{2.3a}$$

Hence any $\hat{\lambda} \in \lambda(\hat{x})$ satisfies $\hat{\lambda} > 0$ and there exists a $\hat{\delta} > 0$ such that

$$f(\hat{x} + \hat{\lambda} h(\hat{x})) - f(\hat{x}) = -\hat{\delta} < 0 . \tag{2.3b}$$

Since $h(\cdot) = -\nabla f(\cdot)$ is *continuous* by assumption, the function $f(x + \hat{\lambda} h(x)) - f(x)$ is continuous in x and hence there exists an i_0 such that for all $i \in K$, $i \geq i_0$,

$$f(x_{i+1}) - f(x_i) \le f(x_i + \hat{\lambda} h(x_i)) - f(x_i) \le -\hat{\delta}/2 \,. \tag{2.3c}$$

Now, by construction, $\{f(x_i)\}_{i=0}^{\infty}$ is monotone decreasing and $f(x_i) \overset{K}{\to} f(\hat{x})$ as $i \to \infty$ by continuity of $f(\cdot)$. Therefore, we must have that $f(x_i) \to f(\hat{x})$ as $i \to \infty$. But this contradicts (2.3c). Hence $\nabla f(\hat{x}) = 0$ must hold. ■

Next let us examine the method of steepest descent geometrically, which requires the following notation. Given any function $g : \mathbb{R}^n \to \mathbb{R}$, we shall denote its level sets by $\mathbf{Lg}(\alpha)$, i.e., $\mathbf{Lg}(\alpha) \triangleq \{ x \in \mathbb{R}^n \mid g(x) \le \alpha \}$. Now, given a point x_i, we see that the method of steepest descent approximates the continuously differentiable function $f(x)$ by the quadratic function

$$q(x;x_i) \triangleq f(x_i) + \langle \nabla f(x_i), (x - x_i) \rangle + \tfrac{1}{2}\|x - x_i\|^2 , \tag{2.4a}$$

and its level set

$$\mathbf{Lf}(f(x_i)) \triangleq \{ x \in \mathbb{R}^n \mid f(x) \le f(x_i) \} , \tag{2.4b}$$

by the "disk"

$$\mathbf{Df}(f(x_i);x_i) \triangleq \{ x \in \mathbb{R}^n \mid q(x;x_i) \le f(x_i) \} , \tag{2.4c}$$

which is tangent to $\mathbf{Lf}(f(x_i))$ at the point x_i. A minimizer of (2.1), $\hat{x}$, defines a "center" of $\mathbf{Lf}(f(x_i))$; $x_i - \nabla f(x_i)$, minimizes $q(x;x_i)$ and is the center of $\mathbf{Df}(f(x_i);x_i)$. The method of steepest descent treats the point $x_i - \nabla f(x_i)$ as an approximation to the point $\hat{x}$. Since this approximation is rather poor, the method of steepest descent performs a line search along the line passing through x_i and $x_i - \nabla f(x_i)$, according to (2.2c) to obtain a somewhat better approximation to $\hat{x}$, x_{i+1}, defined by (2.2d).

We can now consider the simplest minimax problem:

$$\min_{x \in \mathbb{R}^n} \max_{j \in \mathbf{m}} f^j(x), \tag{2.5a}$$

where the functions $f^j : \mathbb{R}^n \to \mathbb{R}$ are continuously differentiable and $\mathbf{m} \triangleq \{ 1,2,\ldots,m \}$. Let $\psi(x) \triangleq \max_{j \in \mathbf{m}} f^j(x)$, then (2.5a) becomes

$$\min_{x \in \mathbb{R}^n} \psi(x) \,, \tag{2.5b}$$

which is a nondifferentiable optimization problem. We need the following results [4].

THEOREM 2.2 : (a) For all $x,h \in \mathbb{R}^n$, the function $\psi(\cdot)$ has directional derivatives at x in the direction h which are given by

$$d\psi(x;h) \triangleq \lim_{t \downarrow 0} \frac{\psi(x + th) - \psi(x)}{t} = \max_{j \in \mathbf{I}(x)} \langle \nabla f^j(x), h \rangle , \tag{2.6a}$$

where $\mathbf{I}(x) \triangleq \{j \in \mathbf{m} \mid f^j(x) = \psi(x)\}$.

(b) If $\hat{x}$ is a local minimizer for (2.5b), then the following *equivalent* statements hold:

(*i*) $\quad d\psi(\hat{x};h) \ge 0, \ \forall \ h \in \mathbb{R}^n$, **(2.6b)**

(*ii*) $\quad 0 \in \mathrm{co}_{j \in \mathbf{I}(\hat{x})} \{ \nabla f^j(\hat{x}) \}$, (2.6c)

(iii) $\quad 0 \in G\psi(\hat{x}) \triangleq \underset{j \in \mathbf{m}}{\text{co}} \left\{ \begin{bmatrix} f^j(\hat{x}) - \psi(\hat{x}) \\ \nabla f^j(\hat{x}) \end{bmatrix} \right\},$ (2.6d)

where "co" denotes the convex hull of the set in question. ■

Next we make the observation that the level sets of the function $\psi(\cdot)$ are the intersection of level sets of the functions $f^j(\cdot)$, i.e.,

$$L\psi(\alpha)) \triangleq \{ x \in \mathbb{R}^n \mid \psi(x) \leq \alpha) \} = \bigcap_{j \in \mathbf{m}} Lf^j(\alpha)) . \tag{2.7a}$$

Proceeding by analogy with the geometric interpretation of the method of steepest descent, given a point $x_i \in \mathbb{R}^n$, we approximate the level set $L\psi(\psi(x_i))$ by the intersection of the discs $Df^j(\psi(x_i))$ which approximate the level sets $Lf^j(\psi(x_i))$, i.e., by the set

$$D\psi(x_i, \psi(x_i)) \triangleq \bigcap_{j \in \mathbf{m}} Df^j(\psi(x_i))) = \{ x \in \mathbb{R}^n \mid q^j(x; x_i) \leq \psi(x_i) \}, \tag{2.7b}$$

and we approximate the "center" $\hat{x}$ which solves (2.5) by the "center" $(x_i + h_i)$ of $D\psi(x_i, \psi(x_i))$ which solves the problem

$$\min_{x \in \mathbb{R}^n} \max_{j \in \mathbf{m}} q^j(x; x_i). \tag{2.8}$$

Adding a line search, we obtain the following extension of Algorithm 2.1.

MINIMAX ALGORITHM 2.2 :

Step 0 : Select a $x_0 \in \mathbb{R}^n$ and set $i = 0$.

Step 1 : Compute the *search direction*

$$h_i = \arg \min_{h \in \mathbb{R}^n} \max_{j \in \mathbf{m}} \{ f^j(x_i) + \langle \nabla f^j(x_i), h \rangle + ½\|h\|^2 \} . \tag{2.9}$$

Step 2 : If $h_i = 0$, stop. Else compute the *step size* $\lambda_i \in \arg\min_{\lambda \geq 0} \psi(x_i + \lambda h_i)$.

Step 3 : Set $x_{i+1} = x_i + \lambda_i h_i$, replace i by $i + 1$, and go to Step 1. ■

The search direction finding problem (2.9) is obviously much more difficult to solve than (2.2c). It is easiest to solve it in dual form. First, it is obvious that

$$\bar{\theta}(x_i) \triangleq \min_{h \in \mathbb{R}^n} \max_{j \in \mathbf{m}} \{ f^j(x_i) + \langle \nabla f^j(x_i, h \rangle + ½\|h\|^2 \}$$

$$= \min_{h \in \mathbb{R}^n} \max_{\mu \in \Sigma} \{ \sum_{j=1}^{m} \mu^j \{ f^j(x_i) + \langle \nabla f^j(x_i, h \rangle + ½\|h\|^2 \} , \tag{2.10a}$$

where $\Sigma \triangleq \{ \mu \in \mathbb{R}^m \mid \sum_{j=1}^{m} \mu^j = 1, \mu \geq 0 \}$. Next, making use of the von Neumann minimax theorem [2], we can interchange the min and max operations in (2.10a), to obtain that

$$\bar{\theta}(x_i) = \max_{\mu \in \Sigma} \min_{h \in \mathbb{R}^n} \sum_{j=1}^{m} \mu^j \{ f^j(x_i) + \langle \nabla f^j(x_i, h \rangle + ½\|h\|^2 \} . \tag{2.10b}$$

Eliminating h from (2.10b) by unconstrained minimization, we obtain that

$$\bar{\theta}(x_i) = -\min_{\mu \in \Sigma} \{ \sum_{j=1}^{m} -\mu^j f^j(x_i) + ½\|\sum_{j=1}^{m} \mu^j \nabla f^j(x_i)\|^2 \} . \tag{2.10c}$$

It now follows from the von Neumann minimax theorem that if $\mu_i \in \Sigma$ is *any* solution of (2.10b), then

$$h_i = -\sum_{j=1}^{m} \mu_i^j \nabla f^j(x_i) \tag{2.10d}$$

is the *unique* solution of (2.9). Since (2.10c) is easily solved by modern quadratic programming algorithms, such as [6], we see that the search direction h_i is readily computed.

Since it is easily shown that the search direction h_i is continuous [8], the following theorem can be proved by repeating the arguments for Theorem 2.1.

THEOREM 2.3 : If $\{x_i\}_{i=0}^{\infty}$ is an infinite sequence constructed by Algorithm 2.2, then every accumulation point $\hat{x}$ of $\{x_i\}_{i=0}^{\infty}$ satisfies $0 \in G\psi(\hat{x})$. ■

We are now ready to consider the general minimax problem

$$\min_{h \in \mathbb{R}^n} \max_{y \in Y} \phi(x,y) \,, \tag{2.11}$$

where $\phi : \mathbb{R}^n \times \mathbb{R}^s \to \mathbb{R}$ and both $\phi(\cdot,\cdot)$ and $\nabla_x\phi(\cdot,\cdot)$ are continuous and $Y \subset \mathbb{R}^s$ is compact. First, we state an extension of Theorem 2.2.

THEOREM 2.4 : (a) For all $x, h \in \mathbb{R}^n$, the function $\psi(x) \triangleq \max_{y \in Y} \phi(x,y)$ has directional derivatives at x in the direction h which are given by

$$d\psi(x;h) \triangleq \lim_{t \downarrow 0} \frac{\psi(x+th) - \psi(x)}{t} = \max_{j \in Y(x)} \langle \nabla_x \phi(x,y), h \rangle \,, \tag{2.12a}$$

where $Y(x) \triangleq \{y \in Y \mid \phi(x,y) = \psi(x)\}$.

(b) If $\hat{x}$ is a local minimizer for (2.11), then the following *equivalent* statements hold:

(*i*) $$d\psi(\hat{x};h) \geq 0, \ \forall\, h \in \mathbb{R}^n \,, \tag{2.12b}$$

(*ii*) $$0 \in \underset{y \in Y(\hat{x})}{\mathrm{co}} \{ \nabla_x \phi(\hat{x},y) \} \tag{2.12c}$$

(*iii*) $$0 \in G\psi(\hat{x}) \triangleq \underset{y \in Y}{\mathrm{co}} \left\{ \begin{bmatrix} (\phi(\hat{x},y) - \psi(\hat{x})) \\ \nabla_x \phi(\hat{x},y) \end{bmatrix} \right\}. \quad ■ \tag{2.12d}$$

If, as we have just done, we redefine $\psi(\cdot)$ by $\psi(x) = \max_{y \in Y} \phi(x,y)$, the formal extension of Algorithm 2.2 to this case is obvious and must be as given below.

MINIMAX ALGORITHM 2.3 :

Step 0 : Select $x_0 \in \mathbb{R}^n$ and set $i = 0$.

Step 1 : Compute the *search direction*

$$h_i = \arg \min_{h \in \mathbb{R}^n} \max_{y \in Y} \{\phi(x_i,y) + \langle \nabla_x \phi(x_i,y), h \rangle + \tfrac{1}{2}\|h\|^2\} \,. \tag{2.13}$$

Step 2 : If $h_i = 0$, stop. Else compute the *step size* $\lambda_i \in \arg\min_{\lambda \geq 0} \psi(x_i + \lambda h_i)$.

Step 3 : Set $x_{i+1} = x_i + \lambda_i h_i$, replace i by $i+1$, and go to Step 1. ■

Referring to [8], we see that the search direction h_i defined by (2.13) is continuous. Hence the following theorem can be proved by identical arguments as for Theorem 2.3.

THEOREM 2.5 : If $\{x_i\}_{i=0}^{\infty}$ is an infinite sequence constructed by Algorithm 2.3, then every accumulation point $\hat{x}$ of $\{x_i\}_{i=0}^{\infty}$ satisfies $0 \in \mathbf{G}\psi(\hat{x})$. ■

The main question to be resolved is whether the search direction h_i can be computed. To this end, we begin by relating (2.10c) to (2.6d). We note that for any $x \in \mathbb{R}^n$, $\mathbf{G}\psi(x) \subset \mathbb{R}^{n+1}$, and that $\bar{\xi} = (\xi^0, \xi)$ is an element of $\mathbf{G}\psi(x)$ if and only if there exists a $\mu \in \Sigma$ such that $\xi^0 = \sum_{j=1}^m \mu^j [f^j(x) - \psi(x)]$ and $\xi = \sum_{j=1}^m \mu^j \nabla f^j(x)$. Hence (2.10c) can be rewritten in the form

$$\bar{\theta}(x_i) = -\min_{\bar{\xi} \in \mathbf{G}\psi(x_i)} \{\xi^0 + \tfrac{1}{2}\|\xi\|^2\}, \tag{2.14a}$$

and the search direction h_i, defined by (2.2), is given by $h_i = -\xi(x_i)$, where $\xi(x_i)$ consists of the last n elements of the $(n+1)$ dimensional vector

$$\bar{\xi}(x_i) = (\xi^0(x_i), \xi(x_i)) \doteq -\arg\min_{\bar{\xi} \in \mathbf{G}\psi(x_i)} \{\xi^0 + \tfrac{1}{2}\|\xi\|^2\}. \tag{2.14b}$$

If we define $\mathbf{G}\psi(x_i)$ by making use of (2.12d), and $\bar{\xi}(x_i)$ as in (2.14b), then the formula $h_i = -\xi(x_i)$ is also valid for Algorithm 2.3. The importance of this observation lies in the fact that $\bar{\xi}(x_i)$ can now be computed by means of a proximity algorithm. These algorithms are descendants of the Gilbert algorithm [7], see e.g., [10]. They depend on our ability to compute tangency points to the sets $\mathbf{G}\psi(x)$, which are defined as solutions of the *contact* problem

$$\min\{ \langle \bar{v}, \bar{\xi}\rangle \mid \bar{\xi} \in \mathbf{G}\psi(x) \}, \tag{2.15}$$

where $\bar{v}$ is any given direction.

The computation of these tangency points in structural design problems does not appear to pose any serious difficulty (see [8] for details). For the sake of completeness, we now state the simplest of these proximity algorithms for solving the problem (2.14a).

PROXIMITY ALGORITHM 2.4 [7]

Step 0: Select a $\bar{\xi}_0 = (\xi_0^0, \xi_0) \in \mathbf{G}\psi(x_i)$ and set $k = 0$.

Step 1 : Set $\bar{v}_k = [\partial/\partial\bar{\xi}](\xi_k^0 + \tfrac{1}{2}\|\xi_k\|) = (1, \xi_k)$.

Step 2 : Compute $\bar{\eta}_k \in \mathbf{G}\psi(\bar{\xi}_i)$ such that $\langle \bar{v}_k, \bar{\eta}_k\rangle = \min\{ \langle \bar{v}_k, \bar{\xi}\rangle \mid \bar{\xi} \in \mathbf{G}\psi(x_i) \}$.

Step 3 : Compute $(\xi_{k+1}^0, \xi_{k+1}) = \arg\min_{\lambda \in [0,1]} \xi_k^0 + \lambda(\eta_k^0 - \xi_k^0)\|^2 + \tfrac{1}{2}\|\xi_k + \lambda(\eta_k - \xi_k)\|^2$.

Step 4 : Replace i by $i+1$ and go to step 1. ■

THEOREM 2.6 : The sequence $\{\xi_k\}_{k=0}^{\infty}$ constructed by Algorithm 2.4, converges to the search direction vector $-h_i$. ■

Algorithm 2.4 tends to converge very slowly. The version in [10] is considerably more complex, but it converges considerably faster.

Finally we turn to algorithms for the solution of constrained minimax problems, of the form

$$\min\{\ \psi^0(x)\mid \psi^j(x)\le 0,\ j\in \mathbf{m}\ \},\tag{2.16}$$

where, for $j = 0,1,2,\dots,m$, $\psi^j(x) = \max_{y_j\in \mathbf{Y}_j}\phi^j(x,y_j)$, $\phi^j : \mathbb{R}^n\times\mathbb{R}^{s^j}\to\mathbb{R}$, and both $\phi^j(\cdot,\cdot)$ and $\nabla_x\phi^j(\cdot,\cdot)$ are continuous and the subsets $\mathbf{Y}_j\subset\mathbb{R}^{s^j}$ are compact.

Let $\delta>0$, let $\psi_+(x)\triangleq\max_{j\in\mathbf{m}}\{0,\psi^j(x)\}$ and, for any $\bar{x}\in\mathbb{R}^n$, and let the parametrized function $F_{\bar{x}}(x)$ be defined by

$$F_{\bar{x}}(x)\triangleq\max\{\psi^0(x)-\psi^0(\bar{x})-\delta\psi_+(x),\psi^j(x),\ j\in\mathbf{m}\}\ .\tag{2.17}$$

We see that if $\hat{x}$ is a local minimizer for (2.16), then it must also be a local minimizer for the function $F_{\hat{x}}(x)$. Hence we deduce from Theorem 2.4 (b) the following result.

THEOREM 2.7 [4]: If $\hat{x}$ is a local minimizer for (2.16), then

$$0\in\text{co}\left\{\bigcup_{j=0}^{m}\ \underset{y_j\in\mathbf{Y}_j}{\text{co}}\left\{\left[\begin{array}{c}\tilde{\phi}^j(\hat{x},y)\\ \nabla_x\phi^j(\hat{x},y)\end{array}\right]\right\}\right\},\tag{2.18}$$

where $\tilde{\phi}^0(x,y)=\phi^0(x,y)-\psi^0(\hat{x})$, and $\tilde{\phi}^j(x,y)=\phi^j(x,y)$ for all $j\in\mathbf{m}$. ■

Just as we obtained algorithms for unconstrained minimax problems by geometric extension of the method of steepest descent, we obtain algorithms for the solution of (2.16) from the following phase I - phase II generalization of the Huard *conceptual* method of centers (see [9]), below (c.f. (2.17)), where $\delta>0$ is given.

CONCEPTUAL METHOD OF CENTERS 2.4 :

Step 0 : Select $x_0\in\mathbb{R}^n$ and set $i=0$.

Step 1 : Compute $x_{i+1}=\arg\min_{x\in\mathbb{R}^n}F_{x_i}(x)$.

Step 2 : Set $i=i+1$ and go to Step 1. ■

It is easy to prove convergence of the above algorithm under the following simplifying assumption.

ASSUMPTION: (a) For every $\bar{x}\in\mathbb{R}^n$, the level sets $\mathbf{LF}_{\bar{x}}(\alpha)$ are compact. (b) for every $\bar{x}\in\mathbb{R}^n$ which is not a local minimizer of (2.16), $\gamma(\bar{x})\triangleq\min_{x\in\mathbb{R}^n}F_{\bar{x}}(x)<0$. ■.

THEOREM : If $\{x_i\}_{i=0}^{\infty}$ is an infinite sequence constructed by the conceptual method of centers, then every accumulation point $\hat{x}$ of $\{x_i\}_{i=0}^{\infty}$ is a local minimizer for (2.16).

PROOF : First, referring to [8], we conclude that $\gamma(\cdot)$ is continuous. Next, suppose that $\{x_i\}_{i=0}^{\infty}$ is an infinite sequence constructed by the conceptual method of centers, and that there exists an i_0 such that $\psi(x_+(x_{i_0}))\le 0$. Then $\psi(x_+(x_i))\le 0$ for all $i>i_0$ and $\psi^0(x_{i+1})-\psi^0(x_i)\le\gamma(x_i)<0$. If $\{x_i\}_{i=0}^{\infty}$ has an accumulation point $\hat{x}$ which is not a local minimum of (2.16), then $\gamma(\hat{x})<0$ and hence, by continuity, there exists an $i_1\ge i_0$ such that for all $i>i_0$ and the elements x_i of the subsequence which converges to $\hat{x}$,

$\psi^0(x_{i+1}) - \psi^0(x_i) \leq \frac{1}{2}\gamma(\hat{x}) < 0$, which leads to the contradictory conclusion that $\psi^0(x_i) \to \infty$ and not to $\psi^0(\hat{x})$. Hence $\hat{x}$ must be a local minimizer.

The case where $\psi_+(x_i) > 0$ for all i, is dealt with similarly, using the function $\psi(x) \triangleq \max_{j \in \mathbf{m}} \psi^j(x)$. ■

To implement the conceptual method of centers, we simply replace Step 1, with a single iteration of Algorithm 2.3, and substitute in the convergence theorem the statement that accumulation points are local minima, by the statement that accumulation points are stationary.

3. CONCLUSION

We have presented a brief survey of the simplest minimax algorithms for engineering design. For a complete exposition as well as further references, the reader should consult [8].

ACKNOWLEDGEMENT

This research was supported by the National Science Foundation grant ECS-8517362; the Air Force Office Scientific Research grant 86-0116; and the Office of Naval Research contract N00014-86-K-0295.

REFERENCES

[1] R. J. Balling, V. Ciampi, K. S. Pister and E. Polak, 'Optimal design of seismic-resistant planar steel frames', *University of California, Berkeley, Earthquake Engineering Research Center* Report No. UCB/EERC-81/20, Dec. 1981.

[2] Berge, C., *Topological Spaces* Macmillan Co., N.Y., 1963.

[3] Bhatti, M. A., E. Polak, K. S. Pister, 'Optimization of control devices in base isolation systems for seismic design', *Proc. International IUTAM Symposium on Structural Control*, University of Waterloo, Ontario, Canada, North Holland Pub. Co., Amsterdam, pp 127-138, 1980.

[4] Clarke, F. H., *Optimization and Nonsmooth Analysis*, Wiley-Interscience, New York, N.Y., 1983.

[5] Clough R. W. and J. Penzien, *Dynamics of Structures*, McGraw-Hill New York, 1975.

[6] Gill, Ph. E., W. Murray, M. A. Saunders and M. H. Wright, Programming', *Stanford University Technical Reports*, SOL 84-6 and SOL 84-7, Sept. 1984.

[7] Gilbert, E. G., 'An iterative procedure for computing the minimum of a quadratic form on a convex set', *SIAM J. Control*, Vol. 4, No. 1, pp 61-79, 1966.

[8] E. Polak, 'On the mathematical foundations of nondifferentiable optimization in engineering design', *SIAM Review*, pp. 21-91, March 1987.

[9] E. Polak, *Computational Methods in Optimization: A Unified Approach*, Academic Press, 1971.

[10] Wolfe, Ph., 'Finding the nearest point in a polytope', *Math. Programming*, Vol. 11, pp 128-149, 1976.

[11] Yang, Reng-Jye and M. J. Fiedler, 'Design modeling for large-scale three-dimensional shape optimization problems', *General Motors Research publication* GMR-5216 May 1986.

DISCRETE-CONTINUOUS STRUCTURAL OPTIMIZATION

Ulf Torbjörn Ringertz
Department of Aeronautical Structures and Materials
Royal Institute of Technology, 100 44 Stockholm, Sweden

ABSTRACT. Methods for structural optimization problems having discrete variables, or a mix of discrete and continuous variables, are treated. The original problem is replaced by a sequence of approximate, sub-problems of a more favourable structure. A method using a generalized Lagrangian function is presented. The primal minimization over the discrete variables is done using neighbourhood searches. The performance of the method is demonstrated on numerical examples. A branch and bound algorithm is used, for comparison, to obtain global minima for a convex structural optimization problem.

1. INTRODUCTION

It is often desirable to mix discrete and continuous variables when formulating structural optimization problems. In a mix of thickness and shape variables, the thickness variables may assume only discrete values, while the shape variables may be continuous. Another example could be a membrane or shell structure where orthotropy orientation variables are continuous while the thickness variables are discrete.

Recently, Abrahamsson [1] and Svanberg [2] studied a method for discrete structural optimization. The original problem was replaced by a sequence of approximate subproblems. Each subproblem was solved using a neighbourhood search technique to minimize a nondifferentiable penalty function over the discrete variables. Ringertz [3] also used neighbourhood search technique, but to minimize a generalized Lagrangian function over the discrete variables. The dual function was maximized using a subgradient method. For comparison, global minima were obtained using a branch and bound algorithm.

The mixed discrete-continuous problem is rarely treated. An important exception is the work by Schmit and Fleury [4]. Schmit and Fleury considered a sequence of separable approximate subproblems. Since the ordinary Lagrangian function is separable for separable subproblems, it is easy to perform the primal minimization over the discrete variables. The dual function is continuous but nonsmooth. Schmit and Fleury used a subgradient algorithm to maximize the dual function. There are some difficulties with this approach. Even though the subproblem is convex in the continuous case, the introduction of

G. I. N. Rozvany and B. L. Karihaloo (eds.), Structural Optimization, 257–264.

discrete variables makes the problem nonconvex. The dual maximum may correspond to a large set of possible primal solutions, and the true discrete optima may not be a member of this set. Schmit and Fleury used explicit enumeration to obtain an approximate solution.

This paper introduces a method where the original problem is replaced by a sequence of approximate subproblems. First, the structure of the mixed problem is discussed. Then follows a description of the suggested method. Finally, the performance of the algorithm is compared with a branch and bound method using numerical examples.

2. DISCRETE PROBLEMS ARE NOT CONVEX

Consider the structural optimization problem in Eqs (1-4).

$$P: \quad \min f(x) \tag{1}$$

$$g_i(x) \le 0 \qquad i = 1, \dots, m \tag{2}$$

$$x_j^{min} \le x_j \le x_j^{max} \qquad j \in I_C \tag{3}$$

$$x_j \in X \qquad j \in I_D \tag{4}$$

Here, $f(x)$ could represent the structural weight, and $g_i(x)$ could be stress and displacement constraints. The set I_C contains the indices corresponding to the continuous variables, while I_D contains the indices corresponding to the discrete variables. One way of solving (P) is to replace (P) by a sequence of convex and separable subproblems (P_k).

$$P_k: \quad \min \bar{f}(x) \tag{5}$$

$$\bar{g}_i(x) \le 0 \qquad i = 1, \dots, m \tag{6}$$

$$x_j^{min} \le x_j \le x_j^{max} \qquad j \in I_C \tag{7}$$

$$x_j \in X \qquad j \in I_D \tag{8}$$

This approach has been used by many authors [5-7] for both continuous and discrete problems, with several choices of the approximate functions $\bar{f}(x)$ and $\bar{g}_i(x)$. Even though the subproblem (P_k) is convex in the continuous case, the introduction of discrete variables will make (P_k) nonconvex. The lack of convexity makes it difficult to use the ordinary Lagrangian function Eq. (9).

$$l(x,\lambda) = \bar{f}(x) + \sum_{i=1}^{m} \lambda_i \bar{g}_i(x) \tag{9}$$

Since (P_k) is not convex when discrete variables are present, the solution to the dual problem

$$\max_{\lambda\geq 0} \min_{x} l(x,\lambda) \tag{10}$$

may not correspond to the correct primal minimum, giving a duality gap [8]. This can be illustrated by the simple equality constrained example in Eqs (11-12).

$$\min 1/x_1 + 4/x_2 \tag{11}$$

$$x_1 + x_2 - 2.8 = 0 \tag{12}$$

Consider the primal perturbation function

$$p(u) = \min_{x} \{ 1/x_1 + 1/x_2 \mid x_1 + x_2 - 2.8 = u \} \tag{13}$$

The graph of p(u) is represented in Fig. 1a for the case when x_1 and x_2 are continuous variables. As clearly seen, p(u) is convex and is suppported at u = 0 by the affine function with slope $p'(0) = -\lambda$, where λ is the optimal Lagrange multiplier of Eqs (11-12). If x_1 is discrete, taking either the value 0.6 or 1.4, the perturbation function is no longer convex as illustrated in Fig. 1b. The nonconvex perturbation function can not be supported by a linear function for u = 0, giving a duality gap δ, defined as the difference between p(0) and the dual maximum in Eq. (10).

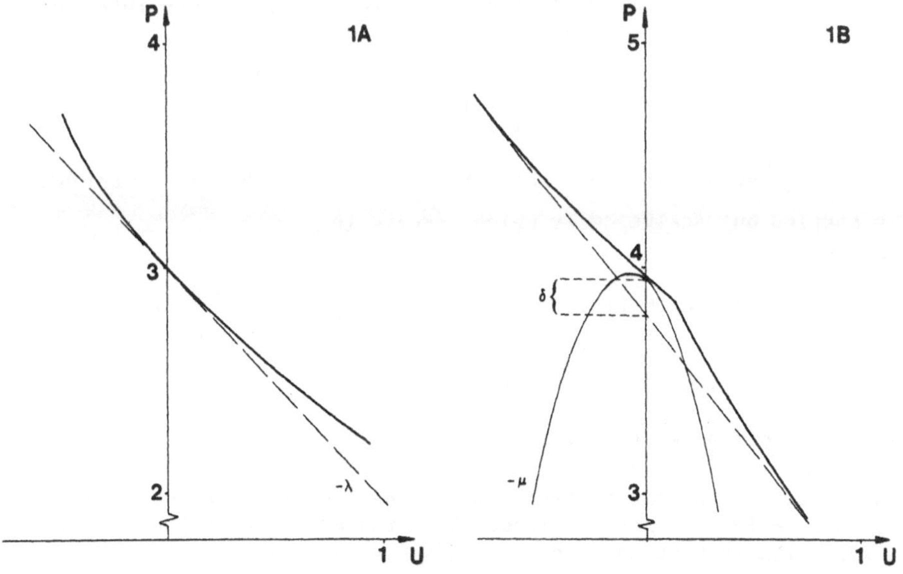

Figure 1. The primal perturbation function for the two cases.

To eliminate the duality gap, a penalty term is added to the ordinary Lagrangian function Eq. (9) giving the generalized Lagrangian function in Eq. (14).

$$l(x,\lambda,\mu) = \bar{f}(x) + \sum_{i=1}^{m} \lambda_i \bar{g}_i(x) + \mu \sum_{i=1}^{m} \bar{g}_i(x)^2 \tag{14}$$

Several authors (e.g. [9-10]) have developed different generalized, or augmented, Lagrangian functions for continuous nonlinear programming (NLP) problems. The addition of the penalty term corresponds to replacing the linear support function in Fig. 1b by a concave support function with curvature μ, as illustrated in Fig. 1b. As clearly seen in Fig. 1b, the concave support function will, for sufficiently large μ, support p(u) at u = 0. The duality gap is reduced to zero and, consequently, the dual solution will correspond to the correct primal minima. The advantage of using the generalized Lagrangian function is that the duality gap can be eliminated. The drawback is that the primal minimization of Eq. 14, over the discrete variables, is much more difficult.

3. A GENERALIZED LAGRANGIAN METHOD

The proposed method combines the use of a generalized Lagrangian function with heuristic neighbourhood searches to perform the primal minimization over the discrete set. A dual function is maximized over the Lagrange multipliers and the penalty parameter.

To treat inequality constraints in Eq. (14), one can introduce slack variables $v_i \geq 0$ giving

$$l(x,v,\lambda,\mu) = \bar{f}(x) + \sum_{i=1}^{m} \lambda_i(\bar{g}_i(x)+v_i) + \mu \sum_{i=1}^{m} (\bar{g}_i(x)+v_i)^2 \tag{15}$$

Following Rockafellar [9], the minimization over the slack variables v_i is carried out analytically which results in

$$l(x,\lambda,\mu) = \bar{f}(x) + \sum_{i=1}^{m} \frac{1}{2\mu} [(\max\{0,\lambda_i + \mu\bar{g}_i(x)\})^2 - \lambda_i^2] \tag{16}$$

The dual function $\Phi(\lambda,\mu)$ is defined as

$$\Phi(\lambda,\mu) = \min_x l(x,\lambda,\mu) \tag{17}$$

The penalty parameter is also regarded as a dual variable, as proposed by Rockafellar [9] for continuous NLP problems. The subgradient of $\Phi(\lambda,\mu)$ with respect to λ is

$$\partial_{\lambda_i} \Phi(\lambda,\mu) = (1/\mu)[\max(0,\lambda_i + \mu \bar{g}_i(x)) - \lambda_i] \tag{18}$$

which is available as $\Phi(\lambda,\mu)$ is evaluated. The dual function is concave and continuous, but the discrete primal variables give rise to occasional discontinuities in the gradient of $\Phi(\lambda,\mu)$. The dual problem

$$\max_{\lambda,\mu} \Phi(\lambda,\mu) \tag{19}$$

is solved in a similar way as suggested by Powell [10] for continuous NLP problems.

Given initial values for λ and μ, compute $\Phi(\lambda,\mu)$ and $\partial_\lambda \Phi(\lambda,\mu)$. A step is taken in the subgradient direction using μ as the step length.

$$\lambda_i^{k+1} = \lambda_i^k + \mu\, \partial_{\lambda_i} \Phi(\lambda^k,\mu^k) \tag{20}$$

The dual variable μ is increased according to

$$\text{if} \quad ||\partial_\lambda \Phi(\lambda^k,\mu^k)||_\infty > ||\partial_\lambda \Phi(\lambda^{k-1},\mu^{k-1})||_\infty / 5 \tag{21}$$

$$\text{then} \qquad \mu^{k+1} = 5\, \mu^k$$

The numerical factor 5 is somewhat arbitrary, numerical experiments suggest that any number in the range [2,10] is satisfactory.

The most difficult part of the algorithm is the primal minimization performed when evaluating the dual function in Eq. (17). The minimization is performed in a sequence of two step minimizations as illustrated by Eqs (22-23).

$$\text{Step 1.} \quad \min_{x_i} l(x,\lambda,\mu) \quad , i \in I_C \tag{22}$$

$$\text{Step 2.} \quad \min_{x_i} l(x,\lambda,\mu) \quad , i \in I_D \tag{23}$$

The minimization in step 1, followed by the one in step 2, is repeated until convergence, which is usually attained in a few steps. The minimization over the continuous variables is performed using Newton's method, since it is in most cases easy to obtain second derivatives of all the functions in the subproblem Eqs (5-8).

The minimization over the discrete set is performed using neighbourhood searches as introduced by Reiter and Sherman [11]. The problem is to minimize an unconstrained function over a discrete set X. A set N(x), called the neighbourhood of x, is defined. Given an initial point, the next point is obtained so that x^{k+1} minimizes $l(x,\lambda,\mu)$ over $N(x^k)$. The algorithm terminates when there is no point $x \in N(x^k)$ such that $l(x,\lambda,\mu) < l(x^k,\lambda,\mu)$. In this work, the set N(x) is defined as all points obtained by taking one discrete step in two variables from x, or one step in one variable. Hence, the size of N(x) is proportional to n_D^2, n_D being the number of discrete variables.

4. NUMERICAL EXAMPLES

This section presents the results obtained by using the algorithm described in the previous section. For the first small scale example, the results are compared with using a branch and bound algorithm to solve each subproblem. The branch and bound algorithm used is a slightly modified version of the one used in [3]. Each subproblem is created using the inverse variable approximations, as suggested by Schmit and Fleury [7], but any convex approximation could be used. Structural gradients are obtained using the pseudo load technique [11]. Initial values are obtained by rounding off the continuous solution. The criterion for convergence is

$$|f(x^k) - f(x^{k-1})|/f(x^k) + || \max(0,g_i(x^{k-1})) || < 10^{-4} \quad (24)$$

4.1. Example 1

Svanberg [12] has proved that minimizing the weight of a truss subject to a symmetric displacement constraint for each loading condition, represents a convex optimization problem. A displacement constraint is said to be symmetric if it may be written as

$$q^T u \leq u^{max} \quad \text{with} \quad q \;//\; p \quad (25)$$

where u is the displacement vector and p is the load vector. Minimizing the weight of the 36 bar truss in Fig. 2 subject to symmetric displacement constraints represents a convex problem.

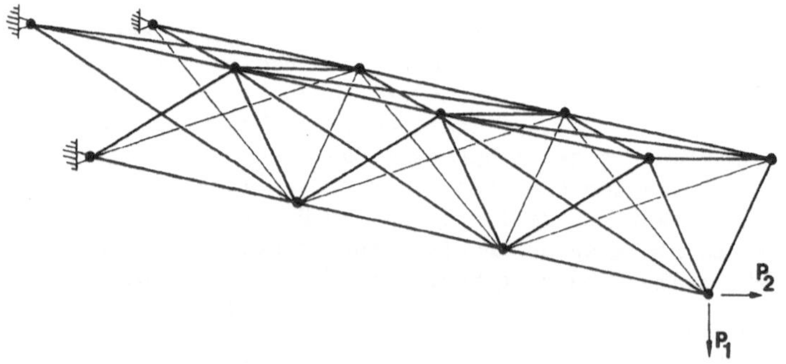

Figure 2. The 36 bar truss.

Detailed information on this example can be found in [3]. There are 21 variables, with variables 1-12 being discrete. For a convex problem it is possible to obtain a global minimum using a branch and bound method where the original problem Eqs (1-4) is treated in each step. The results obtained using the generalized Lagrangian method, and using the branch and bound method to solve each subproblem, is compared with the global minima. As seen in Table I, the use of branch and bound on each subproblem obtains the global minima, which can not always be expected. Even for this small problem, only 12 discrete variables, one can observe the significant difference in computational effort (on a VAX/FPS164 system) needed for the different methods.

TABLE I

Example 1, The convex 36 bar truss. X = (5.0,10.0,...,125.0)

	Initial values	Global minima	Branch & Bound	Generalized Lagrangian
W (lbs)	36351	35845	35845	35913
Feasible	yes	yes	yes	yes
Num. of iter	-	-	7	2
CPU (s)	-	183	434	8.5

4.2. Example 2

The weight of a 63 bar truss is minimized subject to stress constraints and a torsional rotation constraint. Detailed description of this problem, as well as results from a continuous optimization run, are given in [7]. There are 63 variables of which 54 are discrete. Only variables 1,2,3,4,23,42,43,44, and 45 are continuous. The number of discrete variables is too large for using the branch and bound method. The results obtained using the generalized Lagrangian method are given in Table II.

TABLE II

Example 2, The 63 bar truss. X = (0.01,1.0,2.0,...,60.0)

	Initial values	Generalized Lagrangian
W (lbs)	6305.1	6163.3
Feasible	yes	yes
Num. of iter	-	3
CPU (s)	-	2853

5. DISCUSSION

It is possible to obtain a global minimum for each subproblem, using a branch and bound method, for small scale problems (< 20 discrete variables). This may, however, not mean that the global minima will be obtained even though the problem to be solved Eqs (1-4) is convex. For medium size problems (20-50 discrete variables) the computational effort becomes too large to use branch and bound methods. The proposed generalized Lagrangian method makes it possible to efficiently solve medium size problems. If the problem is large scale (> 50 discrete variables), the computational effort necessary to perform the primal minimization in Eq. (17) may become excessive. One possibility to treat large scale problems could be to consider a smaller discrete neighbourhood N(x) then the one used in this study. It could also be possible to improve the results for small scale problems by increasing the size on N(x).

6. REFERENCES

[1] T. Abrahamsson, 'A Method for Discrete Structural Optimization,' Thesis, Dept. of Mathematics, Royal Institute of Technology, Stockholm, (1987).

[2] K. Svanberg, private communication, (1986).

[3] U. Ringertz, 'On Methods for Discrete Structural Optimization', to appear in Eng. Opt.

[4] L. Schmit and C. Fleury, 'Discrete-Continuous Variable Structural Synthesis Using Dual Methods', AIAA J., 18, 1515-1524, (1980).

[5] K. Svanberg, 'Method of Moving Asymptots - A New Method for Structural Optimization', IJNME., 24, 359-373, (1987).

[6] C. Fleury and V. Braibant, 'Structural Optimization, a New Dual Method Using Mixed Variables', Report SA-115, Aerospace Laboratory of the University of Liege, Belgium, 12 pp, (1984).

[7] C. Fleury and L. Schmit, 'Dual Methods and Approximation Concepts in Structural Synthesis', NASA Contractor Report 3226, (1980).

[8] M. Minoux, Mathematical Programming Theory and Algorithms, Wiley, (1986).

[9] R. Rockafellar, 'Augmented Lagrange Multiplier Functions and Duality in Nonconvex Programming', SIAM J. Control, 12, 268-285, (1974).

[10] M. Powell, 'A Method for Nonlinear Optimization in Minimization Problems', in Optimization, (R. Fletcher, ed.), Academic Press, New York, (1969).

[11] S. Reiter and G. Sherman, 'Discrete Optimizing', J. Soc. Indust. Appl. Math., 13, 864-889, (1965).

[12] K. Svanberg, 'On Local and Global Minima in Structural Optimization', In New Directions in Optimum Structural Design, (Atrek, Ragsdell, Gallagher, Zienkiewicz, Eds.), Wiley, (1984).

OPTIMALITY CRITERIA AND LAYOUT THEORY IN STRUCTURAL DESIGN: RECENT DEVELOPMENTS AND APPLICATIONS

G.I.N. ROZVANY
FB 10, Universität GH Essen
4300 Essen 1, FRG

ABSTRACT. After outlining current difficulties in structural optimization and a way of overcoming them by employing *optimality criteria* and *layout theory*, the above concepts are discussed in greater detail and illustrated by simple examples. Finally, a brief review of recent developments in these fields is presented.

1. Introduction

The significance, basic features and early history of optimality criteria and optimal layout theory were summarized in the author's opening address (p. xvii) and detailed recent reviews of these fields can be found elsewhere [1-4]. For this reason, only certain aspects of the above topics will be discussed in this short contribution. The importance of layout optimization via optimality criteria can be better understood if we consider some of the present difficulties in structural optimization (Fig. 1).

Optimization problems involving large structural systems are usually non-convex and highly non-linear; hence the number of local minima is unknown and increases exponentially with the number of variables. For this reason, we are often not sure, if the solution found by a *mathematical programming* (MP) method is a global (and not only a local) minimum. The above problem can be avoided to some extent by using *optimality criteria* (OC) methods. However, in the case of several design requirements (e.g. stress, displacement, stability, dynamic and geometrical constraints) the solution splits up into "regions" governed by different optimality criteria (Fig. 1). Since the layout of these regions is not known prior to the optimization procedure, certain difficulties may arise but these can be removed by employing layout theory, as was demonstrated in the case of grillages [5-6].

Further pitfalls in structural optimization were revealed recently by Kohn and Strang [7] and Cheng and Olhoff [8] who found that (a) shape optimization of continua usually results in (a theoretically infinite number of) *internal boundaries* and (b) least-weight solid plates contain a dense system of *rib-like formations* (Fig. 1). Moreover, the optimized shape of thin shells of constant thickness is likely to contain *corrugations* of theoretically infinitesimal spacing (Fig. 1). Layout theory has been found useful in optimizing the configuration of such holes and ribs [2-4]. Obviously, solutions with an infinite number of cavities or ribs are unpractical but they can easily be approximated, without a significant loss of economy, by designs containing a *finite* number of such formations. On the other hand, numerical methods are often unsuitable for handling these problems because the optimal cost is highly dependent on the number of elements used.

G. I. N. Rozvany and B. L. Karihaloo (eds.), Structural Optimization, 265–272.

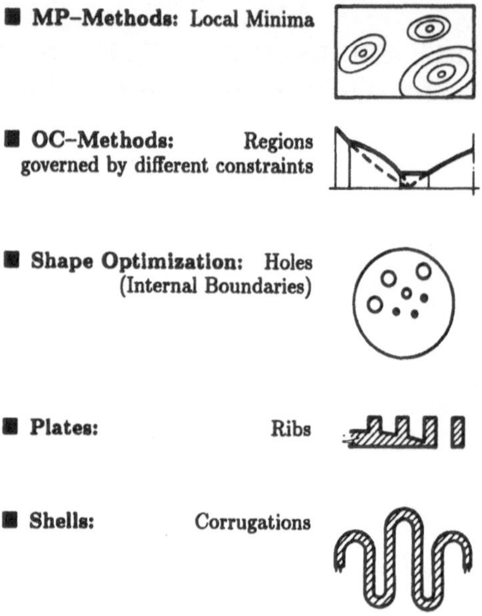

Fig. 1 Current difficulties in structural optimization

2. Static-Kinematic Optimality Criteria

Using once again Prager's terminology, the basic variables of structural mechanics consist of *generalized* stresses **Q** (including stress resultants, e.g. bending moments), strains **q** (including, for example, curvatures), loads **p** and displacements **u**. The earliest general, continuum-type optimality criteria were introduced in the context of ***plastic design*** for a single load condition by Prager and Shield [9] and can be explained as follows: If plastic design is based on the lower bound theorem of limit analysis, then it is sufficient to find a ***statically admissible*** stress field $\mathbf{Q}^S$ which for the given loads **p** satisfies equilibrium (static-continuity) on the entire structural domain D and static constraints (static boundary conditions) on some subset S_1 of the domain D. In order to make the design then *safe* (resisting at least the load **p** without plastic collapse), it is sufficient to adopt cross-sections which for the statically admissible stresses $\mathbf{Q}^S$ satisfy the yield inequality $Y(\mathbf{Q}) \leq Y_0$ on the entire structural domain D (Fig. 2).

While the solution of the above design problem is usually non-unique, we can achieve uniqueness if we also stipulate that (a) all cross-sections are required to be at yield $[Y(\mathbf{Q}^S) = Y_0]$ and a certain quantity called total cost Φ is to be minimized. The latter can be expressed by integrating the so-called specific cost ψ (cost per unit length, area or volume, e.g. cross-sectional area) over the structural domain D. As the specific cost depends, via the yield equality condition, on the generalized stresses **Q**, the above procedure (termed *optimal plastic design*), can be represented in a manner shown in Fig. 3.

Although the above problem can be solved by discretization and direct minimization (e.g. by finite element methods and mathematical programming), a more efficient approach termed *optimality criteria method* is shown in Fig. 4. In the latter the minimality condition

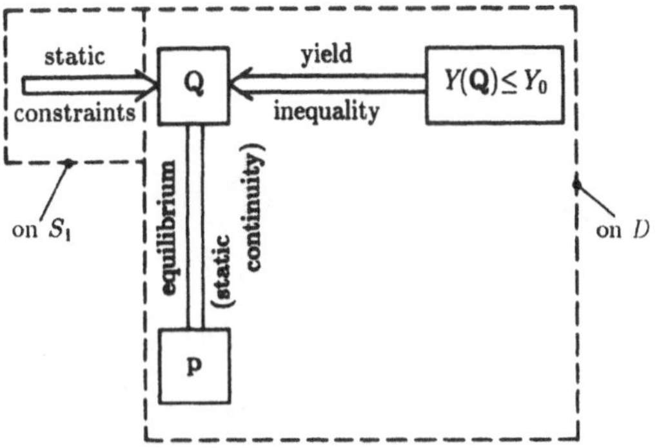

Fig. 2 Basic relations for plastic design

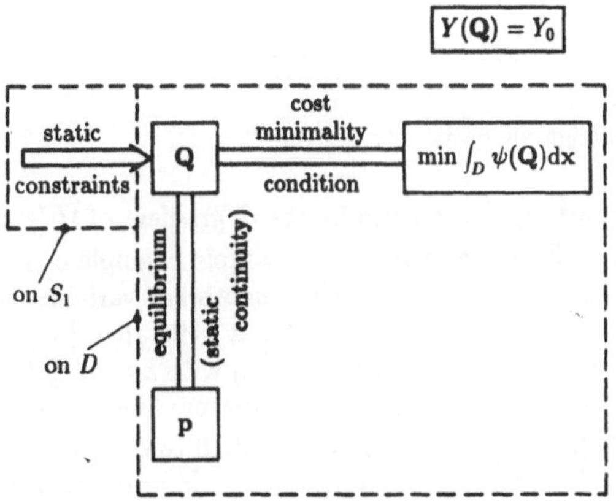

Fig. 3 Basic relations for optimal plastic design – direct cost minimization

is converted in such a way that certain terms in it can be interpreted as ("Pragerian" or "adjoint") strains $\overline{\mathbf{q}}$ and displacements $\overline{\mathbf{u}}$ which are required to satisfy certain "optimality criteria" expressed in the form of strain-stress relations (Fig. 4) and kinematic constraints (kinematic boundary conditions) on the adjoint displacements $\overline{\mathbf{u}}$. As an additional optimality criterion, compatibility (kinematic continuity) of the adjoint fields ($\overline{\mathbf{q}}$, $\overline{\mathbf{u}}$) must be fulfilled. This implies that the adjoint fields must be ***kinematically admissible*** ($\overline{\mathbf{q}}^K$, $\overline{\mathbf{u}}^K$).

The optimal strain-stress relation (static-kinematic optimality criteria) in Fig. 4 follows from the Prager-Shield condition [9] in its generalized form [5] which states that the adjoint strains $\overline{\mathbf{q}}^K$ must equal the G-gradients ($\mathcal{G}$) of the specific cost ψ with respect to the "real" stresses $\mathbf{Q}^S$. For differentiable functions, the G-gradient is simply a vector containing the partial derivatives $\mathcal{G}[\psi(\mathbf{Q})] = [\partial\psi/\partial Q_1, \ \ldots \ , \partial\psi/\partial Q_n]$, but for non-differentiable and discontinuous functions it has an extended meaning [5, 10, 11].

The kinematic boundary conditions (kinematic optimality criteria) for $\overline{\mathbf{u}}$ in the case of zero reaction cost are the same as for rigid supports of the real structure. However, if the cost of the reactions $\mathbf{R}$ is given by a non-zero "reaction cost function" Ω, then the

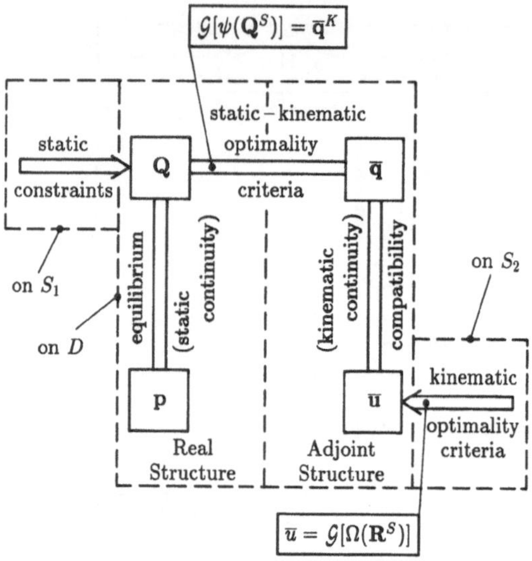

Fig. 4 Optimal plastic design via optimality criteria

displacements along supports (S_2) are given by the G-gradient of Ω (Fig. 4, [5]).

The above procedure is illustrated with a very simple example concerning the optimal *plastic design* of a clamped beam (Fig. 5a) of given depth but variable width (Fig. 5g) with a central point load. If all cross-sections are to be at yield then $|M| = \sigma_y bd^2/4$ and the specific cost function becomes $\psi = bd = k|M|$ (Fig. 5e) with $k = 4/\sigma_y d$. The corresponding static-kinematic optimality condition, i.e. optimal curvature/moment relation (Fig. 4) given by $\bar{\kappa}^K = -\bar{u}'' = \mathcal{G}[\psi(M^S)]$ is the one shown in Fig. 5f. If the cost of supports is zero, then the kinematic optimality (boundary) condition for $\bar{u}$ is $\bar{u}' = 0$ at both ends of the beam (cf. Fig. 4). The adjoint displacement field satisfying all the above optimality conditions is shown in Fig. 5c, and the corresponding optimal moment-diagram in Fig. 5b.

If the beam has (a) rigid (i.e. rigidly clamped) supports and (b) zero reaction cost, then the above solution is also valid for *elastic design* with a stress constraint. This conclusion follows from the fact that for a rectangular beam of given depth d but variable width b the flexural stiffness $S = Ebd^3/12$ is proportional to the moment capacity $M_c = \sigma_0 bd^2/6$ where E is Young's modulus and σ_0 is the permissible stress; hence the stress constraint for a fully stressed design can be expressed as $S \geq a|M|$ with $a = Ed/2\sigma_0$. Since the cross-sectional area $A = bd$ is also proportional to the stiffness ($A = cS$ with $c = 12/Ed^2$), we can minimize the quantity $\Phi = \int_0^L S\,dx$ where $\psi = S(M) = a|M|$ becomes the specific cost function if all cross-sections are fully stressed. This means that the specific cost function, optimal moment-curvature relation and adjoint displacement field $\bar{u}$ in Figs. 5e, f and c are still valid for optimal elastic design (with $k \rightarrow a$) so long as the optimal fully stressed elastic solution gives a kinematically admissible real displacement field (u^K). The moment capacity M_c and stiffness diagrams for the above design are shown in Fig. 5d and the corresponding moment diagram in Fig. 5b. It follows that the real displacement field u is also given by Fig. 5c within a constant factor (ka). As the latter is clearly kinematically admissible, the optimal moment capacity diagram M_c in Fig. 5d is valid for optimal elastic stress design as

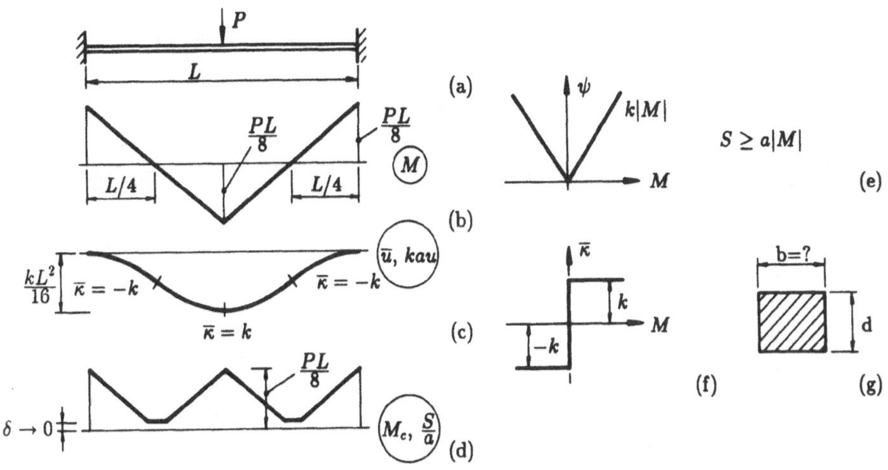

Fig. 5 Optimal plastic design and optimal elastic design with a stress constraint – an example

well as for optimal plastic design. It is assumed in this problem that concentrated rotations (i.e. hinges) do not occur at cross-sections with locally zero area. This can be assured by prescribing a minimum cross-section ($M_c \geq \delta$) and then carrying out a limiting process ($\delta \to 0$, Fig. 5d).

It is a common misconception, however, that optimal elastic stress design always results in a fully stressed design. If the supports in the problem in Figs. 5a and g have the cost $\Omega = \alpha|M_E|$, where M_E is the end moment in the beam then the kinematic optimality criterion in Fig. 4 will give a slope of $|\overline{u}'| = \alpha$ at both beam ends for the adjoint field which results in a negative moment region that is shorter than $L/4$. This means that the corresponding real displacement field u for a fully stressed beam would not satisfy kinematically admissibility which can only be restored by including some understressed segments in the positive moment region (for details see [12]).

Moreover, an optimal elastic stress is, in general, not fully stressed if *selfweight* is taken into consideration (Fig. 6). In the case of optimal plastic design with allowance for selfweight, the static-kinematic optimality criteria in Fig. 4 are replaced by [13]

$$\overline{\mathbf{q}}^K = (1 + \overline{\mathbf{u}}^K)\, \mathcal{G}[\psi(\mathbf{Q}^S)] \tag{1}$$

which gives the adjoint curvatures $\overline{\kappa}$ and displacements $\overline{u}$ shown in Fig. 6c where z, a and b are nondimensionalized (e.g. $z = k^{1/2}\tilde{z}$ where $\tilde{z}$ is the dimensional coordinate and k is the cost factor in the specific cost function $\psi = k|M|$). The above field gives a simple relationship (framed in Fig. 6c, see solutions by E. Becker in Fig. 6f) for the location of the zero moment points which has also been verified by integrating the moment fields in Fig. 6b and subsequently differentiating with respect to b (see [13]).

Considering now optimal elastic strength design with selfweight for the problem in Fig. 6a, we find that the elastic curvatures at fully stressed cross-sections have a constant absolute value and hence the real (elastic) deflection field u for a fully stressed design (continuous lines in Fig. 6d) would not be kinematically admissible. It can be shown readily that two understressed segments occur in the optimal solution [broken line, $\overline{M}$ in the corresponding

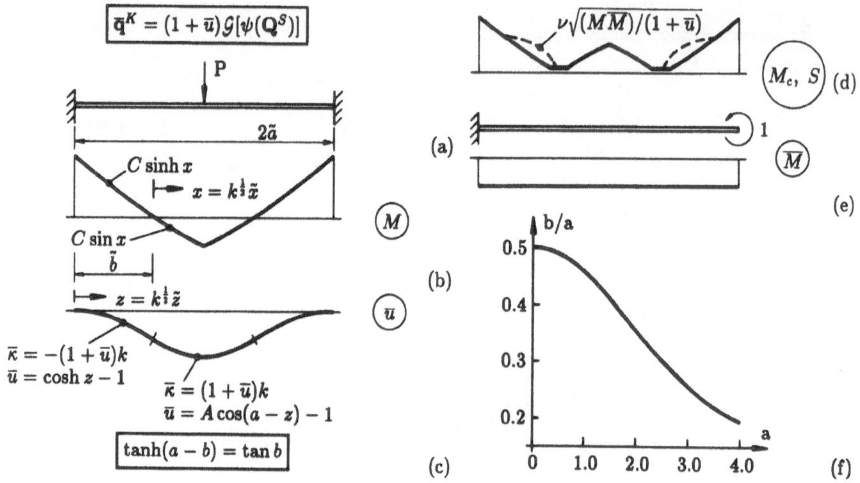

Fig. 6 Optimal beam design with allowance for selfweight

formula is the virtual moment assuring zero rotation at the ends via the work-equation $\int_0^{2a}(\overline{M}M/S)dz = 0$, see Fig. 6e].

Apart from allowance for selfweight ([13], 1977) the author and his associates have extended the Prager-Shield condition [9] to multi-constraint design (1970), alternate loads (1972), multi-component systems (1972), non-convex specific cost functions (1973), partially prescribed cost distribution (segmented structures, 1973), discontinuous-specific cost functions (1974), allowance for cost of supports (1974), optimization of support location (1975), allowance for cost of connections (1975), generalized cost functions (1976), bounded spatial gradients (Niordson-constraints, 1984) and continuous, segment-wise linear cost distribution (1988) in *plastic design*, as well as optimal segmentation (1975), optimal support location (1975), allowance for compliance and deflection constraints (1976), optimization for buckling load (1977), stress constraints (1977), Niordson-constraints (1987), and a combination of deflection and stress constraints (1987) in *elastic design*. For a review of these optimality conditions, the reader is referred to the author's forthcoming book [11] in which natural frequency constraints are also discussed. The use of optimality criteria for elastic design with deflection constraints is shown in Fig. 7.

The Prager-Shield condition was extended to dynamic problems in 1968 by Prager and Taylor [14]. The latter optimality criterion was applied to thin-walled vibrating beams already in 1972 in a paper of historic importance by de Boer [15].

2. Layout Optimization in Structural Design

General aspects of layout optimization were discussed already in the author's opening address (p. xvii). *Classical layout theroy* has been applied to pin-jointed frames (Michell-structures), grillages (or beam-systems), arch-grids and cable-nets of least weight (Prager-structures). Of these, the study of minimum weight grillages represents one of the most remarkable developments, in so far as they constitute the first class of truely two-dimensional

$$\mathcal{G}_{,z}[\psi(\mathbf{z})] + \Sigma_j \cdot \nu_j \cdot \overline{\mathbf{Q}}_j^{SK} \cdot \{\mathcal{G}_{,z} \cdot [\mathbf{F}]\} \cdot \mathbf{Q}^{SK} = 0$$

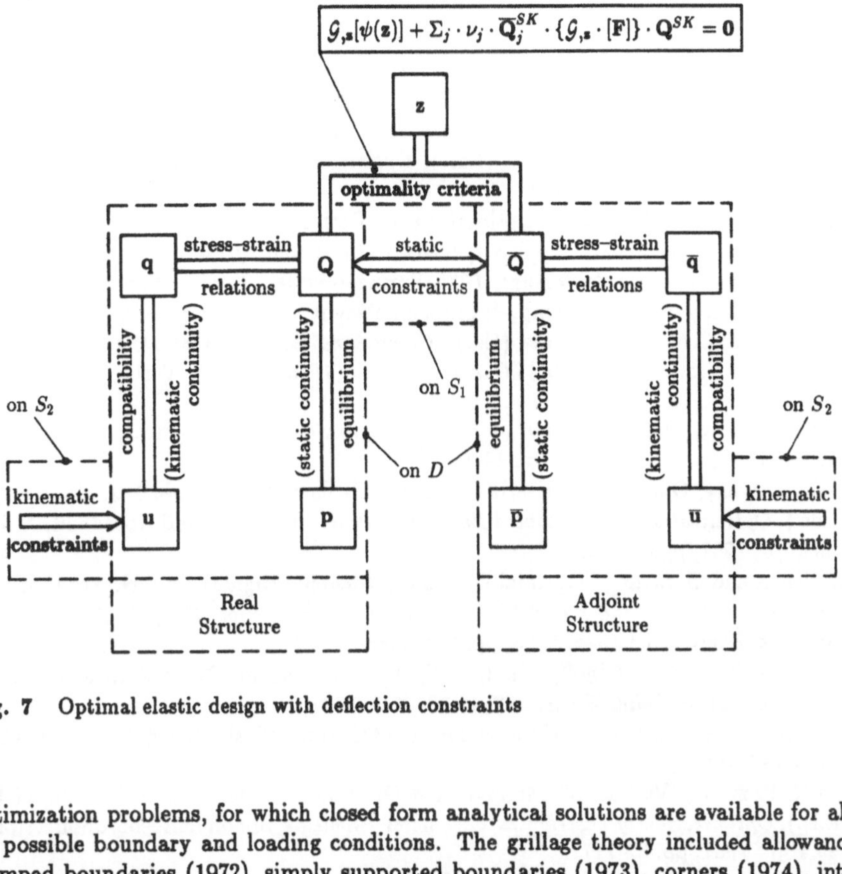

Fig. 7   Optimal elastic design with deflection constraints

optimization problems, for which closed form analytical solutions are available for almost all possible boundary and loading conditions. The grillage theory included allowance for clamped boundaries (1972), simply supported boundaries (1973), corners (1974), internal simple supports (1976), cost of supports (1976), partial discretization (1976), free edges (1977), beam supported edges (1978), bending- and shear-dependent cost (1979), non-uniform depth (1981), upper limit on the beam density (1983), selfweight (1984), selfweight with bending- and shear-dependent cost (1986) for *plastic grillages*, as well as deflection constraints (1986) and a combination of deflection and stress constraints (1987) for *elastic grillages*. Moreover, a computer algorithm was developed for generating analytically and plotting optimal beam layouts (1978). Analytical solutions are also available for Prager-structures with any arbitrary plane, axisymmetric or "quasi-axisymmetric" loading and for additional selfweight. For a review of the above developments see [2, 5, 11].

Applications of *advanced layout theory* included the design of *plastic solid plates* in which regions with ribs of infinitesimals spacing occur in the solution. More recent research revealed that in optimal solutions for *perforated and composite elastic plates* with either compliance or deflection constraints some regions contain a system of intersecting ribs of first and second order infinitesimal width, respectively, in the principal directions. For a detailed discussion of these problems the reader is referred to [3, 4, 11, 16].

The optimal microstructure for elastic plates was discussed by Murat and Tartar, Lurie, Cherkaev and Fedorov as well as Kohn and Strang (see, for example, [17] and [18]).

References

1. M. Save and W. Prager, *Structural Optimization I – Optimality Criteria*, Plenum Press, New York, 1985.
2. G.I.N. Rozvany, 'Structural Layout Theory – The Present State of Knowledge', Chapt. 7 in E. Atrek, R.H. Gallagher *et. al.* (Eds.), *Optimum Structural Design*, John Wiley & Sons, Chichester, 1984, pp. 167-196.
3. G.I.N. Rozvany and T.G. Ong, 'Optimal Plastic Design of Plates, Shells and Shellgrids', in L. Bevilacqua, R. Feijóo *et. al.* (Eds.), *Inelastic Behaviour of Plates and Shells, Proc. IUTAM Symposium in Rio de Janeiro, 1985*, Springer-Verlag, Berlin, 1986, pp. 357-384.
4. G.I.N. Rozvany and T.G. Ong, 'Minimum-Weight Plate Design via Prager's Layout Theory', in C.A. Mota Soares (Ed.), *Computer-Aided Optimal Design: Structural and Mechancial Systems, Proc. NATO ASI in Troia, 1986*, Springer-Verlag, Berlin, 1987, pp. 165-180.
5. G.I.N. Rozvany, *Optimal Design of Flexural Systems*, Pregamon Press, Oxford, 1976; Russian version: Stroiizdat, Moscow, 1980.
6. G.I.N. Rozvany, Optimality Criteria for Grids, Shells and Arches, in E.J. Haug and J. Cea (Eds.), *Optimization of Distributed Parameter Problems*, Sijhoff and Noordhoff, Alphen aan den Rijn, 1981, pp. 112-151.
7. R.V. Kohn and G. Strang, 'Optimal Design for Torsional Rigidity, in S.N. Atluri & R.H. Gallagher (Eds.), *Hybrid and Mixed Finite Element Methods, Proc. Conf. in Atlanta, 1981*, John Wiley & Sons, Chichester, 1983, pp. 281-288.
8. Cheng, K.-T. and N. Olhoff, 'An Investigation Concerning Optimal Design of Solid Elastic Plates, *Int. J. Solids Struct.* **17**, 3, 305-323, 1981.
9. W. Prager and R.T. Shield, 'A General Theory of Optimal Plastic Design', *J. Appl. Mech.* **34**, 1, 184-186, 1967.
10. G.I.N. Rozvany, 'Variational Methods and Optimality Criteria', in E.J. Haug and J. Cea (Eds.), *Optimization of Distributed Parameter Problems*, Sijhoff and Noordhoff, Alphen aan den Rijn, 1981, pp. 82-111.
11. G.I.N. Rozvany, *Structural Design via Optimality Criteria*, Kluwer Academic Publishers, Dordrecht, 1988.
12. G.I.N. Rozvany, 'Plastic vs. Elastic Optimal Strength Design', *J. Engrg. Mech. ASCE*, **103**, EM1, 1977, 210-214.
13. G.I.N. Rozvany, 'Optimal Plastic Design: Allowance for Selfweight', *J. Engrg. Mech. ASCE*, **103**, EM6, 1977, 1165-1170.
14. W. Prager and J.E. Taylor, 'Problems of Optimal Structural Design', *J. Appl. Mech.*, **35**, 1, 102-106, 1968.
15. R. de Boer, 'Optimierung von Stabschwingern mit dünnwandigem Querschnitt', *Der Stahlbau*, **8**, 1972, 245-249.
16. G.I.N. Rozvany, T.G. Ong, W.T. Szeto, R. Sandler, N. Olhoff and M.P. Bendsøe, 'Least-Weight Design of Perforated Elastic Plates I and II', *Int. J. Solids Struct.* **23**, 4, 1987, 521-550.
17. K.A. Lurie, A.V. Cherkaev and A.V. Fedorov, 'Regularization of Optimal Design Problems for Bars and Plates', *J. Optimiz. Theory Appl.*, **37**, 4, 499-543.
18. R.V. Kohn and G. Strang, 'Optimal Design and Relaxation of Variational Problems', *Comm. Pure Appl. Math*, **39**, 113-182, 353-377.

Shape Optimization: Creating a Useful Design Tool

E. Sandgren
School of Mechanical Engineering
Purdue University
West Lafayette, IN 47906, U.S.A.
Mohamed El Sayed
Mechanical & Aerspace Engineering
University of Missouri-Columbia, Missouri U.S.A.

ABSTRACT. Several new areas of research are presented which seek to eliminate some of the barriers to achieving a useful design optimization tool. These areas include consideration of geometrical and topological optimization as well as crossectional optimization, design for latitude uncertainty in loading conditions and problem formulation, member insertion and deletion, new elements for optimization and fundamental new approaches to combine analysis and optimization. The role new computing hardware will play in the implementation of the next generation design tool is investigated. Work in several of these areas will be described. Results for shape optimization using the boundary integral method with substructuring, zero-one integer programming for decision support and a means of performing the analysis within the shape optimization formulation are presented.

1. INTRODUCTION

The factors which keep structural optimization from reaching its full potential are in many ways the same factors that inhibit computer aided design in general. Analysis is an important but also small portion of the design process. Structural optimization has matured primarily from the analysis side. As our ability to perform better analysis has progressed so should have our ability to perform structural optimization. This has not been the case and one prime factor in this respect is that optimization has never been an integral part of the process. Simply coupling a finite element code with a nonlinear programming code has not produced the results desired. Even with analytic design sensitivity information, the problem size and non-linearity simply overwhelm our current capability to solve the problem. Successes in the area of crossectional optimization are overshadowed by our failure to solve practical shape optimization problems. Design criteria such as minimum weight need to be augmented by more realistic factors more in line with function. The issue of uncertainty in the design formulation, particularly in the area of load conditions, must be addressed. The design synthesis problem of considering a set of possible alternatives or to modify the design to better perform its task

G. I. N. Rozvany and B. L. Karihaloo (eds.), Structural Optimization, 273–278.

must be implemented. In short, the current state of the art in structural optimization is a long way from producing a truly useful design tool.

Extending or modifying existing analysis and optimization methodology would not seem to be adequate to handle all of the requirements for design. A new approach, particularly one viewed from the optimization point of view, would appear more promising. This probably means developing a method from first principles which is a monumental task. On the positive side, however, is the fact that to utilize the new generation of parallel and distributed computing hardware, this will probably be necessary anyway. Both analysis and optimization as they are currently implemented, are fundamentally sequential processes. Progress has been made in selected areas in parallelizing portions of algorithms but one only has to look at the updating prescription for a new nonlinear programming algorithm to see its sequential nature. The new design point is formed by taking a step in the specified search direction from the previous point. This is a sequential operation in its present form.

The approach taken in the research presented, is one of addressing the issue of design through optimization. This optimization need not be in a conventional form and in fact, it probably will not be due to the demands of design. Optimization is a potentially useful tool because it allows the engineer to work in a manner which closely parallels the design process. Design should proceed from the general to the specific. Optimization allows for the synthesis of a design which best meets the design criteria and in this respect, it forms the basis of a useful design tool. By extending the traditional formulation to include multiple objective criteria, variability in constraint functions, design for latitude and inner and outer noises, the scope of structural optimization can be broadened considerably.

A brief discussion of the additional capabilities required for structural design will be followed by several highly condensed results from work conducted in the past year. These results will be in three areas. The first will be in the area of substructuring for shape optimization. The second include zero-one nonlinear programming for decision support and the last involves a look at how analysis may be performed simultaneously with optimization.

2. THE CHALLENGES

When optimization is performed on a structure, particularly one composed of beam elements, the design variables would consist of the crossectional dimensions and possibly the position of a few of the element nodes. The results achieved can be dramatically improved by taking the broader of crossectional, geometrical and topological optimization. An automotive seat frame constructed of tubular steel was optimized under both static design criteria and dynamic crash loading. The resulting weight of the optimal design for the case of crossectional,geometrical and topological design are presented in Table 1.

Consideration	Optimal Volume of Material
Starting value	297
Crossection	120
Geometrical	114
Topological	60

TABLE 1. Optimal Design of a Seat Frame

These results demonstrate the importance of looking at the shape of a design. Using the design data, the weight of the seat was cut substantially by reducing the crossectional area in noncritical sections. Allowing the geometry to change from the initial design, by moving nodal points in the finite element analysis decreased the weight a few percent. A big decrease was obtained by allowing the topology to change (i.e., connectivities and addition or deletion of elements). An additional 50% of the weight was removed as the design was allowed to take an unsymmetrical shape to better withstand the applied loading.

The results of the seat from optimization show that significant weight savings can be achieved through shape optimization. Although it was a fairly small problem (variable linking was used to reduce the problem to 19 design variables) even a problem of this size is a computationally expensive task. The difficulty with the results is that they are highly dependent on the specified loading conditions. A change in magnitude or direction of a few percent in one of the four applied loading conditions can alter the results significantly. This is a result of the solution of an exact mathematical problem which is used to model a design with a large uncertainty factor. Structural optimization when performed with a standard nonlinear programming algorithm, will always be guided by the constraints. If there were no constraints, the optimal weight would be zero. Stress, deflection buckling and frequency constraints, then are the determining factors in the optimization. A more realistic approach would be to use weight as one criteria but to also include keeping stresses low, maintaining the natural frequency within a specified band and staying far removed from a buckling condition. This formulation is closely related to design for latitude, that is keeping the design insensitive to changes in factors that cannot be controlled. If the final design is not at the intersection of a set of constraints, it will be less affected by a change in loading conditions. This approach produces what might be termed as more robust design. Multi-criteria design optimization methods exist and some alternatives also exist in the opertions research arena such as goal programming.

Decision support as for the seat frame in removing elements is a difficult area to address. Expert systems and artificial intelligence will begin to play a role for specific applications but for general design, optimization can also play a role. If decision making is thought of as selecting design alternatives at various levels in the design, then it can be represented in a tree structure. By traversing the tree with each decision treated as a zero-one integer value, the design decisions may be made in a convenient fashion. Coupled with a multi-criteria optimization capability, the ability to perform design is greatly enhanced.

The final area of research considered is that of developing new approaches to perform analysis and optimization that not only work well together but can take advantage of the new computer architectures available. While computational speed has increased dramatically over the past twenty years, the pace has slowed considerably now that the roadblock is not manufacturing capability but the physical limitation of the speed of light. Parallel and distributed processing appear to be the future for increasing performance and the software must be adapted to this hardware to take advantage of the speed. This will be a slow process but one area of investigation which may yield immediate results is substructuring. Substructuring has been available in many forms for quite some time but as yet, it has not been implemented where it may well have its greatest impact for parallel processing. The effective use of parallelism in both the analysis and optimization will allow the solution of larger and more complete models. It will also allow for consideration of latitude, multiple design criteria and a more global view of shape optimization.

3. SUBSTRUCTURING FOR SHAPE OPTIMIZATION

The boundary element method is ideally suited for structural shape optimization. Shape optimization in this context refers to the determination of the optimal shape of the boundaries or surfaces of continuous structures and structural components. The boundary element method provides a convenient means of computing accurate stress and deflection information for a component while dealing only with surface information. A surface representation for the structure can be represented by a general shape function and the coefficients of the shape function become design variables in the optimization process. With the boundary element method, the changing surface geometry requires only a surface remeshing rather than a full three dimensional remeshing as would be required with the finite element method. This allows for greater confidence that the search direction for altering the geometry during the optimization is based on the potential driving the shape change rather than the change in accuracy due to remeshing. The addition of a substructuring capability allows the shape change to be isolated to a region of the component which increases the computational efficiency of the optimization and reduces the need to make simplifying assumptions in the problem formulation. This is important for the general three dimensional shape optimization problem is several orders of magnitude more complex than the crossectional optimization of beam and plate elements generally performed.

Substructuring in the boundary element method allows for separate analyses of each substructure with all generalized displacements on common boundaries completely constrained. This is followed by the relaxation of these boundaries to insure equilibrium and the calculation of the interfaced displacements. Naturally, the system equations for the substructuring analysis involve a considerably smaller number of unknowns compared with that of the complete structure without partitioning.

While substructuring has been applied as a natural means of implementing the boundary integral method, it has great potential for isolating the region in which the shape is to be optimized. This technique has been applied to the design of a ladle hook originally proposed by Dieter[1]. A B-spline representation was used for the boundary. Two substructures were used; one with 47 boundary elements and one with 46 elements. The B-spline had 9 control points which resulted in a total of 18 design variables. The generalized reduced gradient code OPT[2] was applied to solve the problem. The substructuring reduced the computational time by a factor of two thirds. The resulting design reduced the volume of the hook considerably without violating the stress constraint. Additional details may be found in Reference [3].

4. DECISION SUPPORT

By applying a zero-one capability to the optimization algorithm, the design tool is enhanced considerably. The zero-one capability can be implemented through the branch and bound method[4], in a nonlinear programming algorithm. If several design alternatives are to be evaluated, one choice would be to optimize each choice independently. As the number of choices increases, particularly in a multilevel design task, the total number of choices involved becomes unmanageable. An alternative approach is to assign each decision a variable which must take on the value of 0 or 1. By requiring the sum of all variables representing a particular choice be one, a decision support formulation has been achieved. Examples involving the optimal design of a load carrying beam and for a three bar planar truss, have been successfully solved[5]. The load carrying member was optimized under a selection of materials and attachment possibilities. The three bar truss was optimized under varying crossectional forms under stress, displacement buckling and frequency constraints. In both cases, the computational time for the zero-one formulation was considerably less than that required to optimize all of the alternatives. Heuristic procedures may be used to prune the decision tree to reduce the time even further.

5. COUPLING OPTIMIZATION AND ANALYSIS

A multitude of options are available to perform the task of analysis. Finite element, boundary integral, finite difference and spectral methods are just a few of the choices. By forming the analysis problem through spectral methods with the optimization variables included with the undetermined coefficients of the approximation function, a structural optimization tool can be created. This method allows the optimization to be carried out simultaneously with the analysis. The analysis is only approximate in the early stages, but it becomes more accurate as the process continues. This is ideal as the most accurate analysis information is available as the solution is approached. A general form (i.e., polynomial or Fourier series) is selected to represent the field variable. The undetermined coefficients are handled with the optimization along with the design variables specifying the shape of the boundary. A multi-criteria objective is used to minimize the error over the domain in the partial differential equation defining the field variable and to minimize the design objective (i.e., weight). Boundary conditions may be satisfied by selecting the appropriate form of the equation for the field variable or by treating them as constraints in the nonlinear programming problem.

An example involving the shape optimal design of a flywheel was solved[6] using this approach. The method performed quite well although it was much more reliable when the boundary conditions were met with the approximating function rather than through the optimization problem as constraints. Since an equation is available for the field variable, the error or residual term may be used to determine if subdivision of the domain is required. The subdivision, if done in an appropriate fashion, may allow for parallel processing.

6. SUMMARY AND CONCLUSIONS

The examples presented show the discrepancy of what we can do now versus what we would like to be able to do in the area of shape optimization. Performing each individual task is easier than performing all tasks simultaneously. Integration of the various design features is the next step. In order to implement the desired elements for a useful design tool, some dramatic changes must be made in the way we formulate and solve problems. This is particularly true if we are to take advantage of new computing architectures. The time is here to take a fresh look at what we are doing. The move from analysis to design will not happen overnight, but it must happen eventually.

REFERENCES

1. Dieter, G.E., Engineering Design: A Materials and Processing Approach, McGraw-Hill, 1983.
2. Gabriele, G.A., and Ragsdell, K.M., OPT: A Nonlinear Programming Code in Fortran 77, Users Manual, Purdue Research Foundation, 1976.
3. Sandgren, E., and Wu, S.J., 'Shape Optimization Using the Boundary Element Method with Substructuring', to appear, International Journal of Numerical Methods in Engineering.
4. Gupta, O.K., and Ravindran, A., 'Nonlinear Integer Programming and Discrete Optimization', Progress in Engineering Optimization, ed R.W. Mayne and K.M. Ragsdell, ASME, 1981.
5. Chang, J.F., 'An Application of Integer/Discrete/0-1 Programming for Nonlinear, Conceptual Design Optimization', M.S. Thesis, University of Missouri-Columbia, May, 1987.
6. Mohamud, O.M., 'Coupling of Analysis and Optimization for Engineering Design Problems', PhD Dissertaion, University of Missouri-Columbia, May, 1987.

OPTIMAL SHAPE OF PENDULUM LINKS

Werner O. Schiehlen
Institute B of Mechanics
University of Stuttgart
7000 Stuttgart 80
Federal Republic of Germany

ABSTRACT. Pendulums with a few links are found in many technical applications like robot arm assemblies and torsional vibration absorbers. Such pendulum devices are characterized by instationary motions ranging from the equilibrium position under gravity forces via irregular chaotic motions to steady-state rotations under centrifugal forces. The method of multibody systems is well qualified for the analysis of such motions as well as for the estimation of corresponding stresses. An optimization of the shape will result in a more homogeneous stress distribution and lower weight. A method for the shape optimization of instationary moving pendulum links is proposed and some results will be shown in detail.

1. Introduction

Advanced engineering problems include not only the analysis but also the synthesis of mechanical systems. In particular, the motion and the strength of mechanical structures is expected to be optimal. Usually, optimal motions are found by parameter variation of discrete elements like springs, dampers, servomechanisms. Moreover, optimal control theory is often applied to improve the dynamical behavior of mechanical systems, see e.g. Haug [1]. The optimal strength of mechanical systems is mainly obtained under static load or small structural vibrations, respectively. Due to the limited stress in material bodies, the optimal strength results in an optimal shape of the parts of the system, see e.g. Eschenauer [2], Karihaloo [3], Rožvany [4], Stadler [5]. An optimal strength of mechanical systems featuring large nonlinear working motions requires a combined consideration of motions and stresses. The variation of the shape of a part results in changes of inertia and stress. In this paper the multibody system approach is applied to the optimal design of pendulum links. The more general treatment is illustrated by a simple example. It is shown that the optimal design depends not only on static gravity but also on the initial conditions of the large dynamic oscillation.

G. I. N. Rozvany and B. L. Karihaloo (eds.), Structural Optimization, 279–288.

2. Modeling of Multibody Systems

Engineering systems featuring large displacement nonlinear motions as mechanisms, vehicles, robots and spacecrafts can be modeled as multibody systems. Such systems are characterized by rigid bodies with inertia as well as springs, dampers and actively controlled servomotors without inertia. The bodies are interconnected by rigid joints or any other kind of supports, Fig. 1, and they are subject to additional applied forces and torques. The method of multibody systems is well qualified for the modeling of mechanical systems with complex geometry resulting in motions with frequencies less than 50 Hz. Structural vibrations of engineering systems may be included in the model by a sufficiently large number of rigid bodies but the structural vibrations of the bodies themselves are neglected.

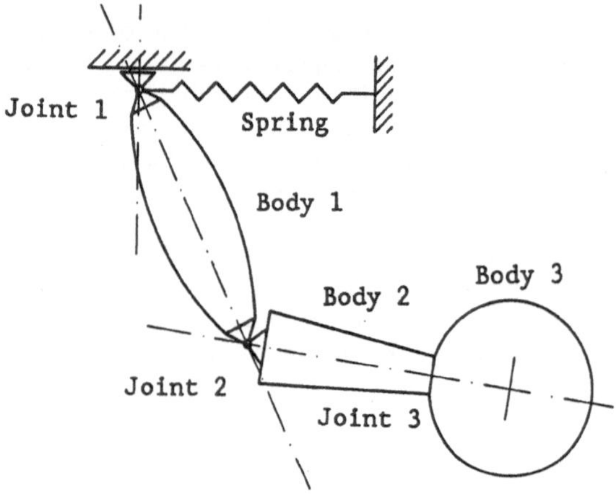

Figure 1. Multibody system

The method of multibody systems has been very well developed during the last two decades by researchers all over the world. The state-of-the-art is represented by proceedings edited by Magnus [6], Haug [1] and in Ref. [7]. General multibody systems include holonomic and nonholonomic constraints as well as proportional, differential and integral applied forces. With respect to the optimal design problem only ordinary multibody systems restricted to holonomic constraints and differential-proportional forces will be treated. Moreover, the multibody systems will be considered ideal, i.e. friction phenomena are neglected.

A holonomic system of p bodies and q holonomic, scleronomic constraints due to rigid joints results in $f = 6p-q$ degrees of freedom. A spherical joint, e.g., is represented by $q=3$ constraints.

The configuration of a holonomic multibody system is given by the 3x1-translational vector $r_i(t)$ of the center of mass and the 3x3-

rotational tensor $S_i(t)$, i=1(1)p, as shown in detail in Ref. [8]. Introducing the fx1-position vector y(t) of the generalized coordinates, the constraints of the system read explicitly as

$$r_i = r_i(y),\ S_i = S_i(y),\ i = 1(1)p. \tag{1}$$

The translational and rotational velocities are found as

$$v_i = J_{Ti}(y)\ \dot{y}(t),\ \omega_i = J_{Ri}(y)\ \dot{y}(t),\ i = 1(1)p, \tag{2}$$

where J_{Ti} and J_{Ri} are the 3xf-Jacobian matrices of the system configuration.

Further, the translational and rotational accelerations are given by

$$\begin{aligned} a_i &= J_{Ti}(y)\ \ddot{y}(t) + \bar{a}_i(y,\dot{y}), \\ \alpha_i &= J_{Ri}(y)\ \ddot{y}(t) + \bar{\alpha}_i(y,\dot{y}), \end{aligned} \tag{3}$$

where $\bar{a}_i$ and $\bar{\alpha}_i$ are local accelerations including coriolis phenomena.

Newton's and Euler's equations read for each body in the inertial frame as

$$m_i\ a_i = f_i^e + f_i^r\ ,\ i = 1(1)p, \tag{4}$$

$$I_i\ \alpha_i + \tilde{\omega}_i\ I_i\ \omega_i = l_i^e + l_i^r\ . \tag{5}$$

The inertia of the system is represented by the scalar mass m_i and the 3x3-inertia tensor I_i of each body. The applied forces f_i^e and l_i^e are due to gravity as well as springs, dampers and serromotors.

The constraint forces are first of all due to the joints and other kinds of supports. According to the holonomic, scleronomic constraints the qx1-vector $g_o(t)$ of the generalized reaction forces can be introduced. Then, it yields

$$f_i^r = F_i(y)\ g_o(t),\ l_i^r = L_i(y)\ g_o(t), \tag{6}$$

where F_i and L_i are the 3xq-distribution matrices.

Using matrix notation the Newton-Euler equations are obtained from (2) to (6) as

$$\bar{\bar{M}}\, \bar{J}\, \ddot{y}(t) + \bar{q}^c = \bar{q}^e + \bar{Q}\, g_o(t), \tag{7}$$

where the global matrices $\bar{\bar{M}}$, $\bar{J}$ and $\bar{Q}$ as well as the global vectors $\bar{q}^c$, $\bar{q}^e$ are used representing inertia, kinematical and dynamical constraint phenomena as well as coriolis and applied forces and torques. Eqs.(7) have dimension 6p with f generalized coordinates and q generalized constraint forces as unknowns.

However, each rigid body of the system can be partitioned, see Fig. 2. The cross-section cutting the body in two parts, the left and the right body, is characterized by a material coordinate u_1. The partitioned system has p+1 bodies and q+6 constraints but an unchanged number of degrees of freedom, f = 6p-q.

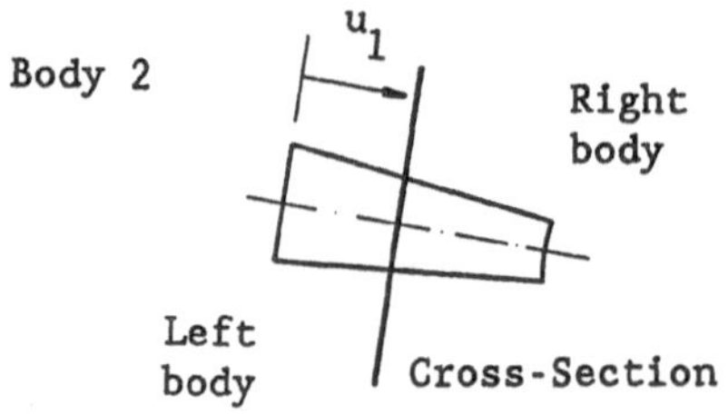

Figure 2. Partitioning of rigid body

This results in extended Newton-Euler equations of dimension 6p+6 as

$$\bar{\bar{M}}(u_1)\, \bar{J}(u_1)\, \ddot{y}(t) + \bar{q}^c(u_1) = \bar{q}^e(u_1) + \bar{Q}(u_1)\, g_p(t). \tag{8}$$

In particular, the (q+6)x1-vector of reactions has to be used

$$g_p = [g_o^T(t) \quad f^{rT}(u_1,t) \quad l^{rT}(u_1,t)]^T, \tag{9}$$

where f^r, l^r represent the 3x1-vectors of the reactions in the cross-section of the partitioned body.

3. Equations of Motion and Reaction

The Newton-Euler equations (7) or (8), respectively, are algebraical-differential equations. The numerical solution of such equations is cumbersome since nonstandard algorithms have to be applied. Using the result of analytical mechanics, the Newton-Euler equations can be separated in purely differential equations of motion and purely algebraical equations of reaction.

The equations of motion may be obtained by Lagrange's equations of the second kind. But this involves a lot of computational work and information on the reactions is not available. Therefore, the most economic way is the application of D'Alembert's principle. It states that the virtual work of the reactions is vanishing for the global system or, more precisely in matrix notation, $\bar{J}^T \bar{Q} = 0$. This orthogonality condition will be applied to the Newton-Euler equations. Premultiplication of (7) or (8), respectively, by the transpose global Jacobian matrix $\bar{J}^T$ results in the equations of motion

$$M \ddot{y}(t) + k = q. \tag{10}$$

The fxf-inertia matrix reads as

$$M = \bar{J}^T \bar{\bar{M}} \bar{J} = \bar{J}^T(u_1) \bar{\bar{M}}(u_1) \bar{J}(u_1) \tag{11}$$

and the fx1-vectors k and q represent the generalized coriolis and applied forces.

On the other hand, premultiplication of (8) by the transpose global distribution matrix $\bar{Q}^T$ and the inverse inertia matrix $\bar{\bar{M}}^{-1}$ leads to the equations of reaction

$$N_p g_p(t) + \hat{q}_p = \hat{k}_p. \tag{12}$$

The (q+6)x(q+6)-reaction matrix of the partitioned system is given as

$$N_p = \bar{Q}^T(u_1) \bar{\bar{M}}^{-1}(u_1) \bar{Q}(u_1) \tag{13}$$

and $\hat{q}_p$, $\hat{k}_p$ are (q+6)x1-vectors of applied and coriolis forces affecting the reactions.

For the computer aided derivation of the equations of motion and the equations of reaction the formalism NEWEUL [9] can be used. NEWEUL generates symbolical and/or numerical equations of motion and numerical equations of reaction.

The equations of motion (10) can be solved by any standard integration code. For the equations of reaction (12) each code using the Cholesky decomposition can be applied. Further, efficient numerical methods for the computation of generalized reaction forces have been developed by Schramm [10].

4. Estimation of Stresses

For the optimal design of the shape of a rigid body, the stresses in the body are most important. However, a rigid body is a statically undetermined system and the stresses cannot be evaluated at all. An elastic body, at the other hand, requires a finite element model with a huge number of degrees of freedom and, correspondingly, huge and often stiff equations of motion. As a compromise, a strength estimation for rigid bodies will be used as presented in Ref. [11].

Under the assumption of a continuous distribution of inertia and applied forces and a linear stress distribution within the rigid body, in each material point of the cross-section the following normal and shear stresses are obtained

$$\sigma_{11}(t) = \frac{f_1^r}{A} + \frac{l_2^r}{J_2} u_3 - \frac{l_3^r}{J_3} u_2,$$

$$\tau_{12}(t) = \frac{f_2^r}{A} - \frac{l_1^r}{J_p} u_3, \quad \tau_{13}(t) = \frac{f_3^r}{A} - \frac{l_1^r}{J_p} u_2. \tag{14}$$

The material point is characterized by the coordinates u_1, u_2, u_3, A is the area and J_2, J_3, J_p are the corresponding second moments of the cross-section considered.

Using the maximum distortion-energy hypothesis, for example, a scalar characteristic stress is obtained

$$\sigma_c(u_1, u_2, u_3, t) = \sqrt{\sigma_1^2 + 3(\tau_{12}^2 + \tau_{13}^2)}. \tag{15}$$

As a matter of principle, any scalar characteristic stress depends on space and on time. The time-variability is the main difficulty in dynamic problems compared to static ones.

5. Optimal Design of Rigid Bodies

The shape of a rigid body exercises influence on its inertia and on its strength. However, an increasing area of the cross-section results in increasing inertia forces as well as increasing strength. Therefore, there will not be available a straightforward strategy for the evaluation of an optimal shape. Moreover, the shape of a body in any machine or structure is time-invariant during operation. The design has to be based with respect to the worst case with respect to the time history of motion.

The following steps may be used during the optimal design process of a multibody system.

i) Modeling of the multibody system with unknown design variables.

The rigid body representing the structure will be substituted by a finite or infinite number of small or infinitesimal rigid sections, respectively. The finite sections may be chosen prismatical or pyramidal, Fig. 3.

The cutting procedure introduced by (8) has to be repeated, the design variables are a finite number of cross-sections A_j. In the case of infinitesimal sections design functions have to be used.

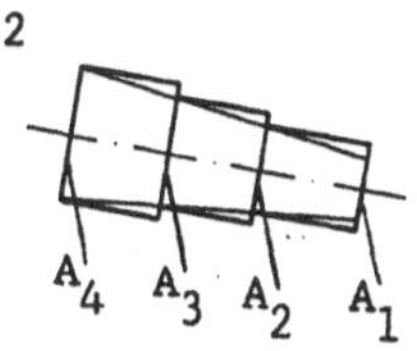

Figure 3. Rigid body substituted by prismatical sections

ii) Solution of equations of motion with some initial design variables.

Symbolical equations of motion (10) indicate directly the influence of the design variables. The generation of symbolical equations is necessary only once. Initial design variables A_j have to be chosen for the numerical solution as well as the initial conditions.

iii) Computation of reaction forces of the motion.

The numerical equations of reaction (12) are the basis for the computation. Efficient methods are due to Schramm [10].

iv) Computation of the maximum stress for some time interval.

The maximum of the characteristic stress (15) has to be determined for some time interval T and all material points for each cross-section A_j. It follows

$$\sigma_{\max j} = \max_{u_1,u_2,u_3} \max_{t\epsilon T} \sigma_c(u_1,u_2,u_3,t). \tag{16}$$

v) Computation of the error with respect to the characteristic stress of material of the structure,

$$e(A_j) = \sqrt{\sum_j (\sigma_{\max j} - \sigma_{\text{material}})^2}. \tag{17}$$

vi) Minimization of the error by variation of the design variables. Evaluation of the optimal design variables.

The mentioned steps result in an optimal design by iteration. In general, conditions of convergence are not available. For economic reasons the number of design variables should be low.

6. Application to Pendulum Link

A very simple example will be treated. Then, the optimization of the shape is achieved completely analytical and the influence of the dynamical phenomena can be clearly observed.

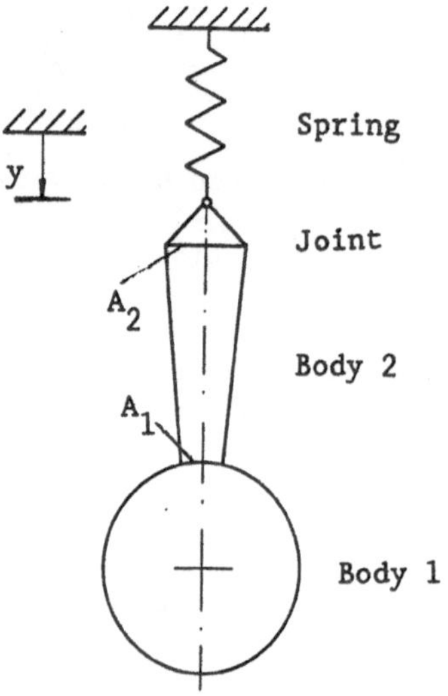

Figure 4. Simple pendulum link

The pendulum, Fig. 4, is given by two bodies and one spring. Body 2 is the structure to be designed. The motion is considered one-dimensional in vertical direction. The system has f=1 degree of freedom and q=1 constraint. The joint and body 2 are separated resulting in another constraint. Then, the extended Newton-Euler equations read as

$$\begin{bmatrix} m_1 & 0 & 0 \\ 0 & m_2 & 0 \\ 0 & 0 & m_3 \end{bmatrix} \begin{bmatrix} 1 \\ 1 \\ 1 \end{bmatrix} \ddot{y}(t) = \begin{bmatrix} m_1 g \\ m_2 g \\ -cy \end{bmatrix} + \begin{bmatrix} -1 & 0 \\ 1 & -1 \\ 0 & 1 \end{bmatrix} \begin{bmatrix} f_1(t) \\ f_2(t) \end{bmatrix} \quad (18)$$

where y is the generalized coordinate, m_1 is the mass of body 1,

$$m_2 = \frac{\rho L}{3} (A_1 + \sqrt{A_1 A_2} + A_2) \quad (19)$$

the mass of body 2, and $m_3 \to 0$. Further, c means the spring coefficient and f_1, f_2 are the generalized reaction forces. The design variables are the areas A_1, A_2 of the cross-sections.

The equations of motion follow as

$$(m_1+m_2)\ddot{y}(t) = -cy(t) + (m_1+m_2)g \tag{20}$$

and for the equations of reaction it remains

$$\begin{bmatrix} m_1+m_2 & -m_1 \\ -m_1 & \frac{m_1 m_2}{m_3} \end{bmatrix} \begin{bmatrix} f_1(t) \\ f_2(t) \end{bmatrix} - \begin{bmatrix} 0 \\ \frac{m_1 m_2}{m_3} cy(t) \end{bmatrix} = 0. \tag{21}$$

The solutions can be given explicitly as

$$y = y_0 \cos \omega t + \frac{1}{\omega^2} g, \quad y_0 > 0, \ \dot{y}_0 = 0, \tag{22}$$

$$f_1 = \frac{m_1}{m_1+m_2} cy, \ f_2 = cy \tag{23}$$

where $\omega = c/(m_1+m_2)$ is the frequency.

For the periodic solution (22) the characteristic normal stresses according to (14) are easily formed as

$$\sigma_{max1} = \frac{1}{A_1} \frac{m_1}{m_1+m_2} [cy_0 + (m_1+m_2)g], \tag{24}$$

$$\sigma_{max2} = \frac{1}{A_2} [cy_0 + (m_1+m_2)g]. \tag{25}$$

The error (17) is completely vanishing for $\sigma_{max1} = \sigma_{max2} = \sigma_m$.

For the evaluation of the optimal design variables, (19) has to be considered. Under the assumption of a constant frequency it yields $c = (m_1+m_2)\omega^2$. Then it follows from (24), (25)

$$A_1 = \frac{m_1}{\sigma_m} (\omega^2 y_0 + g), \quad A_2/A_1 = \frac{2 - \mu^2 + \mu\sqrt{4 - 3\mu^2}}{2(1-\mu)^2} > 0 \tag{26}$$

where $\mu = \rho L A_1 / 3m_1 \ll 1$.

Obviously, in addition to the static case $y_0 = 0$, there is a strong influence, in the dynamic case $y_0 > 0$ even for this very simple example.

References

[1] Haug, E.J.: Computer Aided Analysis and Optimization of Mechanical System Dynamics. Springer, Berlin/..., 1984.

[2] Eschenauer, H.: Rechnerische und experimentelle Untersuchungen zur Strukturoptimierung von Bauweisen. DFG-Forschungsbericht. Universität Siegen, Siegen, 1985.

[3] Karihaloo, B.L.: 'On Minimax Optimum Design of Flexural Members in Presence of Self-Weight.' Mechanics of Structures and Machines 15 (1987) pp. 17 - 28.

[4] Rozvany, G.I.N.; Ong Tiong Guan: 'General Theory of Optimal Layouts for Elastic Structures.' J. Eng. Mech. 112 (1986) pp. 851 - 857.

[5] Stadler, W.: 'Natural structural shapes.' Quart. J. Mech. Appl. Math. 31 (1978) pp. 169 - 217.

[6] Magnus, K.: Dynamics of Multibody Systems. Springer, Berlin/..., 1978.

[7] Bianchi, G.; Schiehlen, W.: Dynamics of Multibody Systems. Springer, Berlin/..., 1986.

[8] Schiehlen, W.: Technische Dynamik. B.G.Teubner, Stuttgart, 1985.

[9] Kreuzer, E.; Schmoll, K.-P.: Programmsystem NEWEUL, Manual AN-18. Institute B of Mechanics, Stuttgart, 1986.

[10] Schramm, D.: Ein Beitrag zur Dynamik reibungsbehafteter Mehrkörpersysteme. VDI-Verlag,Düsseldorf, 1986.

[11] Schiehlen, W.; Kreuzer, E.: 'Strength Estimations in Multibody Systems.' In [7], pp 248 - 259.

AN INTEGRATED KNOWLEDGE-BASED PROBLEM SOLVING SYSTEM FOR STRUCTURAL OPTIMIZATION

K. SCHITTKOWSKI
Mathematisches Institut
Universität Bayreuth
D - 8580 Bayreuth, Germany F.R.

ABSTRACT: LAGRANGE is an interactive integrated programming system supporting the whole life cycle of the solution process of a mechanical structural optimization problem. Providing the geometry of the structure by means of a finite element formulation, the system guides a user to define the optimization problem, e.g. the type of constraints and their numerical data, the variable linking of the design variables, the selection of a suitable optimization routine, the generation of a formatted input file, the automatic execution of the corresponding main program and the evaluation of the results. The system is self-learning and uses rules for proposing a suitable optimization algorithm or for proposing remedies in an error situation.

1. The Structural Optimization Model

The objective function to be minimized, is the weight of the mechanical structure. It is planned to allow also the formulation of other objectives in future, e.g. the reliability of a structure. Design variables can be grouped in the following way:

- cross sectional areas for trusses and beams
- wall thicknesses for menbrane and plate elements
- laminate thicknesses for single layers in composite elements
- angles of layers for composite elements
- nodal coordinates (geometry)
- balance masses

While sizing problems can be solved in the traditional way, special methods were developed for the variation of fibre directions and grid coordinates. Mass balancing is of particular importance for flutter optimization.

A variety of different constraint types is available to

G. I. N. Rozvany and B. L. Karihaloo (eds.), Structural Optimization, 289–297.

allow the modelling of real structures. By using gauge constraints, manufacturing conditions can be satisfied. The suitable combination of restrictions defines the physical model of the structure to be optimized. The following constraints can be treated:

- deformation of nodal points
- strains
- stresses
- buckling
- aeroelastic efficiencies
- flutter speed
- divergence philosophy
- eigenfrequencies
- dynamic responses
- bounds for design variables (gauges)

A couple of large scale structures were optimized by the LAGRANGE system, e.g. the composite vertical stabilizer of the Airbus A310, cf. Kneppe, Krammer and Winkler (1987). In this case, the finite element model consists of 8436 degrees of freedom, 6414 elements and four loading cases. 383 design variables are defined to represent the thickness of the upper and lower skin and, in addition, the thicknesses of stringers and flanges of the inner structure. The resulting optimization problem was solved by a sequential linear programming method and a reasonable answer was found in seven iterations reducing the initial weight to .72 units.

2. System Functions

The main menue of LAGRANGE allows to activate any of the subsequently explained system functions by typing the corresponding initial. By further submenues and questions, the user is guided through the system, so that the learning of a 'language' is not required. Whenever he is not aware of the set of allowed answers, he may type a '?' for getting more detailed information.

a) New design

The command is used to generate a new optimization problem and to include it in the actual data base. First some information on the design to be optimized, must be typed, e.g. the name which is used to identify the design in the data base, some information for documenting the practical situation, and the name of the data file containing the geometrical problem description in a suitable format. After some consistency checks, a window displayes the options that can be chosen to specify certain system facilities, e.g. minimization of bandwidth or output size. A submenue is then displayed to specify the constraint type. The corresponding numerical data which are required to define them numerically e.g. by upper and lower bounds, are inserted in form of a sequence of windows. Whenever possible, the given NASTRAN bulk data deck is investigated by the system e.g. to search for certain cards and element numbers. In addition the design variables must be specified by defining the corresponding element and layer numbers either individually or in form of ranges. Again windows are displayed on the screen in tabular form to facilitate the input as much as possible.

b) Display of design data

All design data and numerical results are kept in an internal data base. On request they can be displayed on the screen individually. Moreover lists of all available problems together with some characteristic data, e.g. number of elements and design variables, and of the optimization history of a special design problem can be shown in form of tables. In the latter case, a user will be informed on all performed solution attempts by indicating the chosen method, the used tolerances and parameters, the initial state, and the most important performance data, i.e. number of function and gradient evaluations and computing time. By offering the possibility to get information on the whole solution process very conveniently, it is hoped that a user will update his own experience very fast, which mathematical algorithm might be preferable in a given situation.

c) Output of the result file

When solving a structural optimization problem interactively, only very limited information is displayed on the screen to inform the user on the current state of the optimization process. Much more detailed information is stored on an output file, which can be investigated on the screen or sent to the line printer on request.

d) Edit design data

In principle all data which are inserted by a user to define a structural optimization problem, may be edited on request, also the data defining the finite element structure. For editing constraints or the variable linking of design variables, the user can select either the operating system editor which is executed automatically, or the same windowing technique which was used before to define the problem data initially.

e) Start optimization run

LAGRANGE requires the generation of a specially formatted input file that contains some control information, the NASTRAN card deck, the definition of constraints and their corresponding data, the variable linking information, and some optimization parameters depending on the chosen solution method. By initiating the start command, this input file is generated automatically either for performing only an analysis at the initial design or to start an optimization run. Since many different optimization routines are available, a user must either define them 'by hand' or he requires a selection made by the system. In this case a rule-based subprocess will send some questions to the terminal and depending on the answers and the information on the design available so far, a heuristic proposal is made. A user may accept the proposed method and parameters or he may choose another code.

If some results are available obtained from a previous run with the same algorithm, it is possible to perform a warm start, i.e. continuation of the iteration which was interrupted before by exceeding the maximum number of iterations. Otherwise a cold start may be activated starting from the last computed iterate or alternatively, a new optimization cycle is initiated starting from the originally given design variables. Then a user determines the computer which is to be used to perform the numerical optimization. If the computer is not changed, the FORTRAN main program is generated automatically with precise dimensioning parameters, compiled, linked only with those modules needed to solve the problem, and executed immediately.

Alternatively the user may prefer the execution in form of a batch job. In this case, solution data are stored on a system file and are sent to the data base when starting LAGRANGE again. Otherwise some control information is displayed on the terminal in each iteration, e.g. objective function, maximum constraint violation, and after finishing the optimization cycle, the main menue can be used then for initiating other system functions, e.g. the output of the achieved solution data which are available now in the data base.

f) Failure analysis

LAGRANGE possesses a very flexible failure system and it is out of the scope of this report to explain all of its features. Severe failures interrupting the optimization, are written to the output file mentioned before, and are sent to the terminal. By activating the failure analysis, a user will see the same failure information again. Subsequently a rule-based, heuristic proposal of a suitable remedy is displayed and the user may accept the proposed action or not.

g) Delete a design

Any structural optimization problem generated by a user, is kept in a data base and can be deleted on request. Since all numerical data defining the geometry, the constraints and the variable linking, are stored internally on files, a user will be asked whether he prefers to delete these files as well.

h) Copy data from one optimization problem to another

In many cases it might be desirable to generate a few different versions of one and the same structural optimization problem or to generate very similar design problems. In this case it is possible to copy data e.g. for defining a special constraint type from one problem to another. If the second problem is not available in the data base, a new design problem is declared automatically where all design data are taken from the first one. They can be edited or modified then in any way.

i) User documentation

The user documentation of the LAGRANGE system is part of the system itself. By choosing a suitable subcommand, a user may require the display of any section of the user documentation on the screen. Alternatively the whole documentation can be printed on request on the available line printer.

j) Sort structural optimization problems

The principle key for identifying a design problem in the data base, is its name defined initially by the user. However the sequence of all available problems in a data base is not sorted at all. Thus a user might wish to sort them according to some information, e.g. name, date, user identification, to get a more readable list when starting the output command. The sorting key selected by a user, does not influence the internal efficiency of the interface.

k) Direct input of operating system commands

When starting the LAGRANGE system, a couple of organisatorial procedures must be initiated, e.g. the internal linking of the data structure, the inclusion of known numerical data, and the preparation of the program to be interpreted. To avoid the necessary waiting time when a few operating system commands are to be executed within a LAGRANGE session, it is allowed to insert them directly. Any operating system command may be typed then and the command is evaluated in form of a separate subprocess.

l) Utilities for the system manager

Only the system manager has additional access rights to perform some actions invisible for all other users of the system. He is allowed to include or edit the mail displayed when starting LAGRANGE, to get information on all user accesses, e.g. calculation times spent, and, in particular, to modify directly the underlying data base. In the latter case, he may use commands of the implementation language SUSY to perform actions that can not be achieved by the predefined system functions. The documentation of the SUSY-language or the data base structure can be displayed on request.

m) Halt

The halt function will save all data again on the data base file that was declared when starting LAGRANGE, and return the control to the operating system.

3. Choice of an Optimization Method

The implementation of the LAGRANGE system is based on the idea to provide as many different optimization methods as available, and to have the possibility to exchange, modify, or extend them whenever it seems to be necessary based on the increasing practical experience. Therefore the whole architecture of the system is highly modular and it is particularly easy to add a new optimization code. The reason for attempting to achieve a fairly large algorithm base is found in the observation that none of the codes tested so far in the frame of the LAGRANGE system, is superior to all other ones with respect to efficiency or reliability. In addition, some codes are applicable only under certain conditions, e.g. feasible starting point or limited number of constraints and design variables.

The forecast, however, which of the available algorithms might be the most suitable one in a given real situation, is very difficult and can be made at most by experts, i.e. those engineers who did work with the system for a long time. Nevertheless they may fail as well and the whole decision process is based on heuristics and some very vague criteria. But since any false decision is, in most cases, cost expensive and delays the design process, there is a strong need to collect the experience of the experts and to implement it in a suitable way, so that their knowledge is available also for 'newcomers' who use the system the first time.

In the present version of LAGRANGE, the following mathematical optimization algorithms are included and may be selected by a user:

IBF Inverse barrier algorithm, i.e. a penalty type method forming a sequence of unconstrained nonlinear minimization problems. All iterates are feasible.

MOM Method of multipliers, i.e. solution of a sequence of unconstrained nonlinear optimization problems minimizing an augmented Lagrangian function.

SLP Sequential linear programming algorithm, i.e. solution of a sequence of linear programming problems obtained by linearizing the objective and constraint functions. The algorithm is stabilized by trust regions, certain bounds on the choice of a search direction that may change dynamically, cf. Kneppe (1986).

SQP1 Sequential quadratic programming algorithm, i.e. solution of a sequence of quadratic programming problems obtained by a quadratic approximation of the Lagrange function and a linear one of the constraints. The algorithm is stabilized by a line search with respect ot an augmented Lagrangian merit function, cf. Schittkowski (1985/86).

SQP2 Sequential quadratic programming algorithm, i.e. solution of a sequence of quadratic programming problems obtained by a quadratic approximation of the Lagrange function and a linear one of the constraints. The algorithm is stabilized by a line search with respect ot an exact penalty function and a 'watchdog' technique, cf. Powell (1982).

GRG Generalized reduced gradient method, i.e. elimination of dependent variables and projection of all iterates onto the feasible domain, cf. Bremicker (1986).

SRM Stress ratio method, i.e. heuristic approach based on optimality criteria and special constraints, e.g. stresses.

Moreover each algorithm requires the definition of certain tolerances and parameters that influence the solution process. There are practical situations, in which any of the mentioned codes might be superior to the other ones. The underlying model structure is quite special, and favours the usage of the one or other code in a certain case. To give an example, consider the squential linear programming method. This code should be used e.g. whenever

(i) the objective function is linear (the only choice in the present version),
(ii) many active constraints are expected to be active at the optimal solution,
(iii) stress constraints dominate.

But the last two items are not known by sure in general, so that the conclusion, to use the SLP-method in this case, depends on the degree of belief whether they are valid or not. Since the implementation of knowledge associated with uncertainties, is hard to achieve in any conventional programming language, it was decided to implement it in form of rules and to use the inference mechanism of the SUSY-language, cf. Schittkowski (1987). Thus one is capable to formalize the known heuristics based on the numerical experience of all previous optimization runs, in form of rules.

The values of the certainty factors range between 0 an 100. When initiating the rule evaluation, the certainty factor obtains a value which depends on the underlying selflearning process of the system and previous solution attempts of the design problem under consideration with the same method. Then this factor is updated according to those rules which premises are satisfied in the actual situation. They are either simple arithmetic expressions depending on available data or depend on other actions, that describe a situation and that are obtained from answers of the user. The resulting certainty values obtained for all included optimization codes, are displayed to facilitate the decision of a user. He is of course allowed to reject the method that got the largest certainty value, and to prefer another one. For more details on the internal inference mechanism, the reader is referred to the SUSY-documentation, see Schittkowski (1987) or to Schittkowski and Zotemantel (1987).

References:

Bremicker M. (1986):
Entwicklung eines Optimierungsalgorithmus der generalisiert reduzierten Gradienten, Anwendung auf Beispiele der Struktu dynamik
Bericht, IMR, Universität Siegen

Kneppe G., Krammer J., Winkler E. (1987):
Structural Optimization of Large Scale Problems Using MBB-LAGRANGE
Report MBB-S-PUB-305, Messerschmidt-Bölkow-Blohm, Munich, Germany F.R.

Kneppe G. (1986):
Direkte Lösungsstrategien zur Gestaltsoptimierung von Flächentragwerken
VDI-Verlag

Powell M.J.D. (1982):
VMCWD: A FORTRAN subroutine for constrained optimization
Report DAMTP 1983/NA4, University of Cambridge, Cambridge, U.K.

Schittkowski K. (1985/86):
NLPQL: A FORTRAN subroutine solving constrained nonlinear§5 programming problems
Annals of Operations Research, Vol. 5, 485-500

Schittkowski K., Zotemantel W. (1987):
LAGRANGE: A Knowledge Based Integrated Structural Optimization System
Bericht, Mathematisches Institut, Universität Bayreuth, Bayreuth, Germany F.R.

Schittkowski K. (1987):
Die Systementwicklungssprache SUSY
Bericht, Mathematisches Institut, Universität Bayreuth, Bayreuth, Germany F.R.

A METHOD OF FEASIBLE DIRECTION WITH FEM FOR SHAPE OPTIMIZATION

Eckart Schnack
Institute of Solid Mechanics
University of Karlsruhe
Kaiserstrasse 12
D-7500 Karlsruhe/German Federal Republic

ABSTRACT. The avoidance of cracks in zones of stress concentrations by minimizing the maximal von Mises stress is very important in practical problems for the industry. In many applications it is necessary to allow for the change of traction vectors in large time intervals. So, one gets as result a multiple loading problem with a finite number of traction vectors, which will be formulated here as a discrete dynamic programming problem. The proposed algorithm is based on the method of feasible directions. It means, in effect, that one has a non-gradient method, combined with FEM, for which the search direction vector is known from physical reasons, see SCHNACK 1979 and SCHNACK and SPÖRL 1986 [1,2].

INTRODUCTION

We consider a homogeneous linear elastic, isotropic material with small deformations. The treatment is restricted to plane and axi-symmetric bodies with Dirichlet and Neumann boundary conditions. A sketch of the problem formulation can be seen in Fig. 1, in which

$$V \subset \mathbb{R}^n, \text{ with } n = 2,3 \tag{1}$$

The Neumann condition is formulated as follows:

$$\vec{t}_\mu := \tau^\mu_{km} \; n_k \; \vec{e}_m \Big|_{\partial V^\mu_K} = \vec{\tilde{t}}_\mu$$

$$\text{with } \partial V^\mu_K \subset \partial V \longrightarrow \mu = 1(1)M \tag{2}$$

and the Dirichlet condition can be presented as,

G. I. N. Rozvany and B. L. Karihaloo (eds.), Structural Optimization, 299–306.

$$\vec{u}_\mu \Big|_{\partial V^\mu_K} = \tilde{\vec{u}}_\mu$$

$$\text{with } \partial V^\mu_K \subset \partial V \longrightarrow \mu = 1(1)M \tag{3}$$

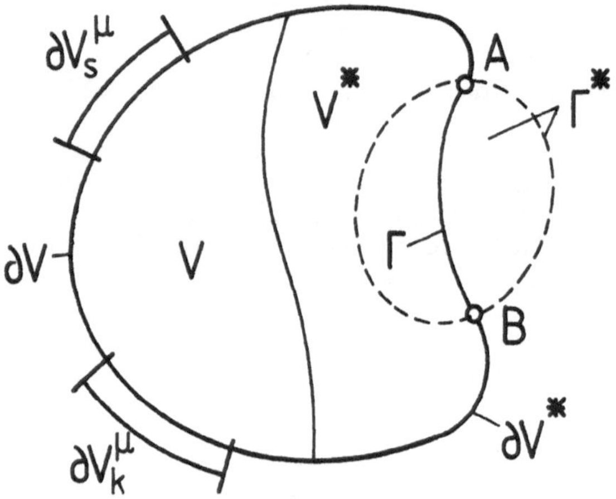

Fig. 1: Problem formulation
$V \subset \mathbb{R}^n$, n = 2,3, domain inside of which the field equations are satisfied
V^*: domain for optimizing the stress field with $V^* \equiv V$ as a special case
Γ^*: variation domain
Γ: boundary for variation between A, B
$\Gamma \subset \partial V^* \cap \partial V$

For the time interval between two of all M loading cases, we have: $\Delta t_\mu \to \infty$. The field equations for a fixed boundary Γ are with the displacement vector u_k, the strain tensor e_{km} and the stress tensor τ_{km}:

$$e_{km} := \frac{1}{2}\left(u_{k,m} + u_{m,k}\right) \tag{4}$$

$$i_{rs} := \varepsilon_{kpr}\, \varepsilon_{mqs}\, e_{km,pq} = 0 \tag{5}$$

$$\tau_{km,k} = 0 \tag{6}$$

$$\tau_{km}: = 2G\left(e_{km} + \frac{\nu}{1 - 2\nu}\, \delta_{km}\, e_{qq}\right) \tag{7}$$

Then, the von Mises equivalent stress value $\bar{\sigma}_{\mu}$ can be computed as follows:

$$\bar{\sigma}^2: = \left(\tau_{kk}\right)^2 - \frac{3}{2}\left[\left(\tau_{kk}\right)^2 - \tau_{km}\,\tau_{km}\right] \tag{8}$$

$$= I_1^2 - 3I_2 \tag{9}$$

The constraints $\Gamma \subset \Gamma^*$ are of geometrical type. The condition that the stress value $\bar{\sigma}_{\mu}$ cannot exceed the upper bound $\tilde{\sigma}$, leads to the following physical constraint:

$$\bar{\sigma}_{\mu} - \tilde{\sigma} \leq 0 \text{ in } V \tag{10}$$

SOLUTION FINDER

Because the problem cannot be solved in an analytical manner, it is formulated as a discrete dynamic programming problem. Then (Fig. 2) one can construct the design vector:

$$\vec{x}^T: = \left[x_1, y_1, \ldots, x_i, y_i, \ldots, x_n, y_n\right] \tag{11}$$

The objective function has the form:

$$\min_{\Gamma}\ \max_{V^*} \left[\bar{\sigma}^1_{\mu}\left(\vec{x}\right), \ldots, \bar{\sigma}^n_{\mu}\left(\vec{x}\right)\right] \qquad \mu = 1(1)M \tag{12}$$

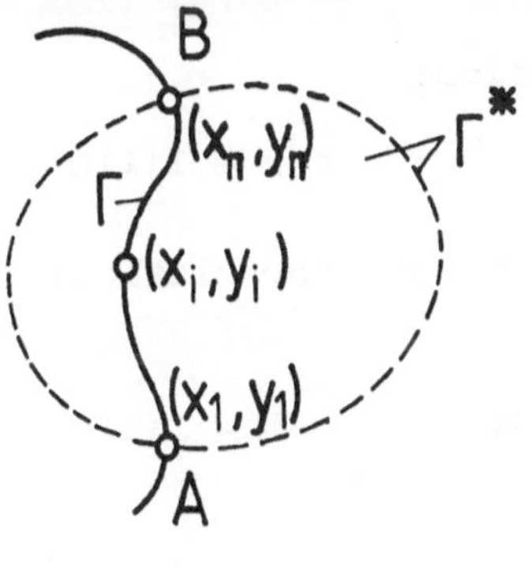

Fig. 2: Discretization of free-boundary Γ

The restrictions are:

$$\left(x_i , y_i\right) \in \Gamma^* \qquad \text{with } i = 1(1)n \tag{13}$$

$$\bar{\sigma}^1_\mu - \tilde{\sigma} \leq 0 \qquad \text{with } i = 1(1)N, \quad \mu = 1(1)M \tag{14}$$

In order to check the second set of constraints, one has to perform the structural analysis by FEM instead of BEM, because if informations must be considered on the interior of the domain too, then BEM is much more time-consuming than FEM. So, one has the following finite element equations:

$$V: = \bigcup_{e=1}^{E} V_e \tag{15}$$

$$\vec{u}_e: = N_e \vec{d}_e \tag{16}$$

$$\vec{\varepsilon}_e: = B_e \vec{d}_e \tag{17}$$

$$k_e: = \int_{V_e} B_e^T D B_e \, dV \tag{18}$$

$$\vec{F}_e^T: = \int_{\partial V_e} \vec{t}_e^T N_e \, dS \tag{19}$$

For the total structure, we have:

$$K_\mu \vec{r}_\mu = \vec{R}_\mu, \qquad \mu = 1(1)M, \tag{20}$$

yielding the stress distribution:

$$\vec{\sigma}_e: = D B_e T_e \vec{r}_\mu \qquad \vec{\sigma}_e: \text{ for } \mu = 1(1)M \tag{21}$$

Accuracy-checks are done by the gap formulation:

$$\text{gap}: = \frac{\max_i \left|\Delta\bar{\sigma}_i\right|}{\max_i \bar{\sigma}_i} 100\% \qquad i = 1(1)N \tag{22}$$

From the elastical perturbation theory (this special class

is called notch stress theory), one has the monotony principle:

$$\bar{\sigma}\ (t\varkappa) \nearrow \qquad \varkappa = \frac{1}{\rho} \quad (t\varkappa\text{: normalized curvature}) \tag{23}$$

This is a special formulation of the Saint-Venant-principle, also called the fade-away theorem of Neuber. In addition, one has the reaction law of the notch stress theory, sometimes called the theory of relieving notch:

$$\bar{\sigma}_{min} \uparrow \longrightarrow \bar{\sigma}_{max} \downarrow \tag{24}$$

From (24) follows the rough description of the solution-idea. By changing the control parameter "curvature", the maximum stress value is reduced:

$$\bar{\sigma}^{j}_{max} \downarrow \tag{25}$$

In addition, the maximum stress value is reduced by increasing the minimum stress value:

$$\bar{\sigma}^{j}_{min} \uparrow \tag{26}$$

In doing this, the geometrical restrictions:

$$\left(x_i^{(j)}, y_i^{(j)}\right) \in \Gamma^* \qquad i = 1(1)n \tag{27}$$

$$\bar{\sigma}_{\mu}^{(j)i} - \tilde{\sigma} \quad \leq 0 \qquad i = 1(1)N,\ \mu = 1(1)M \tag{28}$$

must be satisfied.

Aditionnally, one has to formulate a smoothening process for Γ. From this formulation follows that one needs a hierarchy for statements (25) and (26). This is because a point of $\bar{\sigma}^{j}_{min}$ could be a point of $\bar{\sigma}^{j}_{max}$ for another loading case μ. The result of this description is an iterative change of shape from step j-1 to step j, which can be formulated with the so-called transition function $\vec{f}_j$:

$$\vec{x}_j := \vec{f}_j\ \left(\vec{x}_{j-1},\ \vec{u}_j\right) \tag{29}$$

The transition function can now be explicitly written as follows:

$$
\begin{bmatrix} x_j^i \\ \cdot \\ \cdot \\ \cdot \\ x_j^i \\ \cdot \\ \cdot \\ \cdot \\ x_j^n \\ \hline y_j^1 \\ \cdot \\ \cdot \\ \cdot \\ y_j^i \\ \\ \cdot \\ \cdot \\ y_j^n \end{bmatrix} = \begin{bmatrix} x_{j-1}^i \\ \cdot \\ \cdot \\ \cdot \\ x_{j-1}^i \\ \cdot \\ \cdot \\ \cdot \\ x_{j-1}^n \\ \hline y_{j-1}^1 \\ \cdot \\ \cdot \\ \cdot \\ y_{j-1}^i \\ \cdot \\ \cdot \\ \cdot \\ y_{j-1}^n \end{bmatrix} + \begin{bmatrix} u_j^i \\ \cdot \\ \cdot \\ \cdot \\ u_j^i \left\{ y_{j-1}^i + \gamma_{j-1}^i \left(y_{j-1}^{i-1} - y_{j-1}^i \right) - y_{j-1}^{i+1} \right\} \\ \cdot \\ \cdot \\ \cdot \\ u_j^n \\ \hline u_j^1 \\ \cdot \\ \cdot \\ \cdot \\ u_j^i \left\{ - \left[x_{j-1}^i + \gamma_{j-1}^i \left(x_{j-1}^{i-1} - x_{j-1}^i \right) - x_{j-1}^{i+1} \right] \right\} \\ \cdot \\ \cdot \\ \cdot \\ u_j^n \end{bmatrix}
$$

with $i = 2(1)(n-1)$ (30)

This means that one can define the control or decision vector of the problem:

$$\vec{u}_j^T := \left[u_j^1 \; \ldots \; u_j^i \; \ldots \; u_j^n \right] \tag{31}$$

By formulation of the geometrical constraints, one gets the control space of problem Ω_j:

$$\Omega_j^1 := \Omega_j^n := \{0\} \tag{32}$$

because nodes 1 and n are fixed. It follows that:

$$\vec{u}_j \in \Omega_j \left(\vec{x}_{j-1} \right) := \prod_{i=1}^{n} \Omega_j^i \left(\vec{x}_{j-1} \right) \subset \mathbb{R}^n \tag{33}$$

The theoretical concept of the problem shows the existence of the solution, because from the theorem of *T Y C H O N O V* one has the information, that the domain of the objective function is compact, since the product of compact sets is also compact. Moreover, the theorem of *W E I E R S T R A S S* states that the function has now an extremum, which implies the existence of a solution. This means that one can construct the solution finder by dynamic programming algorithm. The problem can be formulated as follows:

$$\min \sum_{j=1}^{l} g_j \left(\vec{x}_{j-1}, \vec{u}_j\right) \tag{34}$$

$$\text{with:} \quad \vec{x}_j := \vec{f}_j \left(\vec{x}_{j-1}, \vec{u}_j\right) \in \Xi_j \qquad \Xi_j \subset \mathbb{R}^m \tag{35}$$

$$\text{and:} \quad \vec{u}_j \in \Omega_j \left(\vec{x}_{j-1}\right) \qquad \Omega_j \subset \mathbb{R}^r \tag{36}$$

The solution is then found by the feasible direction method. We use the following transformation of the problem:

$$\min x_{2n+1} \tag{37}$$

where x_{2n+1} is a new variable. It follows that one has an extension of the design vector:

$$\vec{x}^{*T} := \left[x_1, \ldots, x_n, y_1, \ldots, y_n, x_{2n+1}\right] \tag{38}$$

The following are the equality restrictions:

$$K_\mu \vec{r}_\mu = \vec{R}_\mu \tag{39}$$

$$\vec{d}^\mu_e = T_e \vec{r}_\mu \qquad h_i = 0 \tag{40}$$

$$\vec{\sigma}^i_\mu = \frac{1}{n_i} \sum_{e=1}^{n_i} D B^i_e \vec{d}^\mu_e, \text{ where } i = 1(1)N \tag{41}$$

and the inequality constraints are as follows:

$$\bar{\sigma}^i_\mu - x_{2n+1} \le 0 \quad i = 1(1)n \tag{42}$$

$$\bar{\sigma}^i_\mu - \tilde{\sigma} \le 0 \quad i = 1(1)N, \qquad \mu = 1(1)M, \qquad g_m \le 0 \tag{43}$$

$$A\vec{x} - \vec{b} \le 0 \tag{44}$$

The first inequality equation (42) is a new one,

coming from the special transformation in (37) and (38). The iteration rule for solving this problem is:

$$\vec{x}^*_j := \vec{x}^*_{j-1} + \alpha_j \vec{s}^*_j \tag{45}$$

One has to check the usability and the feasibility of the search direction vector. By formulating a rezoning process of the finite element mesh, one can avoid the degeneration of elements, due to the changed free boundary Γ. Additionally, in order to improve the efficiancy, one has to formulate the finite element analysis with the total step iteration method; this means that one has to work with an extension of the global stiffness relation (20) by the initial stiffness matrix K_0, in order to get by iteration the corrected displacement vector.

CONCLUSIONS

With the algorithm described above, one can solve many practical problems of the industry, see for example optimal cut-outs, optimization of roundings of gear roots, connecting rod optimization and critical parts of high pressure loading machines.

The main advantages of this algorithm is, that one can avoid gradient computations and consequently one can work with a high number of design variables in solving the optimization problem. Numerical instability problems, which can appear if the gradients are computed from the stiffness relations, are also avoided.

REFERENCES

[1] SCHNACK, E.: An Optimization Procedure for Stress Concentrations by the Finite-Element-Technique. Int. J. Num. Meth. Engng., Vol. 14, No. 1 (1979), pp. 115-124.

[2] SCHNACK, E. and U. SPÖRL: A mechanical Dynamic Programming Algorithm for Structure Optimization. Int. J. Num. Meth. Engng., Vol. 23, No. 11 (1986), pp. 1985-2004.

EXPERIMENTAL DESIGN AND STRUCTURAL OPTIMIZATION

A.J.G. Schoofs
Eindhoven University of Technology
P.O. Box 513
5600 MB Eindhoven
The Netherlands

ABSTRACT. Structural optimization problems are mostly solved iteratively by combining finite element and mathematical progamming methods.

An alternative approach can be used in which the FEM-analyses are planned a priori, both with regard to their number and the values of the design variables, applying techniques used for planning of experiments. From the outcome of the computations an approximate mathematical model is derived by means of regression analysis. This model can in turn be used as a fast analysis module in optimization software. In this paper the use of experimental design in structural optimization is discussed.

The methods have been tested and used extensively for shape optimization of carillon bells, resulting in a new bell which is musically very interesting. They have also been applied successfully to several mechanical engineering problems.

1. INTRODUCTION

The research field of structural optimization is concerned with the design of structures which can meet requirements better, e.g. with respect to the resistance of loads applied to them, or with respect to their performance. The design consists of an iterative procedure of analysis and synthesis. By means of analysis the behaviour of the structure is evaluated, while synthesis is used to modify the structure in a way that it probably meets stated demands better.

We assume that during the optimization process the behaviour of the structure under consideration is determined uniquely by a column matrix $\underset{\sim}{x}$ containing n design variables:

$$\underset{\sim}{x}^T = [x_1 \; x_2 \; \dots \; x_n] \tag{1}$$

G. I. N. Rozvany and B. L. Karihaloo (eds.), Structural Optimization, 307–314.

Then we can write the constrained optimization problem as follows:

Find a set of design variables $\underset{\sim}{x}$ such that a certain objective function $F(\underset{\sim}{x})$ is minimized (2)

The set of design variables generally is subject to:

Inequality constraints: $g_j(\underset{\sim}{x}) \leq 0 \qquad j = 1, \ldots m$ (3)

Equality constraints : $h_k(\underset{\sim}{x}) = 0 \qquad k = 1, \ldots l$ (4)

Side constraints : $x_i^l \leq x_i \leq x_i^u \qquad i = 1, \ldots n$ (5)

For the analytical task in a structural optimization problem mostly the finite element method (FEM) is used for two good reasons. The first reason are its nice and flexible modelling facilities, by which very different structures can be modelled in essentially the same way. Secondly, there is the accuracy of FEM-analysis which can easily be controlled by means of the used element grid. But FEM-analyses of actual engineering structures are often very time-consuming, which is a serious drawback for application of the method during structural optimization because of the iterative character of the optimization process.

If we accept that several FEM-analyses should be spent to solve the optimization problem, another approach can also be used (Schoofs 1987b). With that approach the FEM-analyses are not controlled by iteration steps, but they are planned beforehand, both concerning their number and the values of the set of design variables. From the outcome of the computations approximate, but very efficient mathematical models are derived by means of regression analysis. Such mathematical models can substitute FEM-models and thus can be used as fast analysis modules in optimization software.

For the planning of the FEM-computations and for the analysis of the outcomes, the Experimental Design Theory (EDT) (Montgomery 1984) will be applied. This theory has been developed for the planning and analysis of comprehensive physical experiments. FEM-computations can be regarded as numerical experiments, where the design variables are treated as input quantities. All computable properties of the structure, such as weight, displacements, stresses, etc. can be regarded as response quantities of the numerical experiment. The approximating regression models will be derived for these responses and they can be substituted in the optimization problem given by the equations (1) through (5).

2. EXPERIMENTAL DESIGN THEORY

Advanced scientific and technological research requires comprehensive and expensive experiments and the need for careful planning of the experiments is quite clear. On the one hand it is required to minimize the number of experiments, but on the other it is desired to gather as much information as possible about the relevant aspects of the system under consideration.

The EDT consists of two main parts. The first part, concerns the planning of experiments and ends up with a list of experiments to be carried out. This list is called the experimental design, abbreviated to ED. In the second part the experimental results are analysed and fitted to some mathematical relationship.

In this paper we apply EDT for the planning of deterministic FEM-analyses. Although EDT was developed for stochastic processes, it turns out that much of the common theory may still be used for deterministic processes.

When a structure is determined by n design variables $\underset{\sim}{x}$ according to (1), we may search for m functions describing the response quantities:

$$y_j = y_j(\underset{\sim}{x}) \quad j = 1, \ldots, m \tag{6}$$

in a certain limited area according to (5). In the sequel we will consider only one response y_j and for brevity we omit the index j.

To find the relation $y = y(\underset{\sim}{x})$ we assume a mathematical model. Mostly a linear model of the form will apply:

$$y = \beta_1 f_1(\underset{\sim}{x}) + \ldots + \beta_k f_k(\underset{\sim}{x}) + e \tag{7}$$

in which $\beta_1, \ldots, \beta_k$ are unknown parameters which are being sought; the model is linear in these β's. The functions $f_1(\underset{\sim}{x}), \ldots, f_k(\underset{\sim}{x})$ are known. We can choose both linear and non-linear functions for them; in most cases for (7) a polynomial is chosen. The variable e accounts for the stochastic and/or deterministic model error that is inherent in every model assumption.

The formulation of an ED implies:

1. The choice of discrete values (levels) for all design variables x_i, and
2. the choice of certain combinations of levels of different x_i's. Each combination determines one specific structure.

The ED can be formulated using the so-called 2^n-designs (Montgomery 1984) or by means of the so-called optimal experimental designs (Fedorov 1972, Nagtegaal 1987). We will not discuss this issue here any further.

If we have p combinations as mentioned above in the ED, specified by the column matrices $\underset{\sim}{x}, \ldots, \underset{\sim}{x}_p$, and if we analyse the structure at these p sets of design variables, then according to (7) we can write:

$$y(\underset{\sim}{x}_j) = \beta_1 f_1(\underset{\sim}{x}_j) + \ldots + \beta_k f_k(\underset{\sim}{x}_j) + e_j, \quad j = 1, \ldots, p \tag{8}$$

Equation (8) can be written in matrix notation as:

$$y = X\beta + e \tag{9}$$

where X is a p*k matrix and y, β and e are column matrices representing p observations of y, k unknown parameters and p model errors respectively. Generally p exceeds k and a least-squares solution technique is applied yielding estimates $\hat{\beta}$ of β given by:

$$\hat{\beta} = (X^T X)^{-1} X^T y \tag{10}$$

Subsequently estimates $\hat{y}$ of the function y can be calculated from:

$$\hat{y}(x) = \hat{\beta}_1 f_1(x) + \ldots + \hat{\beta}_k f_k(x) \tag{11}$$

Differentiation of the mathematical model (7) with respect to the design variable x_i gives:

$$\frac{\partial y}{\partial x_i} = \beta_1 \frac{\partial f_1}{\partial x_i} + \ldots \beta_k \frac{\partial f_k}{\partial x_i} + \frac{\partial e}{\partial x_i}, \quad i = 1, \ldots, n \tag{12}$$

In FEM-formulations such sensitivities of y can efficiently be computed and thus (12) can be used with advantage, together with (7), to estimate the parameters β. Furthermore, the accuracy of partial derivatives of the resulting regression models then will be increased, which is advantageous for use of the regression models in optimization algorithms.

In building regression models using EDT the following questions have to be resolved:

1. Which design variables play a role?
2. What is their range of interest?
3. Which functions $f_i(x)$, see (7), must be chosen?
4. How many levels of each design variable are needed?
5. Which ED should be chosen?
6. How can the parameters be estimated and tested?

It is not possible to answer all these questions at once and it is certainly unwise trying to do so. Rather an iterative procedure should be followed. A very valuable aspect of EDT is that it guides model building in a structured way.

3. USE OF EXPERIMENTAL DESIGN IN STRUCTURAL OPTIMIZATION

We have the following reasons for the use of EDT in structural optimization problems:

- The design variable concept appears in both disciplines.
- The need for minimizing the number of expensive elementary operations, be they physical experiments or FEM-analyses.
- FEM-analyses can be regarded as numerical experiments.
- In both disciplines sensitivities can be used with advantage.

- Structural optimization programs involving FEM-analysis are suitable to collect data for experimental designs; for that purpose a powerful computer is needed.
- A regression model can serve as a fast analysis module in a structural optimization program; this can be run interactively even on a personal computer.
- Using such an optimization program an approximate global optimum can be found. It is possible to investigate several objective and constraint functions at low computing costs.
- The approximate global optimal design can be used as a starting point for more accurate optimization, applying direct FEM-analyses in the iteration cycle.

Having discussed the potential of regression models in structural optimization, we must seek the situations in which such models can be used. The main reason for the development of a regression model is the need for a fast analysis model of a certain system which can be described with few variables notwithstanding the complexity of the system and the fact that the desired analysis accuracy is so high that it can only be achieved by elaborate numerical analysis. Formulated in this way, regression models may be developed for a very wide class of applications, even going beyond the field of structural analysis and optimization.

In order to limit the extent of the ED in developing a regression model, the number of design variables should be limited to about 10. The number depends also on the complexity of the assumed mathematical model in the regression. A further necessary condition is a well defined and stable set of design variables and their ranges, by means of which all specimens of the structure can adequately be described. If during the development, changes occur in the set of design variables, it is possible that part of the already collected data is not valuable anymore.

Besides the use of regression models in optimization, they can also be used as stand-alone analysis models in an engineering design office or in a classroom where students attend computer assisted exercises to learn about the real behaviour of systems. The next section begins with an example of such an application.

4. APPLICATIONS

The procedures described in the preceding sections have been applied to several mechanical engineering problems. In this section two applications are presented.

4.1. Three-dimensional bearing problem

Dry running journal bearings are important connecting elements in mechanical engineering. Many practical bearing structures can be modelled according to Fig. 1a. There is a need for a three-dimensional model due to the inclination of the shaft. Wouters (1986) developed for this situation regression models which describe the surface pressure in a number of discrete points of the contact zone between shaft and journal as functions of the design variables indicated in Fig. 1a.

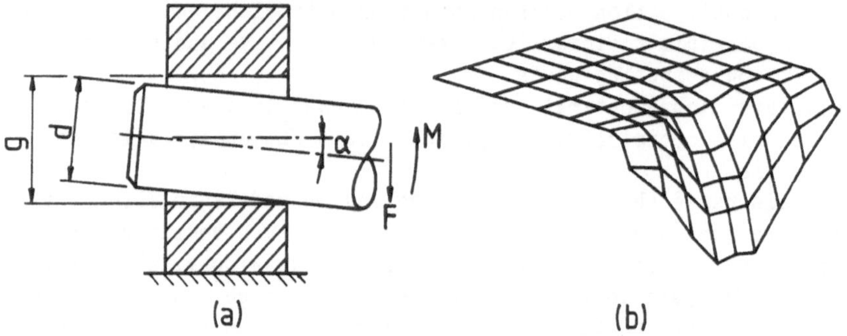

Figure 1. (a) Modelling of a journal bearing indicating the design variables and (b) Qualitative picture of contact pressure for $\alpha = 0.03^{\circ}$.

The required FEM-analyses of 64 different contact problems were carried out using 3D-solid elements and gap elements. Each analysis required 2.5 hours of computing time on a VAX-11/750 computer. Fig. 1b represents a qualitative picture of the contact pressure computed from the derived regression models for a specific inclination angle of the shaft. The resulting regression models proved to be quite accurate. The computing time for the evaluation of Fig. 1b was about 0.5 sec. on the same computer.

4.2 Design of a major-third carillon bell

Some years ago we visited the Royal Eijsbouts Bell Foundry at Asten, The Netherlands, where we were asked the challenging question whether it would be possible to design a tunable major-third bell by means of structural optimization methods. With a major-third bell the frequencies of the five lowest and most important eigenfrequencies should show approximately the ratios 1.0 : 2.0 : 2.5 : 3.0 : 4.0, whereas conventional minor-third bells have approximate ratios of 1.0 : 2.0 : 2.4 : 3.0 : 4.0. Musical reasons for having major-third bells can be found in Lehr (1987). The trial and error approach of several bell founders for more than half a century had rendered little success, in spite of considerable experimental effort.

We solved the problem in the following way (Schoofs et al., 1987a). First we defined a geometrical model of the bell profile (that is the vertical cross-section of the bell) describing it by a set of design variables including radii and wall thicknesses at certain discrete points. The bell profile was subdivided into axisymmetric ring elements subjected to non-symmetric deformation modes, see Fig. 2a. Starting from this initial design some optimization runs were executed based on direct FEM-analyses, resulting in a prototype solution, see Fig. 2b. Around this solution an ED consisting of 128 FEM-analyses was defined using the most important design variables, which are indicated in Fig. 2b.

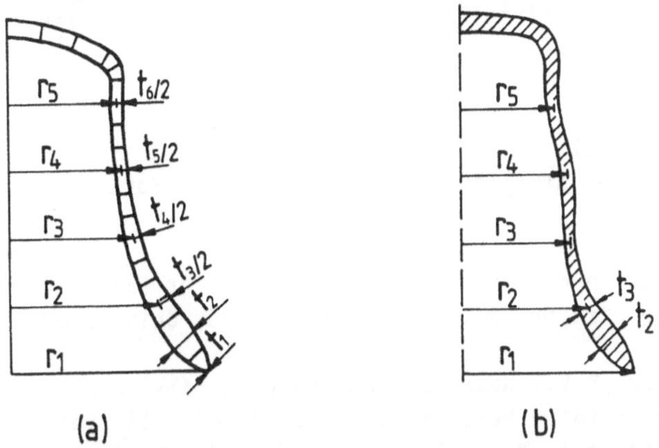

Figure 2. (a) Initial set of design variables for the bell and (b) Prototype solution indicating the set of variables used in the experimental design.

The resulting regression models were used in a simple zero-order optimization algorithm and several approximate solutions for the profile of a major-third bell could be found. From those solutions, the most promising one was selected and the bell foundry cast a bell according to this shape. This bell could be tuned exactly to the desired frequency spectrum. Fig. 3 shows the final geometry of the major-third bell, together with the typical shape of a conventional minor-third one.

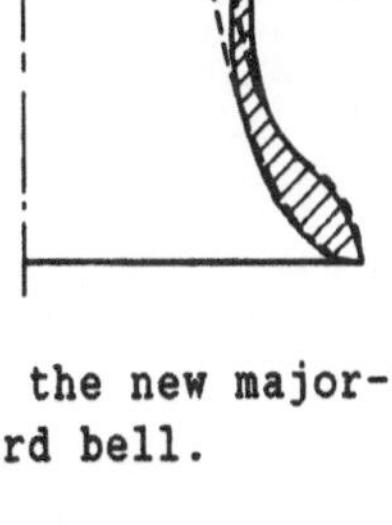

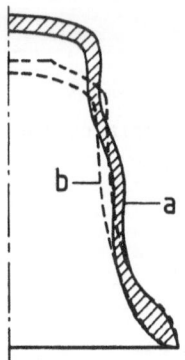

Figure 3. (a) Shape of the new major-thrid bell versus (b) Shape of the conventional minor-third bell.

References

Fedorov, V.V. (1972)
Theory of Optimal Experiments. Academic Press, New York.

Lehr, A. (1987).
'A carillon of major-third bells III: From theory to practice'. Music Perception Vol. 4, No. 3, Spring 1987, Univ. Press, San Diego, California.

Montgomery, D.C. (1984).
Design and Analysis of Experiments. John Wiley, New York.

Nagtegaal, R. (1987).
'Computer Aided Design of Experiments. A program for experimental design and model building' (in Dutch). Report WFW 87.005, Univ. Press Eindhoven.

Schoofs A., Asperen, F.v., Maas, P., and Lehr, A. (1987a)
'A carillon of major-third bells I: Computation of bell profiles using structural optimization'. Music Perception Vol. 4, No. 3, Spring 1987, Univ. Press, San Diego, California.

Schoofs, A.J.G. (1987b).
Experimental Design and Structural Optimization. Ph.D.Thesis, Eindhoven Univ. of Techn., Tiso Press, Enschede, The Netherlands. ISBN 90-71382-19-2.

Wouters, H. (1986)
'The stress distribution in dry running journal bearings' (in Dutch). Report WFW 86.012, Univ. Press, Eindhoven.

Acknowledgement

This research is supported by the Netherlands Technology Foundation (STW).

ON THE SHAPE DETERMINATION OF NON-CONSERVATIVE SYSTEM: A CASE OF COLUMN UNDER FOLLOWER FORCE

Y. Seguchi and S. Kojima
Dept. of Mech. Engineering
Osaka University
Toyonaka, Osaka 560, Japan

ABSTRACT. The shape of a column subject to dissipative and follower forces is determined by the response-based procedure using the space-time finite element scheme. The shape improvement process is derived from the inverse problem of the adjoint variational principle combined with the energy-ratio method. By shifting the time domain included in the functional, the shape is determined in a step-by-step procedure suggesting the possibility of an adaptive shape improvement. The case study shows that the response of improved column is remarkably calm even in unstable region for the uniform column.

1. INTRODUCTION

The shape determination of a structure has been realized to be an attractive and promising field in engineering, not only for the practical structural design, but also from the following reasons: i) the mechanization of the conceptual design process, 2) the exploration of the morphogenetic mechanisms, and 3) the establishment of the new field of applied and computational mathematics. Conventionally, the structure has been optimized in the design parameters such as length, width, thickness, area, and so on, in the preassigned configuration, while the discretization technique such as FEM has developed the shape determination or the structural decision based on the concept of continuum [1], which might be referred to the non-parametric optimization.

The authors and the colleagues have worked in the shape determination problem on the basis of the inverse variational principle (IVP) [2, 3, 4, 5]. The procedure is somewhat different from other approaches [6, 7, 8] in the sense that the objective function consists of only naturally introduced physical functional such as potential energy under the single constraint of volume constancy. The procedure is extended, in this paper, to the shape determination of a non-conservative system. The exemplified system is the dynamical version of the Beck's column which is subject to the follower force and the transverse force at the free end. The problem is formulated through the adjoint variational principle, and the time-varying or adaptive

G. I. N. Rozvany and B. L. Karihaloo (eds.), Structural Optimization, 315–322.

shapes are determined by the combination of the schemes of the space-time finite element and IVP.

2. DYNAMICS OF BECK'S COLUMN

2.1 Basic System and the Adjoint Variational Expression

The column modified by adding the time-varying transverse force $F(\tau)$ is illustrated in Fig. 1. It is clamped at one end and free at another end. Denote the elastic modulus, the cross-sectional area, the moment of inertia and the mass density by E, $A=mA_0$, $I=m^2I_0$ and ρ , respectively, where A_0, I_0 represent the nominal values. Then, the non-dimensional system of basic equation is written as follows [9]:

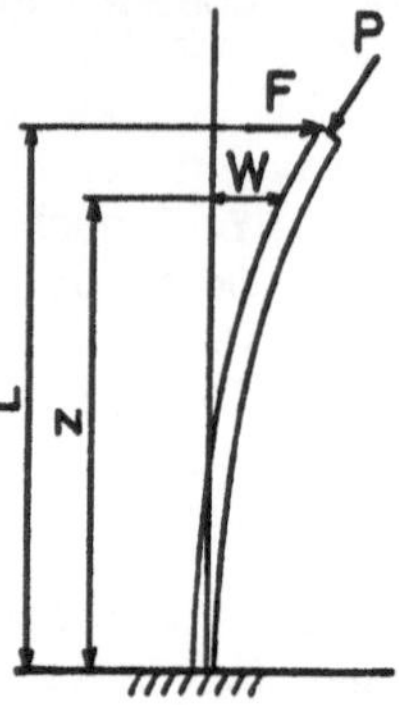

Fig. 1
Beck's column subject to a transverse force

$$(m^2w'')''+pw''+\beta\dot{w}+\gamma(m^2\dot{w}'')''+m\ddot{w}=0, \quad (1)$$

where $()'$, $\dot{()}$ are $\partial()/\partial x$, $\partial()/\partial t$, respectively, and

$$x=z/L,\ w=W/L,\ p=PL^2/(EI_0),$$
$$f=FL^2/(EI_0),\ t=(EI_0/\rho A_0)^{1/2}\tau/L^2,$$
$$\beta=\xi L^2/(\rho EA_0I_0)^{1/2},\ \gamma=\eta(EL_0/\rho A_0)^{1/2}/L^2,\ \tau \text{ is the natural time.} \quad (2)$$

The symbols ξ , η are the external and the internal damping factors, respectively. In this system, we assume the following boundary conditions and the initial conditions:

$$w=0,\ w'=0 \text{ at } x=0, \quad (3\text{-}1)$$
$$m^2w''+\gamma(m^2\dot{w}'')=0,\ (m^2w'')'+\gamma(m^2\dot{w}'')'+f=0 \text{ at } x=1, \quad (3\text{-}2)$$
$$w=0,\ \dot{w}=0 \text{ at } t=0. \quad (3\text{-}3)$$

Now, we consider a problem such that an arbitrary cost function $\int_0^T\int_0^1 C(w)dxdt$ defined in the domain $[0,1]\times[0,T]$ of the distance-time space should be extremum under the subsidiary conditions, Eqs. (1) - (3). According to the Lagrangian principle, the problem is transformed into the unconstrained optimization by introducing a Lagrangian multiplier for each condition. From the stationary condition of the modified functional, i.e. the Lagrangian function, we have again all the equations of the primal system and also of the adjoint system, with all boundary/initial/terminal conditions. To simplify the functional, we eliminate all multipliers of the conditions in Eq. (3), and then carry out the integration by parts by considering the boundary conditions at $x=0$, the initial conditions $t=0$ for the primal system, i.e. Eqs. (3-1), and (3-3), respectively, and the boundary conditions for the adjoint system derived from the stationary condition,

$$y=0,\ y'=0 \text{ at } x=0. \tag{4}$$

Thus, the final simplified functional is written as follows:

$$L_1(w,y)=\int_0^T\Big[\int_0^1(m^2w''y''-pw'y'+\beta\dot{w}y$$
$$+\gamma m^2w''y''+mwy)dx+pw'y(1)-fw(1)-fy(1)\Big]dt,$$

subject to Eqs. (3-1), (3-3) and (4). (5)

where the function $\int_0^1 C(w)dx$ is replaced by $-fw(1)$. The stationary condition of this functional leads to the governing equation of the primal system, Eq. (1) and that of the adjoint system,

$$(m^2y'')''+py''-\beta\dot{y}-\gamma(m^2\dot{y}'')''+m\ddot{y}=0, \tag{6}$$

which is the problem of Reut's column subject to negative damping and a transverse force at the free end, and its natural boundary conditions,

$$m^2y''+py-\gamma m^2\dot{y}''=0 \text{ and } (m^2y'')'+py'-\gamma(m^2\dot{y}'')'=0 \text{ at } x=1,$$
$$(m^2y'')''+\beta y-m\dot{y}=0,\ my=0 \text{ at } t=T. \tag{7}$$

Eqs. (3-2) are again derived as the natural boundary condition. The terminal conditions, i.e. the conditions at t=T of the adjoint system are now explicitly given in Eq. (7). It should be noted that the choice of the cost is arbitrary, and the derived adjoint system depends on it. In this situation, however, it is true that the above-chosen cost is the only possible one which is naturally introduced.

2.2 Finite Element Discretization in Space-Time Domain

Let us consider the shape determination of the column within the framework of discretization. To this end, we adopt the finite element scheme in the space-time domain [10]. Firstly, we assume the simple linear distribution of design variable, m as

$$m(x)=\Phi_A(x)m_A,\ \text{for } x_j\leq x\leq x_{j+1}, \tag{8}$$

where the double indices mean the summation convention. For simplicity, m is not assumed to be time-varying in the formulation. The space-time domain [0, 1]x[0, T] is subdivided into the rectanglar mesh. In each element, the deflection and the adjoint variable, w, y are expressed by the shape functions of the 3rd order polynomial given as (see Fig. 2)

$$w(x,t)=\Phi^e_{Rs}(x,t)w^e_{Rs},$$
$$y(x,t)=\Phi^e_{Rs}(x,t)y^e_{Rs}$$
$$\text{for } x_j\leq x\leq x_{j+1},\ t_j\leq t\leq t_{j+1}, \text{ where}$$
$$w^e_{Rs}=[w_{Rs},w'_{Rs},\dot{w}_{Rs},\dot{w}'_{Rs}]^T,$$
$$y^e_{Rs}=[y_{Rs},y'_{Rs},\dot{y}_{Rs},\dot{y}'_{Rs}]^T. \tag{9}$$

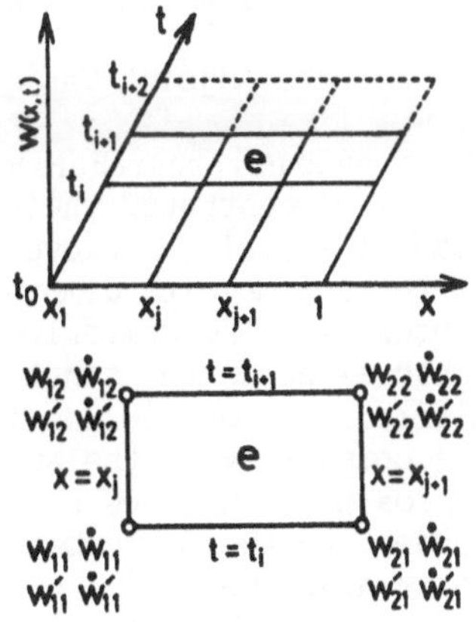

Fig. 2 Space-time element

The external force is also discretized as

$$f(t)=\Phi_A(t)f_A, \text{ for } t_j \leq t \leq t_{j+1}. \quad (10)$$

Substituting Eqs. (8)-(10) into Eq. (5), we have the following descretized version of the functional:

$$L_2=\Sigma L_{2i}=\Sigma[y_i^T[S_i+pQ_i+C_i+M_i]w_i -f_i^T w_i-y_i^T f_i]$$
$$=Y^T[S+pQ+C+M]W-F^T W-Y^T F, \quad (11)$$

where L_{2i} is the functional for the elemental domain $[x_i, x_{i+1}]x[t_i, t_{i+1}]$, w_i and y_i are the element vectors of that domain, W, Y are the nodal vectors of the whole system, S is the stiffness matrix, Q is the load matrix, C is the damping matrix, and M is the mass matrix. From the stationary conditions of Eq. (11) w.r.t. the system vectors, we obtain the following matrix equations governing the system:

$$[S+pQ+C+M]W=F, \quad [S+pQ+C+M]^T Y=F^T. \quad (12)$$

2.3 Computational Procedure for the Response Analysis

In principle, the primal and adjoint systems of Eq. (12) could be solved by the appropriate boundary conditions and the initial/terminal conditions, respectively. Unfortunately, the matrix equations for the total system are highly ill-conditioned, i.e. the small errors from the initial (t=0) or the terminal (t=T) in the primal or adjoint systems, respectively, quickly propagate to cause the divergent oscillation. Thus, the matrix equations of the primal system for example, are solved in the subsystem of each time interval in stepwise process regarding the solution of the sub-terminal of the previous interval as the initial condition of the subsystem. For the adjoint system, the backward scheme is adopted, i.e. the system is solved step by step from the terminal condition. The computation due to the proposed scheme is remarkably stable for any value of the parameters, and also remarkable for memory saving.

3. SHAPE DETERMINATION DUE TO INVERSE ADJOINT VARIATIONAL PRINCIPLE

3.1 Inverse Adjoint Variational Principle

The previous researchers working on the design of the Beck's column have mainly concentrated their efforts on the maximization of the critical load [11]-[13]. Moreover, it might be difficult to find out the design of the column with damping as far as the authors' data base is concerned. The criterion of critical load naturally leads to the shape improvement in the sense of the averaged energy and it is extremely difficult to obtain the compromised solution to every mode because of the complexity of the solution space. Apart from such conventional approaches, thus, we propose the shape determination of the column subject to the disturbance of a conservative transverse force. The shape adjustment will be done in any finite time interval which, if it is done sequentially, leads to the adaptive adjustment of the shape.

The inverse variational principle [2],[3] can be simply extended to the non-conservative system by utilizing the derived functional (11) as follows:

$$L_3=Y^T[S+pQ+C+M]W-F^TW-Y^TF+\mu(V-1),$$
$$V=\int_0^1 m dx. \tag{13}$$

where V is the non-dimensional total volume of the column, and μ is the Lagrangian multiplier assigned to the constraint of volume constancy. The stationary conditions w.r.t.the primal and adjoint variables as well as the design variable m_j $(j=1,2,..,N+1)$, where N is the number of element, are written as

$$[S+pQ+C+M]W=F,\ \ [S+pQ+C+M]^TY=F^T,$$
$$Y^T[\partial S/\partial m_j+\partial C/\partial m_j+\partial M/\partial m_j]W+\mu\partial V/\partial m_j=0,\ \ (j=1,2,..,N+1),$$
$$V=1. \tag{14}$$

The last two equations have been newly introduced for the present shape determination. It is noted that the design variable is dependent not only on the deflection of the column, i.e. the primal variable but also on the adjoint variable.

3.2 Shape Improvement Algorithm

The derived optimality conditions are so highly non-linear that the conventional analytical algorithms such as Newton-Raphson method are not the most promising for the present situation. Thus, the so-called Energy Ratio Method, a class of iterative procedures, which was heuristically found for the conservation system and has been used successfully [3]-[5], are extensively adopted in this case.

Firstly, the initial guess of the shape is given. Then, the dynamical response of the system is solved both for the primal and adjoint systems. For the assumed shape, one can calculate the Lagrangian multiplier from the third of Eq. (14) as

$$\mu_j=-Y^T[\partial S/\partial m_j+\partial C/\partial m_j+\partial M/\partial m_j]W/(\partial V/\partial m_j),\ j=1,2,..,N+1. \tag{15}$$

If the assumed shape is the final one, then the values of μ_j should be identical and have a certain value. Usually, they are not identical so that they may have information to improve the shape. Denote the value of m_j of the kth step by $m_j(k)$ and the median of μ_j by $\bar{\mu}$. Then, the variable m_j is improved by

$$m_j(k+1)^*=m_j(k)+C\log|\bar{\mu}-\mu_j+1|,\ j=1,2,..,N+1,$$
$$m_j(k+1)=m_j(k+1)^*/V(k+1), \tag{16}$$

where V(k+1) is the volume calculated from the variables of the (k+1)th step. The last equation is for adjustment of the volume constancy. The convergence of the sequence is checked by

$$\max|(\bar{\mu}-\mu_j)/\bar{\mu}|<\varepsilon,\ \varepsilon:\text{ small constant.} \tag{17}$$

4. CASE STUDIES AND REMARKS

Three types of the shape improvements are examined. The first one is the shape determination concerning the total time of operation, and the shape determined is not time-varying through the operation, i.e. the functional used is for the domain [0, 1]x[0, T] (Case A). The second shape determination is concerning the domain $[0, 1]x[iT_m,(i+1)T_m]$, $T_m=T/m$, $i=0,1,..,m-1$. The shape is improved time to time sequentially depending on the subsequent functional of the subdomain (Case B). The final case is a modified improving process where the subdomain is chosen as $[0, 1]x[i\Delta, i\Delta+T_m]$, $i=0,1,2,...$ The process is terminated at i=k, if $k\Delta+T_m=T$ (Case C). If the simbol Δ is equal to the time interval of the finite element in the last case, the shape is adjusted in stepwise fashion so that the process might be referred to the adaptive shape improvement. Although there could be many varieties of the improvement process, we just study the above typical cases in this paper. The obtained results are illustrated in Fig. 3 - 5, and listed in Table 1.

For each cases, the initial parameters are chosen as follows:

1. The initial shape is uniform.
2. The time interval, $t_{i+1}-t_i$ is 0.01. The number of subdivision of the column, N is 10.
3. The external and internal damping factors, β, γ are 1.0 and 0.01, respectively.
4. The time discrete of the functional, $T_m=1.0$
5. The follower force, p is 20.0 and the transverse force, f is 0.1cos(t).

Fig. 3 shows the obtained shape of column of Case A compared with that of the uniform column. There can be observed the narrow parts around x=0.2 and the tip of the column. The shape is significantly different from that obtained on the basis of the critical load [11], where only the first and the second modes of the column without damping are considered. As seen in Table 1, from the viewpoint of the critical load, it is somewhat improved into 22.935 from 17.795 of the uniform column, which is known to be 20.052 for no damping. Fig. 4 compares the shapes obtained by the scheme of Case B. The samples are taken t=0, 1, and 4, each obtained from the functional in time domains, [0, 1], [1, 2], and [4, 5], respectively. The oscillatory variation of the shape is observed, which is thought to be due to the time-varying transverse force, though the obtained shape for each time is not so different. The adaptive improvement of the shape (Case C) is shown in Fig. 5, which indicates similar characteristics as Case B within the example shown here. As one sees in Table 1, the remarkable difference is not observed among the critical loads of the three cases.

Finally, Fig. 6 compares the responses of the deflection of the column tip in Case A, and the uniform columns with and without the follower force. The applied transverse force is also illustrated. Naturally, the response of the uniform column with follower force shows divergence because it is beyond the critical load, and the column with improved shape behaves remarkably calmly, while the simple cantilever, i.e. the column without the follower force behaves differently from that with it. It is observed that the magnitude of the transverce

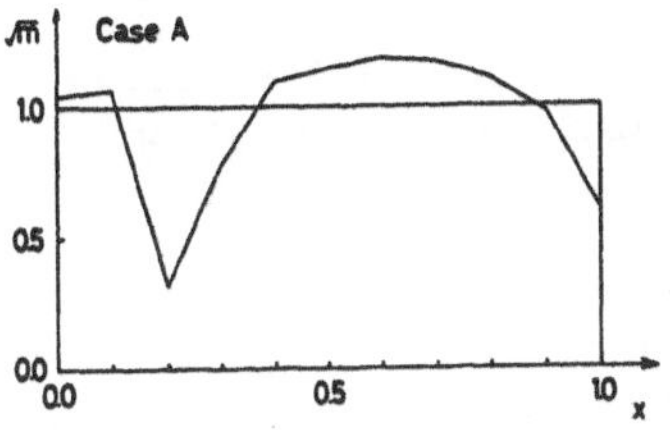

Fig. 3 Improved shape of Case A: The total improvement.

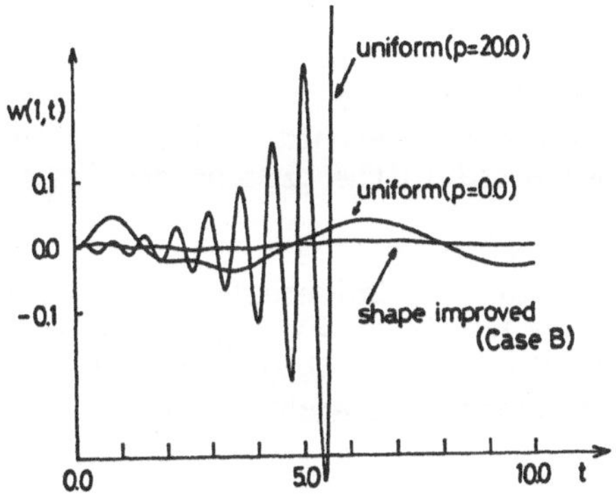

Fig. 6 Comparison of the responses of the columns with the various shapes.

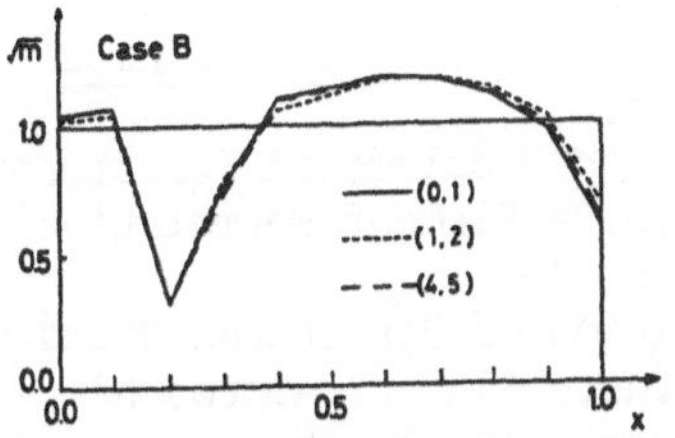

Fig. 4 Improved shapes of Case B: The step-by-step improvement.

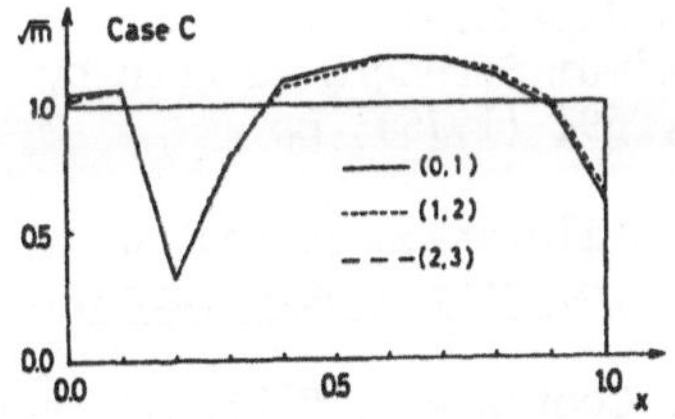

Fig. 5 Improved shapes of Case C: The modified step-by-step improvement.

Table 1 Comparison of the critical loads of the various shapes

		Pcr
uniform (no damping)		20.052
uniform (with damping)		17.795
Case A	(0,10)	22.935
Case B	(0,1)	22.252
	(1,2)	21.275
	(4,5)	22.008
Case C	(0,1)	22.322
	(1,2)	21.729
	(2,3)	22.253

force does not affect the final shape of the column as expected. Actually, this tendency is a basis to propose the shape improvement due to the the response-based approach instead of the critical load-based approach based on the eigenvalues. Although we notice that in the adaptive improvement the time-varying characteristics of the design parameter, m should be included in the governing equations, it is neglected in the present studies.

We would like to thank Dr. M. Tanaka of Osaka University and Dr. Y. Tada of Kobe University for their valuable discussions of the work described in this paper. The study is partly supported by the Ministry of Education, Science and Culture of Japan through Grant-in-Aid for Scientific Research 61850021.

REFERENCES

1. R. E. Richetts and O. C. Zienkiewics, 'Shape Optimization of Continuum Structures' New Directions in Optimum Structural Design (E. Atrek et al. eds.), (1984), J. Wiley & Sons, pp. 137-166
2. V. Horák, Inverse Variational Principles of Continuum Mechanics, (1969), Rozpravy Československé Akad. Věd.
3. Y. Seguchi and Y. Tada, 'Shape Optimization Problems of Structures by Inverse Variational Principle. The Finite Element Formulation', Acta Technica ČSAV, 24, 2 (1979), pp. 139-148
4. Y. Tada and Y. Seguchi, 'Shape Determination of Structures Based on the Inverse Variational Principle/The Finite Element Approach', New Directions in Optimum Structural Design, (E. Atrek et al. eds.),(1984), J. Wiley & Sons, pp. 197-207
5. Y. Tada and Y. Seguchi, 'Structural Shape Determination under Uncertain Loading Conditions', Computational Mechanics '86 (G. Yagawa et al. eds.), (1986), Springer-Verlag, pp. X59-X64
6. O. Pironneau, Optimal Shape Design for Elliptic Systems, (1984), Springer-Verlag
7. Y. Umetani, 'The Shape of Bones and the Growing-Reforming Method', J. JSME, 79, 693 (1976), pp. 749-754 (in Japanese)
8. J. Oda, 'On a Technique to Obtain an Optimum Strength Pattern by the Finite Element Method', J. JSME, 79, 691 (1976), pp. 494-502 (in Japanese)
9. R. C. Kar, 'Stability of a Nonuniform Cantilever Subjected to Dissipative and Nonconservative Forces', Computer & Structures, 11, 3 (1980), pp. 175-180
10. J. T. Oden, Finite Elements of Nonlinear Continua, (1972), McGraw-Hill
11. M. Hanaoka and K. Washizu, 'Optimum Design of Beck's Column', Computer & Structures, 11, 6 (1980), pp. 473-480
12. J. L. Claudon, 'Characteristic Curves and Optimum Design of Two Structures Subjected to Circulatory Loads', Journal de Mecanique, 14, 3 (1975), pp. 531-543
13. C. R. Thomas, 'Stability and Mass Optimization of Non-conservative Euler Beams with Damping', J. Sound and Vibration, 47, 3 (1976), pp. 395-401

A MATHEMATICAL PROGRAMMING APPROACH FOR FINDING THE STOCHASTICALLY MOST RELEVANT FAILURE MECHANISM

L.M.C. SIMÕES
Departamento de Engenharia Civil
Faculdade de Ciências e Tecnologia
Universidade de Coimbra
Portugal

ABSTRACT. In calculating the failure probability of structural systems, the most important operation is the search for the stochasticly most relevant failure mechanism. The nodal and mesh description for the modelling of a flexural frame with fully plastic behaviour and slabs discretized into triangular finite elements whose behaviour conforms the yield line theory are considered. The mathematical programming method arising from these models can be formulated as the minimization of a quadratic concave function over a linear domain.

1. Introduction

Most of the research into the synthesis of structures is based on models with deterministic behaviour. Using these models, it is only possible to guarantee that the structure resists the most unfavorable static loading that is supposed to act during its life for a given behaviour of the constituent materials. Since neither the loading nor the materials are deterministic, research has been conducted to assess.the reliability of structures. To avoid the difficult numerical integration of the probability density functions involved, the reliability index is obtained from the limit state equation using the second moment approximation.

The characteristics of the algorithms used for mathematical programming require the discretization of trusses, frames, plates and shells. The finite element method was choosen owing to its versability in the automatic generation of the stochastically most relevant failure mechanism, that is the mechanism with the smallest reliability index β and the highest probability of failure p_F.

The efficiency in obtaining the mechanism with the highest probability of failure in 2-D structures has been impaired by difficulties in solving the corresponding mathematical programming problem. Even in simple portal frames the algorithms for convex programming used in the evaluation of the stochastically most important failure mechanism may end up with misleading results, because this problem is of nonconvex type. A mathematical programming method that leads to the stochasticly most important failure mechanism is deduced and strategies that give the global optimum (eliminating suboptima) and can be used to enumerate local solutions are presented. This formulation is limited to ductile structures such as steel frames and reinforced concrete slabs.

2. Fundamental Relations

There are two types of formulation available to describe the fundamental relations for the problems to be discussed. One formulation reflects the finite element connectivity at their

G. I. N. Rozvany and B. L. Karihaloo (eds.), Structural Optimization, 323–330.

common corner nodes and it is thereby called nodal description. Alternatively, the finite element connectivity may be reflected through their common sides leading to the formulation of meshes and it is thereby called mesh description.

The suggested finite element model can be regarded as a direct extension to slabs of the finite element modelling of flexural plane frames. Modal deformations are plastic rotations at pre-located critical sections in the case of frames and at pre-located element sides in the case of slabs. Nodes are positioned in between two or more critical sections in the case of frames and at the intersection of two or more element sides in the case of slabs. Futhermore, just as the bars of a frame form meshes, the finite elements modelling the slab can be grouped together forming also meshes as exemplified in Fig.1.

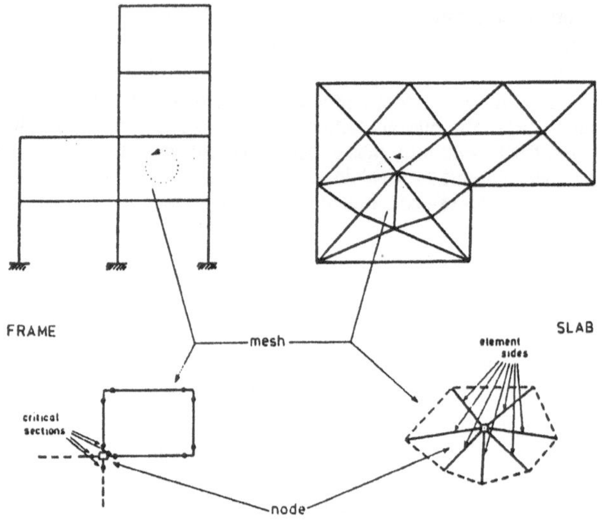

Figure 1

The frame shown in Fig.1 can be represented through a mesh model where an external mesh defines the frame in the two dimensional space. The slab can also be described through a mesh model where the meshes referring to the boundary nodes define the slab in the two-dimensioned space. Therefore, the structure (frame or slab) is replaced by a structural model which can be described in two-dimensional space either through the nodes (Nodal Descrition) defined at the intersection of the finite elements or through the meshes (Mesh Description) defined by chains of finite elements.

2.1. CONSTITUTIVE RELATIONS

The yield criterion for structures with perfectly plastic behaviour imposes bounds on the values of the moments in all critical sections. For example, if the negative and positive plastic resisting moments at the critical section i are m_{*i}^- and m_{*i}^+, respectively, then we have:

$$-m_{*i}^- \leq m_i \leq m_{*i}^+ \qquad (1)$$

Similarly, the yield line method considers a very simple yield criterion involving solely the normal bending moment and the normal angular discontinuity at every element side.The yield conditions impose limit values on the magnitude of the resulting bending moments at the element side. Let the positive and negative bending moment capacities per unit length for the

i^{th} element side be, respectively $\underline{m}_{*i}^{+}$ and $\underline{m}_{*i}^{-}$ and the bending moment per unit length $\underline{m}_i$, with:

$$-\underline{m}^{-}{}_{*i} \le \underline{m}_i \le \underline{m}^{+}{}_{*i} \qquad (2)$$

Now, if such an element side has a length l_i, the yield conditions in terms of total moments for the whole element side are:

$$-m_{*i}{}^{-} = -\underline{m}^{-}{}_{*i}\, l_i \le m_i \le \underline{m}^{+}{}_{*i}\, l_i = m^{+}{}_{*i} \qquad (3)$$

The mechanism deformation can only take place at the element sides where the normal bending moment reaches one of its limiting values. That is to say the angular discontinuity $\Delta\theta_i$ at the element side can only take a positive value $\Delta\theta_i^{+}$ when m_i is equal to m_{*i}^{+} and it can only take a negative value $\Delta\theta_i^{-}$ when m_i is equal to m_{*i}^{-}.

By neglecting the elastic defomations, the deformations of the mechanism are equal to the total deformations of the structure. Provided that at the incipient plastic collapse the displacements are small enough for the plastic analysis to be based on the underformed geometry of the structure, elastoplastic defomations need not to be considered. Clearly, whereas the plastification in frames can be considered to be restricted to pre-located critical sections, in the slabs the plastic behaviour is not necessarly restricted to the sides of the finite elements. Thus, the finite element modelling leads to a correct representation of the frame, but leads to a representation which is only approximate for the slab.

2.2. MESH AND NODAL DESCRIPTIONS

2.2.1. Mesh Description of Statics. The static indeterminacy number (α) of the frame represented in Fig.2 is 3 and the number of critical sections (c_r) that have to be considered is 7. This frame can be reduced to a statically determinate structure in many different ways, one of which is achived by introducing a cut adjacent to critical section 7.

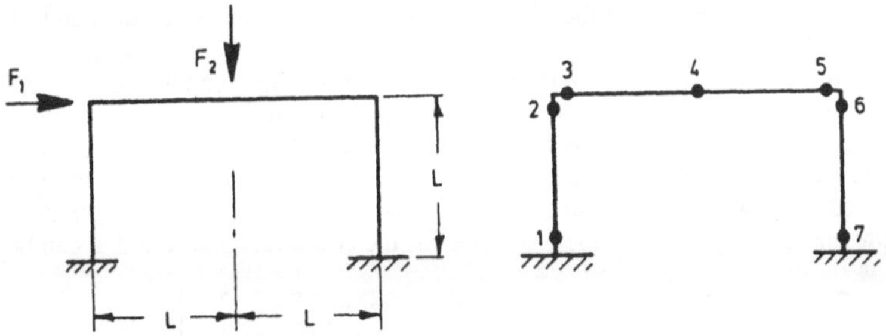

Figure 2

The bending moments at the critical sections can then be expressed in terms of the ordenates of the influence diagrams associated with unit magnitudes of the loads (F_1,F_2) and the indeterminate forces (p_1,p_2,p_3) in the following way:

$$m = B\,p + B_0 F \qquad (4)$$

With more complex frames, the derivation of the basic matrices (B,B_0) becomes a more important and difficult issue. This subject is discussed in Ref.1 where it is seen that the more convenient basis for generating the mesh matrix B cannot generally be derived from physical release systems.

2.2.2. Mesh Description of Kinematics. If a mechanism is built up due to the formation of hinges in the critical sections of the reduced structure, the angular discontinuities Δv can be written in terms of the rotation $\Delta\theta$ from the undeformed geometry of the frame, by means of the coefficients of the static matrix B:

$$\Delta v = B^t \Delta\theta \tag{5}$$

If the structure has a set of compatible displacements, then we have:

$$\Delta v = 0 \quad \Rightarrow \quad B^t \Delta\theta = 0 \tag{6}$$

These equations are valid for every mechanism. Similarly the displacements u corresponding to the loads F can be written in terms of the rotations $\Delta\theta$,

$$u = B_0^t \Delta\theta \tag{7}$$

Then the Kinematic relations for the mesh description in frames become:

$$\begin{bmatrix} 0 \\ u \end{bmatrix} = \begin{bmatrix} B^t \\ B_0^t \end{bmatrix} \Delta\theta \quad \text{or} \quad \begin{bmatrix} 0 \\ u \end{bmatrix} = \begin{bmatrix} B^t & -B^t \\ B_0^t & -B_0^t \end{bmatrix} \begin{bmatrix} \Delta\theta^+ \\ \Delta\theta^- \end{bmatrix} \tag{8}$$

where the rotation in the critical section i is decomposed in the pair of nonnegative variables $\Delta\theta^+$ and $\Delta\theta^-$, as in the mathematical programming algorithms:

$$\Delta\theta = \Delta\theta^+ - \Delta\theta^- \tag{9}$$

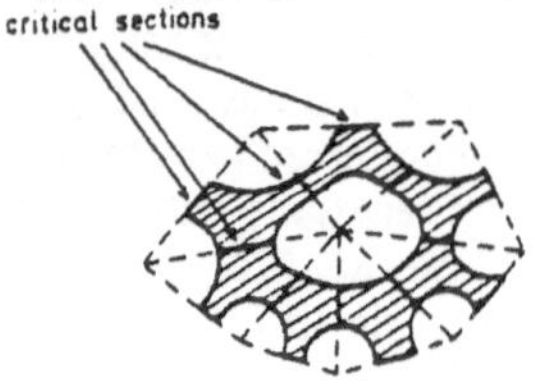

Figure 3

Acording to the yield line theory, the collapse is due to the formation of yield lines along which the slab folds when a mechanism is activated. In the mesh description compatibility conditions for slabs can be established directly, if the finite element meshes are enforced to remain as closed chains of elements during deformation. Such condition of compatibility may be stated for every mesh which can be defined in the discretized slab. A set of linearly independent meshes will be constituted by the meshes defined at each corner with the exception of one. It can be seen from Fig.3 that the static indeterminacy number associated to the finite element modelling of slabs is 2(NC-1), where NC is the number of corner nodes.

Taking one of these meshes, the corresponding compatibility conditions must ensure that when the collapse mechanism is activated, angular discontinuities develop along element sides at yield, maintaining continuity of vertical displacements. That is achieved if the two agebraic sums of the projections in two different directions of those angular discontinuities developing along a fixed sense around the mesh are set to zero. The two compatibility relations for the mesh represented in Fig.4 and for the projection in the two directions X and Y can be stated as follows.

$$\begin{bmatrix} \operatorname{sen}\alpha_1 & \operatorname{sen}\alpha_2 & \operatorname{sen}\alpha_3 & \operatorname{sen}\alpha_4 & \operatorname{sen}\alpha_5 \\ \cos\alpha_1 & \cos\alpha_2 & \cos\alpha_3 & \cos\alpha_4 & \cos\alpha_5 \end{bmatrix} \begin{bmatrix} \Delta\theta_1 \\ \Delta\theta_2 \\ \Delta\theta_3 \\ \Delta\theta_4 \\ \Delta\theta_5 \end{bmatrix} = \begin{bmatrix} 0 \\ 0 \end{bmatrix} \tag{10}$$

where $\Delta\theta_i$ is the total rotation of the element side, that is the sum of the rotations of all element sides that share the same side.

Now, if these conditions are established for each mesh and they are all assembled, the resulting compatibility conditions may be written in the compact form:

$$\Delta V = B^t \Delta\theta = 0 \tag{11}$$

where the vector ΔV has 2 (NC -1) elements, the vector $\Delta\theta$ has NS elements - NS is the number of element sides - and the matrix B is called kinematic transformation matrix.

If all edges are clamped, the moments from which the matrix B_0 can be built are easily determined if the discretized slab is split into cantilever slabs formed by chains of finite elements. If all edges are either clamped or simply supported, the boundary can be considered to be provided by fixed finite elements. The same cannot be said with regard to free edges, but such a difficulty can be overcome if extra boundary finite elements are defined along them. New simply supported edges are thus obtained whilst the original free edges become internal lines along which the normal bending moment capacity is zero. In order to consider supporting columns, either provision is given to account for fixed external constraints or the contour defined by the column in the slab is taken as a clamped edge.

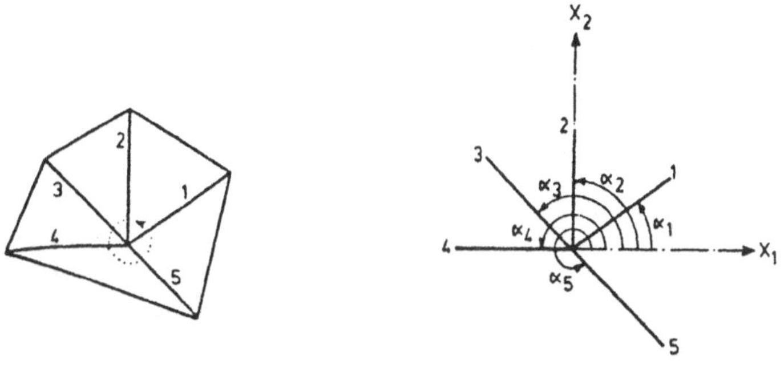

Figure 4

2.2.3. Nodal Description of Kinematics. The mesh description has its origins in the concept of static determinacy and in the expression of any static field in terms of the loading actions and of the mesh forces. The nodal description may for the present case be considered to have its bases on the concept of mechanisms. The elementary mechanisms for a frame are a collection of all the sway and joint mechanisms shown in Fig. 5 .

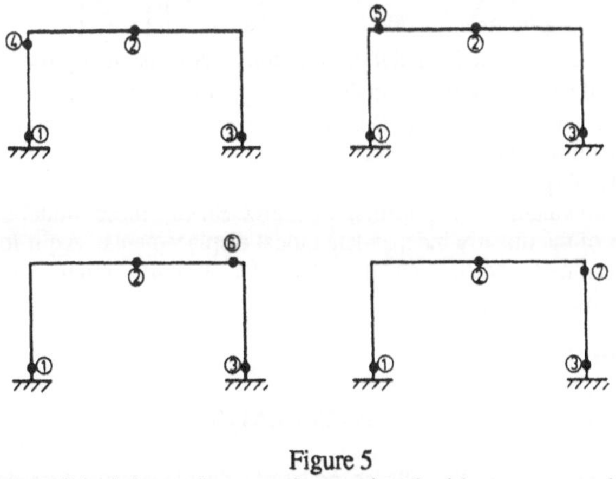

Figure 5

Any set of compatible deformation rates is associated with parameters that play the role of nodal displacements and can be written in the following form:

$$\Delta\theta = C\ \Delta q \qquad (12)$$

Equations similar to these can be written for any number of critical sections c_r and any number of independent mechanisms in the basis $b = c_r - \alpha$.

Similarly the displacement corresponding to the applied loads can be expressed in terms of the nodal displacements:

$$u = C_0 \, \Delta q \tag{13}$$

Hence the nodal kinematic relations become:

$$\begin{bmatrix} \Delta\theta \\ u \end{bmatrix} = \begin{bmatrix} C \\ C_0 \end{bmatrix} \Delta q \qquad \text{or} \qquad \begin{bmatrix} \Delta\theta^{+} - \Delta\theta^{-} \\ u \end{bmatrix} = \begin{bmatrix} C \\ C_0 \end{bmatrix} \Delta q \tag{14}$$

According to the hypothesis of the yield line theory, when the collapse mechanism is activated, the finite elements behave as rigid but angular discontinuities may be generated across the element sides whilst providing for continuity of vertical displacements (Ref.2).

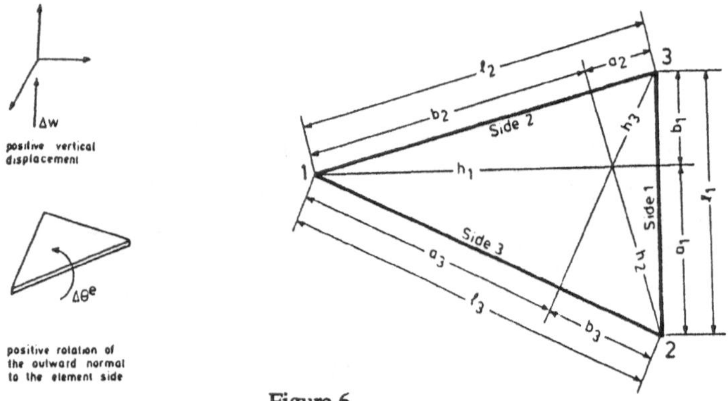

Figure 6

Taking one single finite element of a slab, as represented in Fig.6, the three relations $\Delta\theta_i^e$ of the outward normals to the three sides can then be expressed in terms of the three corner vertical displacements Δq in the following way:

$$\begin{bmatrix} \Delta\theta_1^e \\ \Delta\theta_2^e \\ \Delta\theta_3^e \end{bmatrix} = \begin{bmatrix} -1/h_1 & b_1/(l_1 h_1) & a_1/(l_1 h_1) \\ a_2/(l_2 h_2) & -1/h_2 & b_2/(l_2 h_2) \\ b_3/(l_3 h_3) & a_3/(l_3 h_3) & -1/h_3 \end{bmatrix} \begin{bmatrix} \Delta q_1 \\ \Delta q_2 \\ \Delta q_3 \end{bmatrix} \tag{15}$$

For an interelement side the total angular discontinuity $\Delta\theta_i$ is clearly the algebraic sum of the rotations $\Delta\theta_i^e$ of the two outward normals with respect to the two finite elements sharing such a side. The assemblage of relations (14) for all finite elements is thus readily performed and may be written in the following compact form:

$$\Delta\theta = C \, \Delta q \tag{16}$$

where, C is a (NS x NC) kinematic transformation matrix. Since these modal deformations are obtained as functions of the linearly independent modal displacements Δq it follows that such deformations are necessarily compatible and the rotation/displacement relations may be taken as the compatibility conditions.

3. Problem Statement

3.1. MATHEMATICAL PROGRAMMING FORMULATION

The identification of all the significant collapse modes of a ductile structural system is necessary in the analysis and evaluation of the system reliability, including the evaluation of the corresponding bounds. In this paper only the mathematical programming method corresponding to finding the most important mechanism is analysed. In the case of structures composed of ductile members such as components with elastic-perfectly plastic behaviour, the structural strength would be independent of the failure sequences of the components. It is usual to employ an approximate procedure called second order method (Ref.3) that only requires the mean and loading and structural resistance. For statistically independent random normal variables, the identification of the stochastically more important mechanism consists of finding the position of the limit-state equation closer to the origin of the reduced normal variables. This amounts to minimizing the distance - reliability index - β :

$$\min \beta = \sqrt{(m'_*)^2 + (\lambda'_F)^2} \tag{17}$$

By associating the rotations of the critical sections (or element sides) with the rotations of the members represented by the same random variable through the incidence matrix J_θ and the displacements (or deflections of the triangular elements centroids) corresponding to the loads linked by the same random variable through J_u, we have:

$$\Delta\theta_* = J_\theta \, \Delta\theta^+ + J_\theta \, \Delta\theta^- \quad ; \quad u_* = J_u \, u \tag{18}$$

The limit-state function represents the aptitude of the structure to support the loading equates the external and internal work produced by each mechanism:

$$g(m_*, \lambda) = m_*^T \Delta\theta_* - \lambda_F^T u_* = 0 \tag{19}$$

The relationships linking the reduced normal variables to the normal variables are:

$$m_* = \mu_M + \sigma_M \, m'_* \quad ; \quad \lambda_F = \mu_F + \sigma_F \, \lambda'_F \tag{20}$$

where μ_M, μ_F and σ_M, σ_F are the mean and standard deviation of the random variables m_* and λ_F, respectively. Another bilinear equation is yielded by substituting m_* and λ_F for (20) in the limit-state equation (19):

$$(\sigma_M \, m'_*)^T \Delta\theta_* - (\sigma_F \, \lambda'_F)^T u_* + \mu_M{}^T \Delta\theta_* - \mu_F{}^T u_* = 0 \tag{21}$$

In the limit-state equation (21) the plastic deformations of the mechanism are present as state variables. Since it is required to find the minimum distance from the origin of the reduced normal variates to the yield surface, the values of λ'_F and m'_* given by the minimum norm solution of (21) are:

$$m'_* = \frac{\sigma_M \, \Delta\theta_* \, [\mu_M{}^t \, \Delta\theta_* - \mu_F{}^t \, u_*]}{(\sigma_M)^{2\,t} (\Delta\theta_*)^2 + (\sigma_F)^{2\,t} \, (u_*)^2} \tag{22}$$

$$\lambda'_F = \frac{\sigma_F \, u_* \, [\mu_M{}^t \, \Delta\theta_* - \mu_F{}^t \, u_*]}{(\sigma_M)^{2\,t} (\Delta\theta_*)^2 + (\sigma_F)^{2\,t} \, (u_*)^2} \tag{23}$$

For positive $[\mu_M{}^t \, \Delta\theta_* - \mu_F{}^t \, u_*]$ the identification of the stocastic more relevant mechanism consists of finding:

$$\min \beta = \frac{\mu_M{}^t \, \Delta\theta_* - \mu_F{}^t \, u_*}{\sqrt{(\sigma_M)^{2\,t} (\Delta\theta_*)^2 + (\sigma_F)^{2\,t} \, (u_*)^2}} \tag{24}$$

subject to the linear incidence equations (18), the kinematic relations of the mesh description (9) and the sign constraints:

$$\Delta\theta^+ \geq 0, \; \Delta\theta^- \geq 0, \; u \geq 0, \; \Delta\theta_* \geq 0, \; u_* \geq 0 \tag{25}$$

In the nodal description, (14) is used instead of (9) to represent the kinematic relations and $\Delta q \ngeq 0$ should be added to the sign constraint (25).

The solution of these quadratic fractional programming problems is obtained by minimizing the quadratic concave function:

$$\min - 1/\beta^2 = -(\sigma_M \, \Delta\theta_*)^2 - (\sigma_F \, u_*)^2 \tag{26}$$

subject to (9), (18), (25) in the mesh description - (14), (18), (25) in the nodal description - and

$$\mu_M \, \Delta\theta_* - \mu_F \, u_* = 1 \tag{27}$$

The global optimum of these programs gives the plastic deformations for the stocastic more important mechanism and the random variables can be evaluated using (22)-(23).

3.2. SOLUTION ALGORITHMS

Finding of the global optimum in mathematical programming problems with nonconvex objective function and/or constraints is normally considered a NP-hard problem: The computational effort required grows exponentially with the number of variables in the worst case. One of the few cases for which there are algorithms available that work reasonably well is the minimization of a concave quadratic objective function subject to linear constraints.

3.2.1. Branch and Bound Techniques - The general nonconvex domain is transformed in the branch and bound (B & B) strategy into a sequence of intersecting convex domains by the use of underestimating convex functions. The two main ingredients are a combinatorial tree with appropriately defined nodes and some upper and lower bounds to the final solution associated with each node of the tree.

Each node of the tree in the B & B strategy is associated with a linear programming problem. For the concave function $-x^2$ defined in the interval [a, b] the convex underestimate is:

$$-(a+b)\,x + a\,b \le -x^2 \tag{28}$$

More details on the implementation of this technique can be found in Ref.4.

3.2.2. Cutting Plane Methods - Since the nonconvexity is restricted to the objective function, it is possible to solve this problem through techniques that exploit its special structure: its local solutions are vertices of the domain. In Konnos's algorithm (Ref.5) a local maximum is found and a cut is generated by the Simplex algorithm. In the next iteration, this procedure either generates a point which is strictly better than the last local maximum found, or generates a cut which is deeper until the convergence criteria to a ε-solution is met. This algorithm claims to be more efficient than the cutting plane methods suggested by Tui and Ritter.

4. Discussion

Acording to the yield line theory, the collapse of the slab is due to the formation of yield lines along which the slab folds when a mechanism can be activated. An automatic procedure to derive the trial mechanisms can be devised if the yield line method is formulated as a form of triangular finite element representation in which the yield lines are restricted to element sides. Numerical experience suggests that although cutting plane methods are more efficient than branch and bound techniques when the number of constraints and variables is small, the latter is easier to implement and more reliable. Simplex based algorithms were employed in each iteration of the branch and bound strategy. Since its efficiency is directly related to the number of variables and the cube of the number of constraints, the mesh description should be used for frames and the nodal description for slabs because they lead to smaller problems.

References

1. Simões, L.M.C. 'On the Reliability of Ductile Structural Systems by Mathematical Programming', submitted to *J. Struct. Div.* , *ASCE* , (1988).
2. Munro,J. and Da Fonseca, A.M.A. 'Yield Line Method by Finite Elements and Linear Programming', *The Structural Engineer* , **2** (1978).
3. Ditlevsen, O. 'Generalized Second Moment Reliability Index', *J. Struct. Mech.* , **7** (1979).
4. Simões, L.M.C. 'A Branch and Bound strategy for finding the Reliability Index with Nonconvex Performance Functions', *Structural Safety* , in publication, (1988).
5. Konno, H. 'Maximization of a Convex Quadratic Function under Linear Constraints', *Math. Program.*, **11** (1976).

STRUCTURAL SHAPE OPTIMIZATION IN MULTIDISCIPLINARY SYSTEM SYNTHESIS

Jaroslaw Sobieszczanski-Sobieski
NASA Langley Research Center, MS246
Hampton, Virginia 23669, U.S.A.

ABSTRACT. Structural shape optimization couples with other discipline optimization in the design of complex engineering systems. For instance, the wing structural weight and elastic deformations couple to aerodynamic loads and aircraft performance through drag. This coupling makes structural shape optimization a subtask in the overall vehicle synthesis. Decomposition methods for optimization and sensitivity analysis allow the specialized disciplinary methods to be used while the disciplines are temporarily decoupled, after which the interdisciplinary couplings are restored at the system level. Application of decomposition methods to structures-aerodynamics coupling in aircraft is outlined and illustrated with a numerical example of a transport aircraft. It is concluded that these methods may integrate structural and aerodynamic shape optimizations with the unified objective of the maximum aircraft performance.

Nomenclature.

a -	subscript referring to aerodynamics.
c -	airfoil chord.
C -	cumulative constraint.
$e(Y_*,X)$ -	vector of equality constraints.
$F(X)$ -	system objective function.
$G(X)$ -	vector of system inequality constraints.
h -	airfoil depth.
J_{as} -	$J(U_a,U_s)$, jacobian matrix of partial derivatives of U_a with respect to U_s.
l -	subscript indicating lower bound.
M -	Mach number.
s -	subscript referring to structures.
u -	subscript indicating upper bound.
$U(X)$ -	vector of system behavior variables.
U_i -	vector of behavior variables output from analysis of the i-th discipline.
X -	vector of system design variables.
Y_i -	vector of i-th subsystem design variables.

Introduction.

Structural optimization for shape has a long history of theoretical developments and successful applications, e.g., ref.1. It encompasses both the overall geometry of assembled structures, e.g., ref.2 (Ch.4 and 16), and the shape of individual components as in ref.1 (Ch.9). Its tool inventory includes a rich variety of methods ranging from the classical

G. I. N. Rozvany and B. L. Karihaloo (eds.), Structural Optimization, 331–338.

variational approaches reviewed in ref.2 (Ch.5) to sophisticated finite element-based techniques, e.g., ref.1 (Ch.5 and 6). These developments have reached the point where it becomes practical for an engineer to use shape optimization methods not only in design of a structure treated as an isolated object, but also in design of structure whose shape determines its interaction with the environment.

Aircraft structural design deals with probably the most difficult cases where structures and the surrounding airflow interact through the shape of the structure boundary. Similar examples may also be drawn from naval architecture and automotive industry. Based on aeronautical experience, this paper presents the structural shape optimization as part of the overall aircraft system synthesis in which the structural discipline must interact with aerodynamics and other disciplines in search for the shape that maximizes the flight vehicle performance.

Structures - Aerodynamics Coupling.

EXAMPLES OF COUPLING

One mechanism that couples structures and aerodynamics in near-sonic cruise aircraft is the wave drag which causes the wing drag to rise as the flight Mach number approaches umity. The drag rise shown in Fig.1 shifts toward lower M and becomes steeper as the h/c ratio increases. On the other hand, the structural weight of a bending-dominated wing generally decreases with the increase of h/c. Consequently, a conventional shape optimization performed for minimum weight within the discipline of structures would tend to increase h/c, while an aerodynamic optimization for minimum drag at a constant M would tend to lower h/c. However, both weight and drag are detrimental to flight performance, such as payload delivered over a given range for a constant gross weight at take-off. Thus, a compromise between the two conflicting trends must be sought, by formulating a unified optimization problem, drawing its objective function from the system performance, and including constraints from aerodynamics and structures.

The optimization task is further complicated because the wing structural and aerodynamic behavior are coupled not only through the system performance but also directly via aerodynamic loads whose magnitude and distribution on the wing depend on the wing planform and airfoil shapes and on the wing structural deformations. Dependence of the deformations on the loads and wing shape completes the coupling.

Swept wings provide a second example of the effect of this type of coupling on the metalic wing structural weight plotted in Fig.1 as a function of sweep angle. Lift in a forward-swept wing increases the streamline airfoil angle of attack, that generates more lift, and so on - a positive feedback that may result in a wing divergence. A supercritical airfoil wing is a third example of the structures-aerodynamic coupling, again, through the loads and deformations. As shown in Fig.2, the supercritical airfoil's lift resultant is shifted aft relative to that of a conventional airfoil. As a result, torque on the wing box is increased and this may require more structural material for strength, and also for stiffness necessary to keep the wing from twisting excessively which redistributes the lift inboard with an attendant increase of induced drag.

METHODS FOR COUPLED, STRUCTURES - AERODYNAMIC OPTIMIZATION

The structures-aerodynamics-performance interactions may be conveniently presented in the form of a directed graph (a means widely used in Operations Research (OR) system analysis, e.g., ref.3) shown in Fig.3. Each discipline on the graph represents a disciplinary analysis which from a system's perspective is simply a black box transforming input into output. For the wing design, the inputs and outputs are inscribed by the arrows in the figure. Assuming that structure must fit inside of aerodynamic envelope, the external shape geometry is shown as an input from aerodynamics to structures. The loads and deformations add to the direct coupling of the two disciplines, the remainder of the coupling is through performance analysis. This representation of a wing as a system corresponds to the aspect decomposition defined in OR (ref.4) as one in which the system itself remains an indivisible object but each of its aspects (disciplines) is represented by a black box - as opposed to the object decomposition applicable when a black box representation may be assigned to each physically separable subsystems. Without decomposition, the structural shape optimization and aerodynamic design methods (e.g., ref.5) cannot be used to their full potential in design of a wing as a coupled system because neither design method accounts for the mutual influence of the two disciplines, and, therefore, the individual optimizations may work on cross-purposes. Further, optimization without decomposition requires repetition of the entire system analysis for every design variable perturbation (a multidimensional parametric study) to determine the full effect of the variables on the system performance. In contrast, the system decomposition approach temporarily decouples the disciplines and permits the application of specialized methods inside each black box. This is crucially important to being able to use the disciplinary advanced methods in design of coupled systems.

System Optimization Algorithms.

There are several ways the system optimization algorithm may be organized and the distinguishing element is usually the means of restoring the coupling to the temporarily decomposed system. Two recently developed algorithms are outlined.

HIERARCHAL, TOP-DOWN DECOMPOSITION.

When system optimization by hierarchal decomposition (ref.6, 7, and 8) is applied, the design variables are divided into the system variables that directly affect aerodynamics, structures, and performance, and the local (subsystem) variables which directly affect only aerodynamics, or structures. By these criteria, the wing planform and airfoil shape variables are system variables because they affect all three analyses. The structural weight is also a system variable since it enters the performance analysis. Amplitudes of basis functions representing the wing structural deformation are also variables at the system level due to the influence they have on aerodynamics and structural stiffness. In contrast, the structural cross-sectional dimensions are local variables affecting only the structure, and the variables defining the wing shape outside of the structural box are local variables affecting only aerodynamics.

The system optimization proceeds top down. The optimization problem solved at the system level in the i-th iteration is:

find X, within move limits

$X_l <= X <= X_u$,

such that F(X) is minimum, subject to constraints

$$G(X) <= 0,\ C_a <= 0,\ C_s <= 0. \quad (1)$$

Examples of F(X) and G(X) are the take-off gross weight, and the required climb rate. The constraints C are cumulative constraints formulated using the Kreisselmeier-Steinhauser envelope function (ref.9). They are differentiable functions of the constraints local to aerodynamics and structures, and approximate the largest (most violated) constraint providing a single measure of feasibility of each subsystem design. In eq. 1, the cumulative constraints are approximated as functions of X by linear extrapolation from which the approach takes its name. For example, the extrapolation for C_s is

$$C_s = C_{smin} + \nabla_x C_s \, \Delta X \tag{2}$$

The derivatives forming the $\nabla_x C_s$ are carried from the aerodynamic and structural optimizations executed in the previous iteration. These subsystem optimizations may be generalized by a single definition, using * as a substitute for a and s:

find Y_* such that $C_*(Y_*,X)$ is minimum, subject to constraints

$$e(Y_*,X) = 0, \quad Y_{*l} <= Y_* <= Y_{*u}; \tag{3}$$

The purpose of the optimization is to proportion the wing shape and structure cross-sections to minimize the aerodynamic and structural constraint violations represented by C. The equality constraints, e, enforce compliance with the prescribed structural weight and structural deformation. The aerodynamic loads used in the optimization of the structure in the i-th iteration are computed in iteration i-1. The aerodynamic and structural optimizations are, in effect, decoupled and may be carried out concurrently. Optimum sensitivity analysis using the algorithm described in ref.10 or 11 yields derivatives of C_{*min} with respect to X to form the $\nabla_x C_*$ used in eq.2. Thus, the shape design decisions made at the system level to improve the system performance include an approximate knowledge of their effects on aerodynamic and structural constraints.

NON-HIERARCHAL DECOMPOSITION.

Under this approach, decomposition is introduced only for the purposes of the system behavior sensitivity analysis while optimization is carried out for the entire system. The sensitivity analysis yields the derivatives used in linear extropolations of the system behavior which substitute for system analysis in optimization performed in a piecewise-linear manner. The algorithm for sensitivity analysis by decomposition from ref.12 yields the derivatives of the system response as a solution of a set of simultaneous, linear, algebraic equations

$$\begin{bmatrix} I & -J_{as} \\ -J_{sa} & I \end{bmatrix} \begin{Bmatrix} dU_a/dX_x \\ dU_s/dX_k \end{Bmatrix} = \begin{Bmatrix} \partial U_a/\partial X_k \\ \partial U_s/\partial X_k \end{Bmatrix} \tag{4}$$

The Jacobians in the matrix of coefficients in eq. 4 represent sensitivity of the output from one disciplinary analysis to the inputs received from the other, e.g., partial derivatives of aerodynamic pressure with respect to the structural deformations. Similarly, the partial derivatives on the right hand side measure sensitivity of the output from aerodynamic and structural analyses to the design variables, e.g., partial derivatives of structural deformation to the wing sweep angle as a shape variable. The partial derivatives, by definition, may be computed within each decoupled disciplinary analysis, so that specialized methods of disciplinary sensitivity analysis may be used, such as structural analysis for sensitivity with respect to shape, ref.1 (Ch.3), and the corresponding analysis in CFD formulated in ref.13. The method is new and experiences with its use in optimization is limited to ref. 14.

Numerical Example.

An optimization study in which hierarchal, top-down decomposition of the type outlined in the foregoing was applied to a transport aircraft shown in Fig.4 was reported in ref.15. The objective was to minimize fuel consumption subject to constraints on the aircraft performance, aerodynamics, and structures, including the detailed local buckling constraints of the stiffened panel in the wing covers. The design variables governed the airfoil shape and wing structure cross-sectional dimensions, down to the stringer detail level. Individual cover panel, ribs, and spars were represented in the finite element model. The procedure performance was satisfactory; a typical histogram for the fuel objective and structural weight is shown in Fig.5. It demonstrated that a large number of design variables (1303), including shape variables, and constraints (1950) representing diverse disciplines may be included in a large engineering system optimization, and that the system performance may be linked mathematically to the design details.

Conclusions

When shape optimization is applied to a structure interacting with the surrounding airflow, the optimization task becomes multidisciplinary and has to account for the coupling of structural mechanics and aerodynamics involving the mutual dependence of structural deformations and aerodynamic pressures. Aircraft design for optimum shape is an example of a large-scale engineering problem where the above occurs and where the shape optimization becomes a subtask in a system optimization because both structures and aerodynamics have a strong impact on performance.

Methods for system optimization by decomposition have been developed for such multidisciplinary applications. They allow use of specialized methods for optimization and sensitivity, by separating the disciplinary optimization tasks while accounting for the interdisciplinary couplings. As illustrated by a numerical example for a large transport aircraft, these methods have begun to approach the level of maturity required for effective applications in large scale engineering design.

References.

1. The Optimum Shape, Automated Structural Design ; editors: Bennett, J. A., and Botkin, M. E.; Plenum Press, N.Y., 1986.
2. New Directions in Optimum Structural Design ; editors: Atrek, E.; Gallagher, R. H.; Ragsdell, K. M.; Zienkiewicz, O.C.; John Wiley & Sons, 1984.
3. Steward, D. V.; Systems Analysis and Management ; PBI, N.Y., 1981.
4. Archer, B.:'The Implication for the Study of Design Methods of Recent Developments in neighbouring Disciplines'. International

Conference on Engineering Design, Hamburg, Aug. 1985, Proceedings of, vol.**1**, pp.833-840, publ. Heurista, Zurich.
5. Shankar, V.: 'A Full Potential Inverse Method Based on a Density Linearization Scheme for Wing Design'. AIAA 14th Fluid and Plasma Dynamics Conference, Palo Alto, Cal., June 23-25, 1981.
6. Sobieszczanski-Sobieski, J.: A Linear Decomposition Method for Large Optimization Problems - Blueprint for Development. NASA TM 83248, Feb. 1982.
7. Sobieszczanski-Sobieski, J.; Barthelemy, J. M.; and Giles, G. L.: 'Aerospace Engineering Design by Systematic Decomposition and Multilevel Optimization'; Paper No ICAS-84-4.7.3; 14th Congress of the International Council of the Aeronautical Sciences, Toulouse, France, Sept. 1984.
8. Sobieszczanski-Sobieski, J.; and Barthelemy, J. M.: 'Improving Engineering System Design by Formal Decomposition, Sensitivity Analysis, and Optimization; International Conference on Engineering Design', Hamburg, Aug. 1985; Proceedings of, pp.314-321.
9. Kreisselmeier, G; and Steinhauser, R.: 'Systematic Control Design by Optimizing a Vector Performance Index'; International Federation of Active Control Symposium on Computer-Aided Design of Control Systems, Zurich, Switzerland, Aug. 29-31, 1979.
10. Sobieszczanski-Sobieski, J.; and Barthelemy, J. M.; and Riley, K. M.: 'Sensitivity of Optimum Solutions to Problem Parameters'; AIAA J, Vol.**20**, Sept. 1982, pp. 1291-1299.
11. Sobieszczanski-Sobieski, J.; and Barthelemy, J. M.: 'Optimum Sensitivity Derivatives of Objective Functions in Nonlinear Programing'; AIAA J, Vol.**21**, No.6, June 1983, pp.913-915.
12. Sobieszczanski-Sobieski, J.: On the Sensitivity of Complex, Internally Coupled Systems, NASA TM 100537 January 1988.
13. Yates, E. C., Jr.: Aerodynamic Sensitivities from Subsonic, Sonic,and Supersonic Unsteady, Nonplanar Lifting-Surface Theory; NASA TM 100502, Sept. 1987.
14. Sobieszczanski-Sobieski, J.; Bloebaum, K.; and Hajela, P.: Sensitivity of Control-Augmented Structure Obtained by a System Decomposition Method, NASA TM 100535 January 1988.
15. Wrenn, G. A.; and Dovi, A. R.: 'Multilevel Decomposition Approach to the Preliminary Sizing of a Transport Aircraft Wing'; AIAA/ASME/ASCE/AHS 28th Structures, Dynamics, and Materials Conference, Monterey, Cal., April 6-8, 1987, AIAA Paper No. 87-0714-CP.

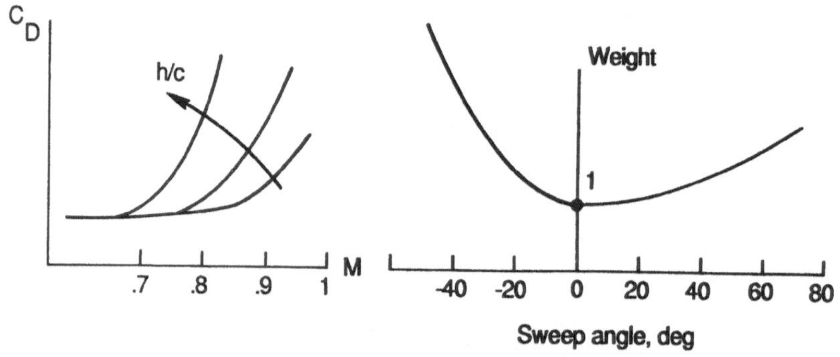

Figure 1. left- Drag rise as a function or M and h/c; right- Structural weight of a wing vs. the wing sweep angle(positive = aft).

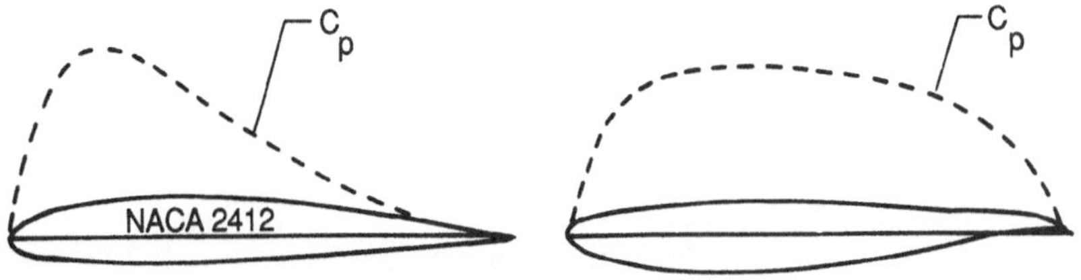

Figure 2. Aft shift of the lift resultant on supercritical airfoil (right) vs. conventional airfoil (left).

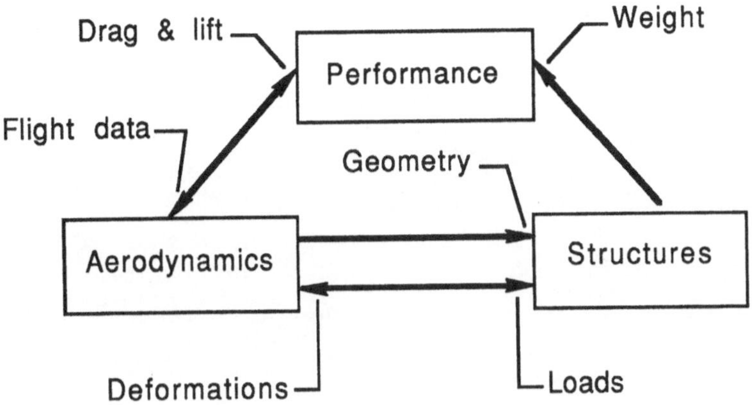

Figure 3. Disciplinary analyses coupled in wing design.

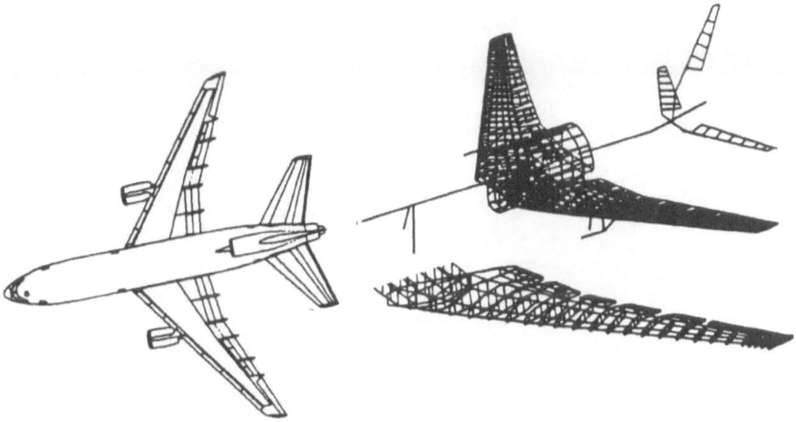

Figure 4. Transport aircraft wing optimization, ref 15.:aircraft, and its finite element model.

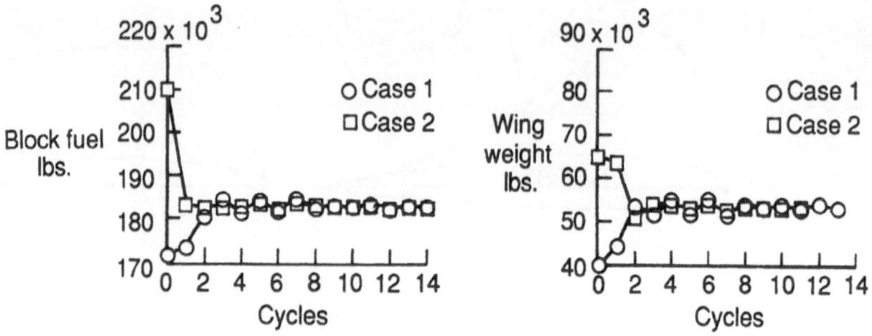

Figure 5. Histograms for the fuel consumption as objective function (left) and wing structural weight (right) for transport aircraft wing (case 1, 2: initial design infeasible and feasible, respectively).

BOUNDARY ELEMENT APPROACH TO OPTIMAL STRUCTURAL DESIGN BASED ON THE INVERSE VARIATIONAL PRINCIPLE

Yukio Tada
Department of Systems Engineering
Faculty of Engineering
Kobe University
Rokkodai, Nada, Kobe 657
Japan

Hideki Taketani
Department of Control Systems
Nagasaki Works
Mitsubishi Electric Corp.
Maruo-6-14, Nagasaki 850-91
Japan

ABSTRACT. This paper proposes the use of the boundary element method in the optimal structural design based on the inverse variational principle. The condition for optimum in the inverse variational shape determination method can be expressed in terms of the boundary data alone. Firstly, using the displacements obtained by BEM, the shape of a body is reformed so that the strain energy density might become uniform everywhere on the designed surface. Volume changes in the iterative process are calculated by the movements of the extreme points of the BEM. Optimum shapes are obtained by a comparable number of iterations with the finite element approach. The boundary element approach has the merit that the initial division into elements and redivision for reanalysis can be more easily performed than in FEM.

1. INTRODUCTION

Shape determination problem is one of categories of optimum structural design, where the optimum shapes of structures are searched for under several conditions. Various schemes for optimum structural design have been proposed already[1]. Authors also have proposed a shape determination method[2] based on the inverse variational principle[3]. Most of shape determination methods, including our methods, make models of the objective bodies using the finite element method for the analysis and design, and create a shape freely by reforming an assumed shape through the movements of nodal points on the outer surface in contrast to 'parameter design'. In the finite element method, however, the whole region of the objective body must be discretized into elements, and the

G. I. N. Rozvany and B. L. Karihaloo (eds.), Structural Optimization, 339–346.

work of making and inputting data is a very cumbersome task. Moreover, in most cases, the shape assumed as the initial solution varies largely with the process of shape reformation, and shapes of elements are also distorted. It is important to keep the uniformity of elements in both shape and size during shape reformation process in order to obtain sufficient accuracy for finite element analysis. It is, however, difficult to set a good discretizing scheme from the outset so that it may give a uniform mesh after shape reformation. Therefore, redivision into elements is necessary.

Incidentally, the boundary element method has recently received remarkable notice as a numerical method for solving several physical problems. It has a great merit that the work of inputting data is carried out easily because it discretizes only the boundary of the body. In the area of optimum structural design, some applications of BEM are observed, some of which aim to obtain uniform stress state on the surface of the body[4].

As the optimality conditions of the inverse variational problem are expressed in terms of the boundary data alone, it is supposed that the boundary element method can make the optimization scheme more effective than by FEM. This paper proposes the use of the boundary element method in the optimal structural design based on the inverse variational principle and shows procedures for the numerical treatment.

2. INVERSE VARIATIONAL PROBLEM

2.1. Inverse Variational Principle

In the inverse variational principle, variations of the energy functionals are considered with respect to variables which represent the shape of a body as well as those which represent behaviors of the body, under the condition that a certain global invariant on structural shape such as volume and surface area is specified. The inverse variational principle or the principle of the minimum potential energy considers such a shape and displacement that make the total potential energy of the body stationary with the subsidiary condition of the volume constancy. When the specified volume is C, the principle is stated as follows,

$$\Omega=\int_V(1/2C_{ijkl}\varepsilon_{ij}\varepsilon_{kl}-K_iu_i)dV-\int_{Sp}P_iu_idS+\lambda(\int_V dV-C)$$

$$\longrightarrow \text{stationary,} \quad (1)$$

where u_i, K_i and P_i are components of displacement, volume force and surface traction, respectively. ε_{kl} is a component of strain tensor and has the relationship between σ_{ij}, a component of stress tensor, which obeys the Hooke's law. The stationary conditions of Eq.(1) are obtained by taking variations of Ω with respect to each variable, that is,

[a]from δu_i --> Equilibrium conditions in the volume region V and boundary conditions on the surface S_p.

[b]from $\delta\lambda$ --> Condition of volume constancy.

[c]from δn --> Equation for optimal shape, that is,

$$(^1/_2)C_{ijkl}\varepsilon_{ij}\varepsilon_{kl}-K_iu_i+\lambda-\partial(P_iu_i)/\partial n+2HP_iu_i=0 \text{ on surface S }, \quad (2)$$

where n is an outward distance in the direction of the normal line from the surface and H is the mean curvature. In a previous report[2], the above problem was solved by formulating it with finite element method and the usefulness of the method was shown. However, as the condition for the optimal shape is described only on the surface [Eq.(2)], it can be said that only quantities on the surface of the body are needed in the evaluation of the optimality for the shape. Therefore, the use of the boundary element method which uses data on the boundary alone is expected in place of finite element method which divides the whole body.

2.2. Representation of stationary condition by boundary data

2.2.1. Equilibrium condition. On considering the variation of Eq.(1) with respect to displacement, if the displacement u_i* of an infinite plate which is subjected to a unit concentrated load is adopted as an admissible displacement field, then the equation which represents the displacement of the point on the boundary can be obtained by a similar procedure to boundary element method. Hence, the equilibrium condition, one of the stationary condition of the functional, is replaced by the boundary integral equation.

2.2.2. Condition of the volume constancy. The constraint of the constant volume is interpreted that the volume of the body must not be changed, that is, change of the volume is zero.

2.2.3. Condition for the determination of shape. If the volume force is disregarded and no surface traction is applied at the designable surface (that is, the surface which can be reformed), then Eq.(2) for the shape determination is reduced to

$$(^1/_2)C_{ijkl}\varepsilon_{ij}\varepsilon_{kl}+\lambda=0 \text{ on S }. \quad (3)$$

This means that the strain energy density is uniform on the surface S.

As said above, all stationary conditions are expressed in terms of quantities on the boundary. Therefore, it becomes possible to solve the inverse variational problem by a boundary element approach.

3. PROCEDURE FOR SOLVING THE INVERSE VARIATIONAL PROBLEM

In this paper, an iterative scheme is adopted which repeats the analysis by BEM and the reformation of shape to obtain an optimal shape, according to the strain energy density and the volume change.

In the boundary element method, the structural analysis is performed by dividing the boundary of a body and its shape is represented by the coordinates of extreme points of each element. Hence, the points are regarded as representative ones for the shape reformation.

3.1. Shape reformation by strain energy density

This paper uses the constant-element in BEM, which assumes constant displacements and surface tractions on each boundary element, so the strain energy density is constant in an element, but is discontinuous at extreme points.

As our purpose is to make the strain energy density uniform on the designable surface, firstly we evaluate strain energy densities element by element and reform the shape by following procedure .

Let the number of boundary elements be n and the strain energy density of the i-th element E_i. Then, Eq.(3) reduces to

$$E_i=-\lambda=\text{constant} \quad (i=1,\dots,n) \ . \tag{4}$$

From the characteristic of the strain energy density, it may be assumed that thin parts of the body have large strain energies and conversely thick parts have small values. Therefore, in order to obtain uniform E_i for each i, a shape reformation scheme in the direction of the outward normal of the surface is considered, as

$$\delta\theta_i=\alpha\cdot\log(E_i/\bar{E}) \quad (i=1,\dots,n) \ , \tag{5}$$

where α is a positive constant depending on a representative size of the objective structure. $\bar{E}$ is the mean of n E_i's, that is,

$$\bar{E}=\sum_{i=1}^{n} E_i/n \ , \tag{6}$$

which is adopted as an estimated value of the actual $(-\lambda)$. Because the strain energy density has discontinuity at extreme points as mentioned before, the amount of movement for the i-th extreme point, $\delta\xi_i$ is derived by a vector addition of $\delta\theta_{i-1}/2$ and $\delta\theta_i/2$ as shown in Fig.1(a).

3.2. Treatment of volume constraint

Design variables ξ_i updated by the previous method may violate the constraint of the volume constancy and have a new volume $V^*(\neq C)$. Therefore, the second reformation

$$\xi^{i+1}=\xi^*+\beta\xi^* \tag{7}$$

is applied to the body in order to make the volume given constant C. The unknown coefficient β is determined by a linear approximation of the volume using the Taylor expansion, that is

$$\beta=(C-V^*)/(\sum_{i=1}^{n} \xi_i^* \ \partial V^*/\partial\xi_i) \ . \tag{8}$$

In this paper, quantities in Eq.(8) are obtained by using only movements

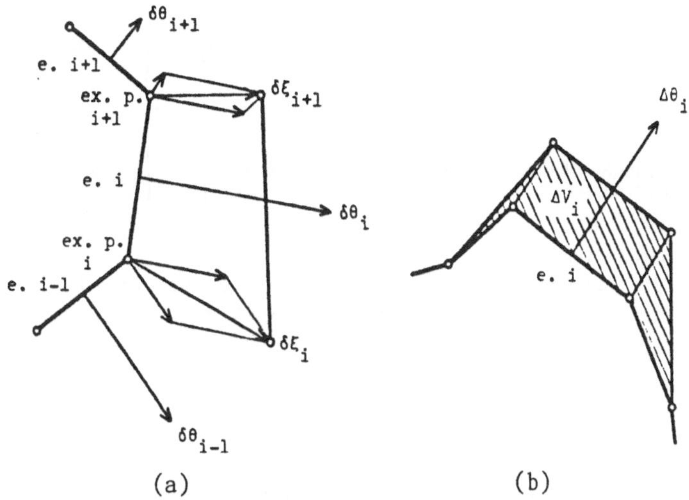

Fig.1 Shape reformation
(a)Reformation by strain energy density
(b)Calculation of $\Delta V/\Delta\theta_i$

of extreme points and not considering the whole body. The change of the volume after the first shape reformation is calculated by summing up those of all elements. As ξ_i is dependent on θ_i and θ_{i-1},

$$\partial V/\partial\xi_i=(\partial V/\partial\theta_{i-1})(\partial\theta_{i-1}/\partial\xi_i)+(\partial V/\partial\theta_i)(\partial\theta_i/\partial\xi_i) \ . \tag{9}$$

$\partial V/\partial\theta_i$ is evaluated as $\Delta V/\Delta\theta_i$ and calculated from the volume change by infinitesimal $\Delta\theta_i$ as Fig.1(b). Similarly, $\partial\theta_{i-1}/\partial\xi_i$ is regarded as $\Delta\theta_{i-1}/\Delta\xi_i$ and $\Delta\xi_i$ is derived from $\Delta\theta_{i-1}/2$ and $\Delta\theta_i/2$ by the second cosine rule.

3.3. Judgment for convergence

The termination of the iterative process is judged by

$$\varepsilon=\max_i\left|(E_i-\bar{E})/\bar{E}\right|\leq\varepsilon_o \ , \tag{10}$$

4. EXAMPLES OF SHAPE DETERMINATION

4.1. Both-end-clamped beam

First example is a beam whose ends are both clamped and which is subjected to a concentrated load at the midpoint of its span. Figure 2(a) shows an initial shape and the distribution of strain energy density for each boundary element due to the concentrated load. In this example, the upper and lower surfaces are designable ones and their shapes are

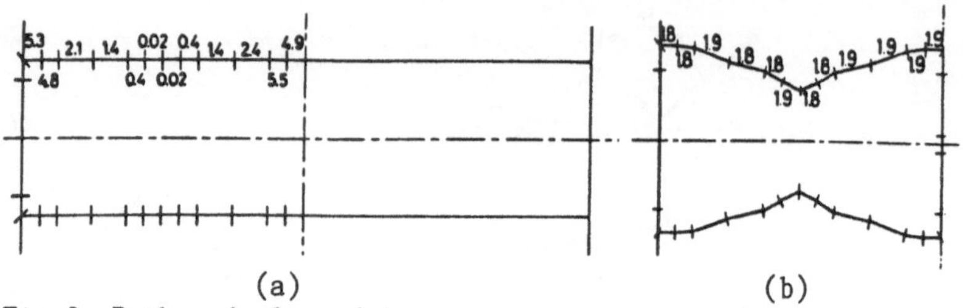

Fig.2 Both-end-clamped beam
(a)Initial shape and strain energy density (x10^{-8})
(b)Convergent shape and strain energy density (x10^{-8})

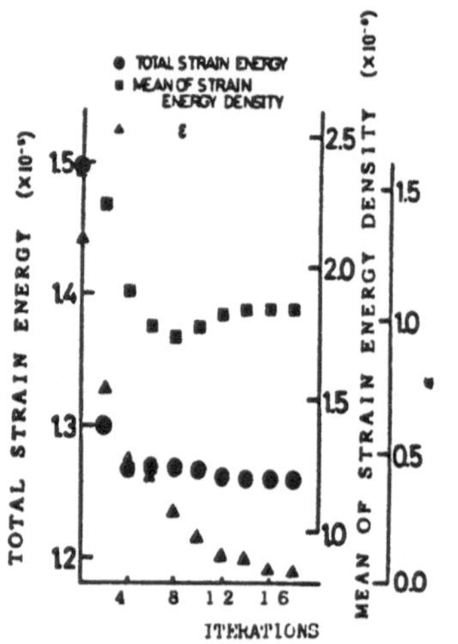

Fig.3 Process to convergence in the shape determination of beam

reformed in the vertical direction. The amounts of movements for extreme points in this case are evaluated similarly to that in the section 3.1. Figure 2(b) shows the convergent shape and the distribution of strain energy density in it for the same load. Only the left hand side is drawn because of the symmetry. The supported and loaded sections, where the strain energy density is high for the initial shape, become thick and the uniformity of strain energy density is accomplished. In Fig.3, the process to convergence is plotted with iteration numbers; each circle shows the total strain energy stored in each body, that is

$$E_T=\int_V(^1/_2)C_{ijkl}\varepsilon_{ij}\varepsilon_{kl}dV \ , \tag{11}$$

which is obtained from the work done by the outer load. Squares are the means of strain energy densities, $\overline{E}$, in Eq.(6). Triangles are ε's in Eq.(10), which represent the degree of the variance among E_i's It is observed from the figure that the total strain energy decreases as the shape reformation proceeds and becomes stationary at its minimum value. This means the stationarity of the total potential energy, which is the original purpose of the inverse variational problem. The behavior of ε in Fig.3 shows the stableness of the present optimization process.

4.2. Plate with a hole

The shape determination problem of a hole in a plate under biaxial loading isconsidered as the second example [Fig.4(a)]. The ratio of the horizontal force to the vertical force is 1/2. The shape of the hole is reformed in the normal direction to the element and a quarter region is analyzed for the symmetry. Figures 4(b) and (c) show the initial and convergent shapes, respectively. The values in the figures are strain energy densities in each shape. The convergent shape has a uniform distribution of the strain energy density by elongating the hole in the vertical direction. The solid line of the left side in Fig.4(c) is an ellipse with the aspect ratio of 2:1, which is known as the optimal shape of a hole in the infinite plate under the same loading condition as this example[5].

4.3. Γ-shaped structure

The object of the third example is a Γ-shaped structure which is subjected to a uniform distributed load at its right shoulder (the part of the protrusion) as shown in Fig.5(a). The designable surface is the left side surface, and it is reformed in the horizontal direction. The angle of the corner is rounded off from the outset in the initial shape in order to avoid the effect of the corner point and because it can be expected that the strain energy density is small in the left corner part in this structure. Fig.5(b) is the convergent shape.

A similar problem was solved by using finite element method[2] and the result is shown in Fig.5(c). As it was difficult in this example to distribute inner nodal points automatically and suitably for a variable shape, a new revised initial shape was given and the body was divided

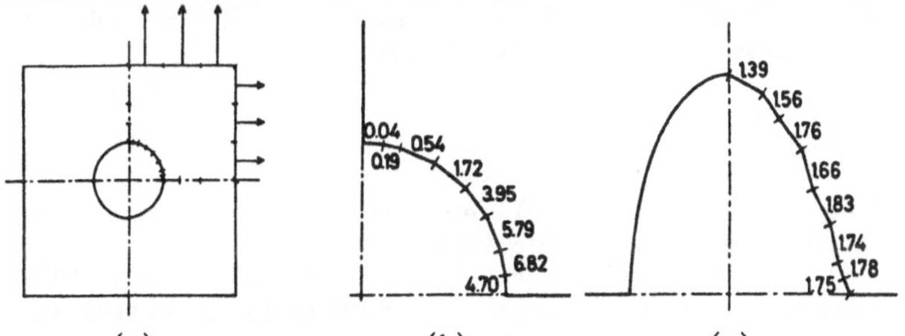

(a) (b) (c)

Fig.4 Plate with a hole
(a)Initial shape (b)Strain energy density (x10^{-4})
(c)Convergent shape, strain energy density (x10^{-4}) and analytical solution of infinite plate

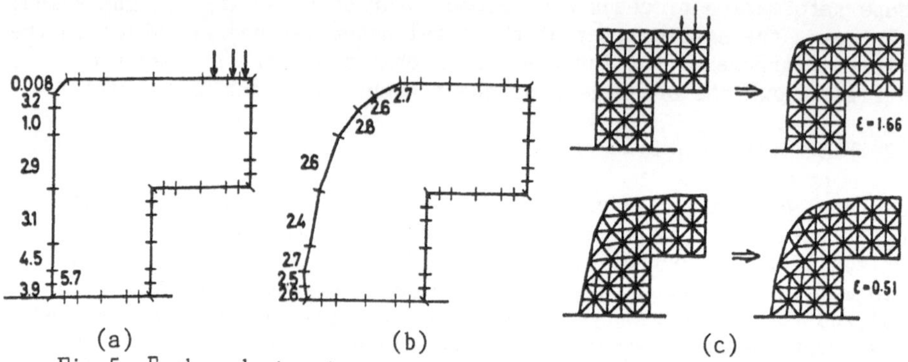

Fig.5 Γ-shaped structure
(a)Initial shape and strain energy density ($x10^{-5}$)
(b)Convergent shape and strain energy density ($x10^{-5}$)
(c)Optimum shape by finite element approach

into elements afresh.

5. CONCLUSIONS

From several numerical examples, the following is concluded;
(1) As the inverse variational problem is the one that is constituted by the combination of the analysis and shape determination, the accuracy of the former has a great influence on the result of the latter.
(2) The present approach can produce similar results to those obtained by finite element approach, with a less work for data such as the mesh division and with the same number of iterations.
(3) In this paper, the mean of the strain energy densities, $\overline{E}$, is adopted as a standard for the calculation of the amounts of the shape reformation[Eq.(5)]. But, leaving out the constraint of the volume constancy, if a certain specified value for the strain energy density is adopted as a denominator in Eq.(5) instead of $\overline{E}$, then the present procedure can be applied to the problem of the minimum weight design under the constraint of the constant strain energy density.

REFERENCES

[1] Atrek, E. et al. (Eds.), *New Directions in Optimum Structural Design*, (1984), 197, John Wiley & Sons.
[2] Seguchi, Y. and Tada, Y., *ACTA TECHNICA ČSAV*, 24-2 (1979), 139.
[3] Horák, V., *Inverse Variational Principles of Continuum Mechanics*, (1969), 25, Rozpravy ČSAV.
[4] Mota Soares, C. A., Rodrigues, H. C., and Choi, K. K., *Trans. ASME, J. Transm. Autom. Des.*, 106-4 (1984), 518.
[5] Kristensen, E. S. and Madsen, N. F., *Int. J. Numer. Methods in Eng.*, 10-5 (1976), 1007.

OPTIMAL SHAPE OF LEAST WEIGHT ARCHES

K.L. Teo
Department of Mathematics
University of Western Australia
Nedlands, W.A. 6009
Australia

C.M. Wang
Dept of Civil Engineering
National University of
Singapore, Kent Ridge
Singapore 0511

ABSTRACT. The paper concerns the shape optimization of plastically designed arches under bending and axial compression. In addition to the arch shape being unknown *a priori*, the problem is further made difficult by a nonsmooth objective functional. A method for solving this class of nonlinearly constrained nonsmooth optimization problem is presented and illustrated with some numerical examples.

1. INTRODUCTION

Shape optimization of fully stressed funicular arches for minimum weight has been investigated by Rozvany and Wang [1,2]. They derived optimality conditions which form the basis for the development of a general procedure for constructing the optimal shape of arches under any given vertical load system with or without selfweight and having supports at the same or different levels. More recent studies include optimization of multispan arches with variable support locations [3] and arches of unknown constant cross-section [4-7]. Note that although Refs [4-7] dealt with cables, but when inverted these cables become arches.

The investigations have not been only confine to isolated arches but also cover archnetworks. Rozvany and Prager [8] have derived the most efficient (minimum-weight) solution for two-way arch systems (or archgrids). Subsequent publications, on archgrids shape optimization, incorporated the effects of selfweight [9], simplified and refined the algorithm for solution as some arches may take on a zero cross-sectional area [10] and extended the original formulation to cover a system of parallel arches running in any number of arbitrary directions [11].

So far, the above arch structures discussed have been subjected to axial forces only. Shape optimization of arches under bending and axial compression has only been briefly treated by Rozvany, *et. al.* [12]. The current work examines this nonfunicular arch problem in greater detail. The optimization problem is challenging because the shape of the arch axis is unknown from the outset and the objective

G. I. N. Rozvany and B. L. Karihaloo (eds.), Structural Optimization, 347–354.

weight function is nonsmooth. To overcome these difficulties, spline functions have been employed to parameterize the unspecified arch axis and a smoothing function used to approximate the nonsmooth function. The resulting standard constrained optimization problem can then be solved by any nonlinear programming method.

2. PROBLEM FORMULATION

Consider the plastic design of an arch of constant depth, d, and whose material obeys the yield condition (see Fig. 1a)

$$\frac{|\bar{M}|}{M_o} + \left(\frac{\bar{N}}{N_o}\right)^{k+1} = 1 \tag{1}$$

in which $\bar{M}$ is the bending moment, $\bar{N}$ the compressive axial force and M_o, N_o the yield moment and yield force with indicator, k, given by

$k = 1,\ M_o = bd^2\sigma_o/4,\ N_o = bd\sigma_o,$	for a rectangular cross-section of variable width, b	(2)
$k = 0,\ M_o = btd\sigma_o,\ N_o = 2bt\sigma_o,$	for an ideal sandwich wide-flange section of variable width, b and thickness, t.	

where σ_o is the yield stress.

In view of Eqs. (1), (2) and denoting $\bar{\beta} = 4/d$, the total weight of the arch is given by

$$\bar{\Phi} = \int_0^L (\gamma/2\sigma_o)[\bar{\beta}|\bar{M}| + \sqrt{k\bar{\beta}^2\bar{M}^2 + 4\bar{N}^2}]\sqrt{1 + (\bar{y}')^2}\, d\bar{x} \tag{3}$$

in which L is the span of the arch; $\bar{x}$ the horizontal distance measured from one end of the arch; $\bar{y}$ the unspecified centroidal axis of the arch, γ the specific weight of arch material and the prime denotes differentiation with respect to $\bar{x}$.

Using the following nondimensional parameters,

$$x = \bar{x}/L;\ y = \bar{y}/L;\ W = \int_0^1 w dx;\ \alpha = W/H > 0;\ M = \bar{M}/WL;\ N = \bar{N}/W;$$

$$\lambda_i = \bar{M}_i/WL,\ i = 1,2;\ \beta = \bar{\beta}L;\ \Phi = 2\bar{\Phi}\sigma_o/\gamma, \tag{4}$$

in which w(x) is the load distribution, H the horizontal force, and $\bar{M}_i$ are the support moments (see Fig. 1b), the optimization problem can then be expressed as

$$\underset{\lambda,\alpha,y}{\text{Min}}\ \Phi = \int_0^1 \{[\beta|M| + \sqrt{k\,\beta^2M^2 + 4N^2}]\sqrt{1 + (y')^2}\}dx \tag{5}$$

where from statical considerations

$$M = \lambda_1(1-x) - x(1-\lambda_2) + (y-x\tan\theta)/\alpha$$
$$+ [x\int_0^x wdx - \int_0^x wxdx + x\int_0^1 wxdx]/W \tag{6}$$

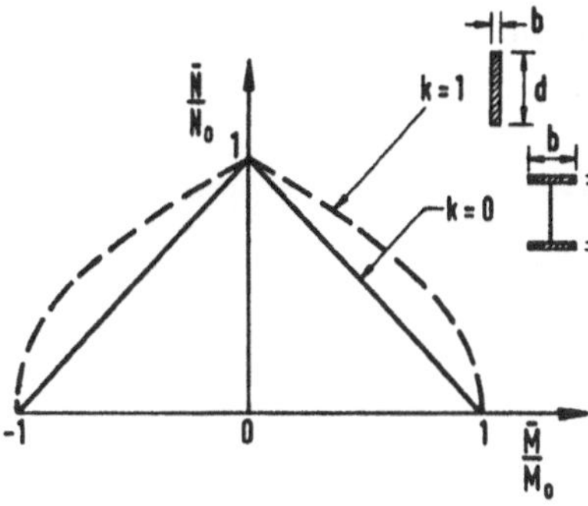

Fig. 1(a) Yield condition

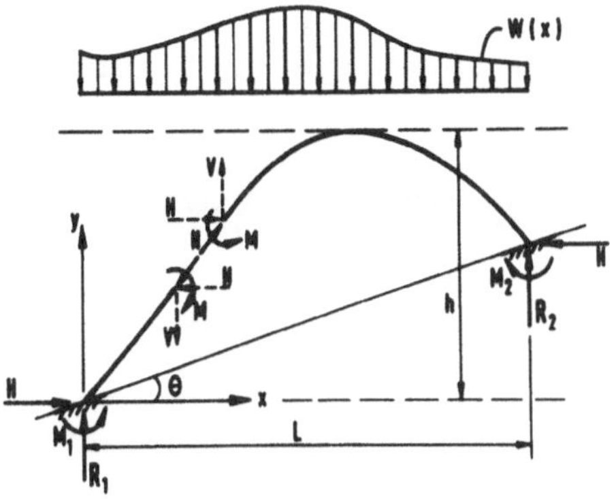

Fig. 1(b) Nonfunicular arch

$$N = \frac{\frac{1}{\alpha} + [1 - \lambda_2 + \lambda_1 + \tan\theta/\alpha - \{\int_0^1 wxdx + \int_0^x wdx\}/W]y'}{\sqrt{1 + (y')^2}} \tag{7}$$

and subjected to the following boundary conditions

$$y(0) = 0 \tag{8}$$

$$y(1) = \tan\theta \tag{9}$$

$$y(x) > x\tan\theta, \ \forall\ x \in [0,1] \tag{10}$$

in which θ is the angle of support inclination to the horizontal. Moreover, if the arch is restricted from exceeding a certain height, h, then a further geometrical constraint is imposed

$$y(x) < h, \ \forall\ x \in [0,1] \tag{11}$$

Note that the λ values depend on the support conditions where for clamped supports at both ends ($\lambda_1 \neq 0$, $\lambda_2 \neq 0$); for clamped support at end A and simple support at B ($\lambda_1 \neq 0$, $\lambda_2 = 0$) and for simple supports at both ends ($\lambda_1 = \lambda_2 = 0$). Alternatively, λ values can be prescribed as applied end moments on the arch.

3. SOLUTION PROCEDURE

Clearly, the aforementioned problem (denoted by (P) for brevity) is a nonlinearly constrained nonsmooth optimization problem. Thus, standard optimization software packages such as MINOS [13] and NLPQL [14] cannot be used. To transform problem (P) into a standard nonlinear programming problem, we shall first parameterize the unspecified arch axis and then employ a smoothing function to approximate the nonsmooth cost functional.

3.1. Parameterizing the arch axis - Adopting a cubic spline parameterization of $\bar{y}$ over a uniform partition $\rho = \{x_0, x_1 \dots x_n\}$ of the interval [0,1], where $x_j = j/n$; $j = 0, 1, \dots, n$ and n is the number of sections of the uniform partition, the arch axis, y(x), is given by

$$y(x) = \sum_{j=-1}^{n+1} C_j \Omega(nx-j) \tag{12}$$

$$C_j = (C_{-1}, C_0, C_1, \dots C_n, C_{n+1})^T \in R^{n+3}$$

where

$$\Omega(s) = \begin{cases} 0 & |s| > 2 \\ -|s|^3/6 + s^2 - 2|s| + 4/3 & 1 < |s| < 2 \\ |s|^3/2 - s^2 + 2/3 & |s| < 1 \end{cases}$$

The parameterized function y(x;C) is differentiable with first derivative given by

$$y'(x;C) = \sum_{j=-1}^{n+1} nC_j \Omega'(nx-j) \tag{13}$$

where

$$\Omega'(s) = \begin{cases} 0 & |s| > 2 \\ (-s^2/2 + 2|s| - 2)\mathrm{sgn}(s) & 1 < |s| < 2 \\ (3s^2/2 - 2|s|)\mathrm{sgn}(s) & |s| < 1 \end{cases}$$

From Ref [15], it is recalled that cubic splines can be used for approximating any continuous function to any degree of accuracy by increasing the number n. Clearly, the function y(x) and its derivative y'(x) are uniquely determined once the parameters C_{-1}, C_0, C_1, ..., C_n, C_{n+1} are given.

With the substitution of y and y' into Eq. (5), the cost functional is to be minimized with respect to the parameter vector $\eta = (C_{-1}, C_0, \dots, C_{n+1}, \alpha, \lambda)$ subject to the constraints given in Eqs. (8) to (11) which are respectively equivalent to

$$C_{-1} = -4C_0 - C_1 \tag{14}$$

$$C_{n+1} = 6\tan\theta - 4C_n - C_{n-1} \tag{15}$$

$$\int_0^1 [\mathrm{Min}\{y(x) - x\tan\theta, 0\}]^2 dx = 0 \tag{16}$$

and

$$\int_0^1 [\mathrm{Min}\{h - y(x), 0\}]^2 dx = 0 \tag{17}$$

3.2. Smoothing the cost functional - To complete the transformation of problem (P) into a nonlinear programming problem, we introduce an appropriate concept to "smooth" out the part $|M(\eta,x)|$ of the cost functional. This can be handled in the following manner.

For each $x \in [0,1]$ and $\varepsilon > 0$, consider the function

$$M^{\varepsilon}(\eta,x) = \begin{cases} |M(\eta,x)| & , \text{ if } |M(\eta,x)| \geq \varepsilon/2 \\ [\{M(\eta,x)\}^2 + \varepsilon^2/4]\varepsilon^{-1}, & \text{if } |M(\eta,x)| < \varepsilon/2 \end{cases} \tag{18}$$

Clearly, the function $M^{\varepsilon}(\eta,x)$ possesses the following properties:

(i) $M^{\varepsilon}(\eta,x)$ is continuously differentiable with respect to η for each $x \in [0,1]$;

(ii) $M^{\varepsilon}(\eta,x) \geq |M(\eta,x)|$ for each $(\eta,x) \in R^n \times [0,1]$; and

(iii) For each $(\eta,x) \in R^n x[0,1]$, $\left|M^{\varepsilon}(\eta,x) - |M(\eta,x)|\right| \leq \varepsilon/4$.

By virtue of these properties, $M^{\varepsilon}(\eta,x)$ is an ideal approximation for the nonsmooth function $|M(\eta,x)|$. By replacing $M^{\varepsilon}(\eta,x)$ for $|M(\eta,x)|$ in the cost functional (5), we obtain

$$\Phi^{\varepsilon}(\eta,x) = \int_0^1 [\beta M^{\varepsilon}(\eta,x) + \sqrt{k\beta^2 M^2(\eta,x) + 4N^2(\eta,x)}]\sqrt{1+y'^2}\, dx \tag{19}$$

where the functional $\Phi^{\varepsilon}(\eta,x)$ is now continuously differentiable with respective to η for each $\varepsilon > 0$. Now, we can approximate the nonsmooth optimization problem (P) by a standard optimization problem where the cost functional (19) is to be minimized subject to the constraints (14) to (17). Let this optimization problem be referred to as (Q_{ε}).

For each $\varepsilon > 0$, let η^*_{ε} be an optimal solution to the approximate problem (Q_{ε}). Furthermore, let η^* be an optimal solution to the original problem (P). Then, how much does $\Phi(\eta^*_{\varepsilon})$ differ from $\Phi(\eta^*)$? Furthermore, what is the relationship between η^*_{ε} and η^*? The next two theorems attempt to answer these two questions.

Theorem 1. Let η^*_{ε} and η^* be respectively, optimal solutions to the problems (Q_{ε}) and (P). Then $0 \leq \Phi(\eta^*_{\varepsilon}) - \Phi(\eta^*) \leq \varepsilon/4$.

Proof. By virtue of property (ii) of the function $M^{\varepsilon}(\eta,x)$, we have

$$\Phi^{\varepsilon}(\eta) \geq \Phi(\eta) \geq \min_{\eta} \Phi(\eta) = \Phi(\eta^*) \tag{20}$$

This, in turn, implies that

$$\min_{\eta} \Phi^{\varepsilon}(\eta) \geq \Phi(\eta^*) \tag{21}$$

and hence

$$\Phi^{\varepsilon}(\eta_{\varepsilon}^{*}) > \Phi(\eta^{*}) \tag{22}$$

Next, from property (iii) of the function $M^{\varepsilon}(\eta,x)$ we have

$$0 < \Phi^{\varepsilon}(\eta^{*}) - \Phi(\eta^{*}) < \varepsilon/4 \tag{23}$$

But

$$\Phi^{\varepsilon}(\eta_{\varepsilon}^{*}) < \Phi^{\varepsilon}(\eta^{*}) \tag{24}$$

Thus, the conclusion of the theorem follows from (22), (23) and (24).

To prove the next theorem, an additional assumption is required:

Assumption 1

$$\Phi(\eta) \to \infty \quad \text{as} \quad ||\eta|| \to \infty \tag{25}$$

where $||\cdot||$ denotes the Euclidean norm.

Theorem 2. Let (η_{ε}^{*}) be a sequence of optimal solutions to the corresponding sequence of approximate problem (Q_{ε}), and let Assumption 1 be satisfied. Then, there exists an accumulation point of the sequence (η_{ε}^{*}) for $\varepsilon \to 0$. Furthermore, any such accumulation point is an optimal solution to the original problem (P).

Proof. From Assumption 1, it follows that (η_{ε}^{*}) is bounded. Thus, there exists an accumulation point $\bar{\eta}$ and a subsequence of the sequence (η_{ε}^{*}), which is again denoted by the original sequence, such that

$$||\eta_{\varepsilon}^{*} - \bar{\eta}|| \to 0 \quad \text{as} \quad \varepsilon \to 0 \tag{26}$$

Since the cost function Φ is continuous, it follows that

$$\Phi(\eta_{\varepsilon}^{*}) \to \Phi(\bar{\eta}) \quad \text{as} \quad \varepsilon \to 0 \tag{27}$$

Now, suppose $\bar{\eta}$ is not an optimal solution to the original problem (P). Then, there exists an optimal η^{*} and a $\delta > 0$ such that

$$\Phi(\bar{\eta}) > \Phi(\eta^{*}) + \delta \tag{28}$$

But, from Theorem 1, we have

$$0 < \Phi(\eta_{\varepsilon}^{*}) - \Phi(\eta^{*}) < \varepsilon/4 \tag{29}$$

This implies that

$$\Phi(\eta^{*}) > \Phi(\eta_{\varepsilon}^{*}) - \varepsilon/4 \tag{30}$$

Hence, (28) is reduced to

$$\Phi(\bar{\eta}) > \Phi(\eta^*_\varepsilon) - \varepsilon/4 + \delta \qquad (31)$$

On this basis, it follows from (27) that

$$\Phi(\bar{\eta}) > \Phi(\bar{\eta}) + \delta \qquad (32)$$

But $\delta > 0$, leading to a contradiction. Thus, the proof is complete.

3.3. Algorithm - We can now propose an algorithm for solving the problem. The algorithm is an iterative technique for the approximate problem (Q_ε) with decreasing values of ε.

Step 0 Set $\varepsilon_1 > 0$ (say, 10^{-4}); set $k = 1$.
Step 1 Solve problem Q_{ε_k} by NLPQL. Denote the solution by $\eta^*_{\varepsilon_k}$.
Step 2 Set $\varepsilon_{k+1} = \varepsilon_k/N$, where N is a prespecified number such as 10 or 100.
Step 3 If $\varepsilon_{k+1} < \delta$ and/or if $||\eta^*_k - \eta^*_{k-1}|| < \Delta$, where δ and Δ are prespecified small numbers depending on the accuracy desired. STOP. Otherwise, set $k = k+1$ and go to Step 1.

4. SOME NUMERICAL RESULTS

Rectangular arches carrying uniformly distributed load and subjected to either (a) applied equal end moments or (b) a maximum height constraint are considered. Using $n = 6$ and $\varepsilon = 10^{-6}$, the optimal solutions are given in Fig. 2 and Fig. 3, respectively.

5. REFERENCES

1. Hill, R.D., Rozvany, G.I.N., Wang, C.M. and Leong, K.H., 'Optimization, Spanning Capacity and Cost Sensitivity of Fully Stressed Arches,' *J. of Struct. Mech.*, **7**, pp. 375-410, 1979.
2. Rozvany, G.I.N. and Wang, C.M., 'On Plane Prager-Structures I & II', *Int. J. of Mech. Sci.*, **25**, pp. 519-541, 1983.
3. Wang, C.M., 'Optimization of Multispan Plane Prager-Structures with Variable Support Locations', *Eng. Struct.*, **9**, pp. 157-161, 1987.
4. Thevendran, V. and Wang, C.M., 'Minimum Weight Design of Cables with Supports at Different Levels', *Int. J. of Mech. Sci.*, **27**, pp. 519-529, 1985.
5. Wang, C.M., Pulmano, V.A. and Lee, S.L., 'Cable Optimization under Selfweight and Concentrated Loads', *J. of Struct. Mech.*, **14**, pp. 191-208, 1986.
6. Teo, K.L., Ang, B.W. and Wang, C.M., 'Least Weight Cables - Optimal Parameter Selection Approach,' *Engineering Optimization* **9**, pp. 249-263, 1986.
7. Goh, C.J., Wang, C.M. and Teo, K.L., 'Optimal Design of Multispan Continuous Cables with General Support Conditions', *Engineering Optimization*, **12**, pp. 299-314, 1987.

8. Rozvany, G.I.N. and Prager, W., 'A New Class of Structural Optimization Problems: Optimal Archgrids', *Comput. Meth. Appl. Mech. Eng.*, **19**, pp. 127-150, 1979.
9. Rozvany, G.I.N., Nakamura, H. and Kuhnell, B.T., 'Optimal Archgrids: Allowance for Selfweight', *Comput. Meth. Appl. Mech. Eng.*, **24**, pp. 287-304, 1980.
10. Alwis, W.A.M. and Wang, C.M., 'On Optimal Archgrids', *Engineering Optimization* **8**, pp. 315-331, 1985.
11. Thevendran, V. and Wang, C.M., 'On the Optimality Criteria for Archgrids', *J. Struct. Eng. ASCE*, **112**, pp. 185-189, 1986.
12. Rozvany, G.I.N., Wang, C.M. and Dow, M., 'Arch Optimization via Prager-Shield Criteria, *J. Eng. Mech. Div., ASCE*, **106**, pp. 1279-1286, 1980.
13. Murtagh, B.A. and Saunders, M.A., 'MINOS 5.0. Userguide', *Technical Report SOL 83-20*, Stanford University, 1983.
14. Schittkowski, K., 'NLPQL: A Fortran Subroutine for Solving Constrained Nonlinear Programming Problems', *Operation Research Annals*, **5**, 1985.
15. Li, Y.S. and Qi, D.X., *Spline Functions Methods*, Science Press, Beijing, China, 1974 (in Chinese).

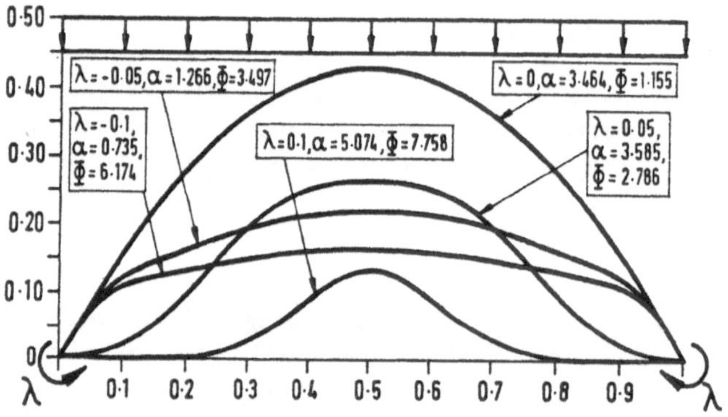

Fig. 2 Optimal shapes of arches for various λ values and β = 150

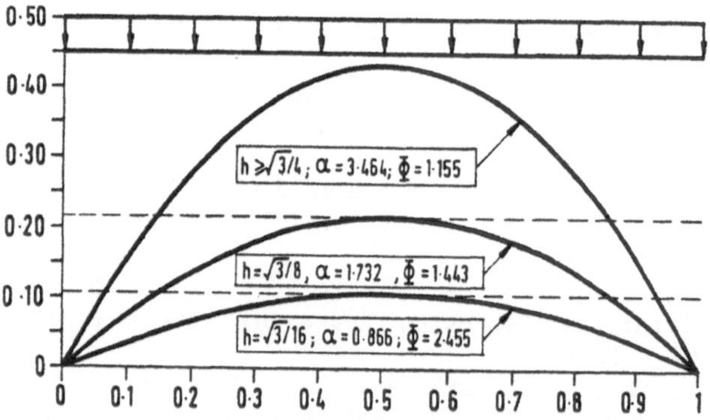

Fig. 3 Optimal shapes of arches for various h values and β = 150

OPTIMIZATION OF CONICAL SHELLS FOR STATIC AND DYNAMIC LOADS

David P. Thambiratnam
National University of Singapore
Department of Civil Engineering
Kent Ridge, Singapore 0511

ABSTRACT. Optimization of conical shells under static or dynamic loads is treated herein. Volume minimization subject to constraints on either allowable stresses or fundamental frequency and frequency maximization subject to a constraint on volume are studied. Piecewise linear variation in shell thickness is allowed in the study. A numerical procedure incorporating the finite element method and direct search optimization techniques is used. Results indicate that considerable saving in material and elevation in frequency are possible.

1. INTRODUCTION

Shell structures find increasing use today as they offer both an economic design and a pleasing appearance. Conical shells are used in several areas of engineering in addition to their use as fluid (or solid) retaining structures. The natural frequencies of a structure must be known in order to avoid the destructive effects of resonance with nearby rotating or reciprocating equipment or other dynamic excitations. The designing of a structure to isolate it from the destructive effects of resonant vibration, rather than strengthening it to withstand the vibration is a recognised fact. This is achieved by elevating the natural frequencies of the structure so that they are sufficiently higher than the frequencies of the possible dynamic excitations in the vicinity. With these in mind, the present study treats the optimum design of conical shells.

In the first part of the study, conical shells used as fluid (or solid) retaining structures are considered and their minimum volume is sought subject to constraints on allowable stresses. The second part of the study focusses its attention on conical shells subjected to vibration. Here, two separate problems are considered. In one the maximum elevation in the natural frequencies is sought for a given volume of material, while in the other, the minimum volume of the shell is obtained for a prescribed frequency of vibration. The objectives of the study are material saving and vibration isolation. The shells are restrained at the base (having the smaller radius) and the inner radius is kept constant while piecewise linear variation in the outer radius is allowed. A numerical procedure incorporating the finite element method and direct search optimization techniques is used.

G. I. N. Rozvany and B. L. Karihaloo (eds.), Structural Optimization, 355–362.

Optimum design of structures has received considerable attention in the literature and are referenced for example in the ASCE publication titled Structural Optimization [1]. However, shell structures have not received much attention probably due to the complexity in their analysis. Using numerical methods, the author along with his colleague has treated cylindrical shells subjected to hydrostatic pressure and vibration [2, 3]. The present work is an extension of that study to conical shells. The parameters used in the study are inner base radius to height ratio, number of slopes of the outer wall and the semivertex angle. Results indicate that by changing the wall shape from uniform thickness to one which varies piecewise linearly, it is possible to achieve considerable saving in material and elevation in frequency.

2. METHOD

2.1 Volume Minimization of Conical Water Tanks, with Constraints on Allowable Stresses

An optimum design of a conical water tank with given internal base radius R, cone angle Ø and height L is sought with the objective of minimizing the volume of concrete used in its construction. The tank is subjected to hydrostatic pressure from the top to the bottom and the design conforms to British Standard BS 5337 [4] which specifies that the hoop and bending stresses (s_1 and s_2) should not exceed 1.44 N/mm² and 2.02 N/mm² respectively for grade 30 concrete (used in this study). Using the penalty function technique this problem can be stated in an unconstrained form as

$$\text{Minimize } P(x,k) = V(x) + \sum_{i=1}^{2} k\,[g_i(x)]^2 \tag{1}$$

where P(x,k) is the objective function, V the volume, x the design variables (thicknesses), k the penalty parameter and g_i are given by

$$g_1(x) = (s_1^1 - s_1) - |s_1^1 - s_1|$$

$$g_2(x) = (s_2^1 - s_2) - |s_2^1 - s_2| \tag{2}$$

where s_i refers to the critical hoop and bending stresses for i=1 and 2 respectively, while s_i^1 refers to the allowable values of these stresses. A value of $k = 10^6$ was found to be satisfactory. Analysis (for stresses) was by the finite element method while optimization was achieved by the Hooke and Jeeve Pattern search technique.

2.2 Minimization of Volume with Constraint on Frequency

Optimization of conical shells for axisymmetric vibration is the essence of the study described in this and the next section. It is appropriate to mention here that as ω_A the lowest frequency corresponding to an axisymmetric mode is elevated during the process of optimization, all the frequencies (corresponding to both axisymmetric and non-axisymmetric modes) are elevated. Optimization with respect to ω_A pertains to an axisymmetric problem and hence is preferred.

Using the penalty function technique the modified objective function to be considered is

$$\text{Minimize } F(x,k) = V(x) + k\,\{\Omega - |\Omega|\}^2 \tag{3}$$

where $\Omega = \omega_A - \omega_A^*$, with ω_A^* being the prescribed value of the frequency. Frequency analysis was by the finite element method while the complex method was used in the optimization for problems in sections 2.2 and 2.3.

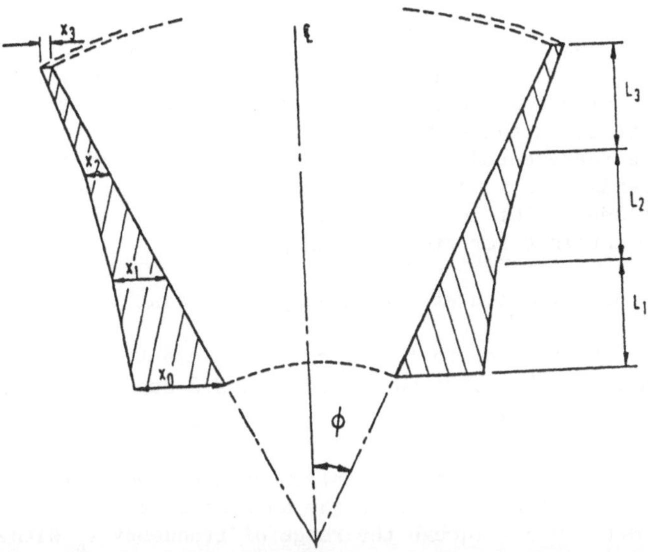

Figure 1 Conical shell with piecewise linearly varying thickness and having three slopes of the outer wall surface.

2.3 Frequency Maximization with Constraint on Volume

The problem statement here is

$$\text{Maximize } \omega_A, \tag{4}$$

subject to $V = V_o$, where V_o is the given volume.

In all three problems the design variables x, which represent the wall thickness at a location were restricted to be within the lower and upper bounds ℓ and u respectively, i.e.

$$\ell \leq x \leq u \tag{5}$$

Figure 1 shows a conical shell with piecewise linearly varying wall thickness, as in the shells used herein, together with the notations.

3. RESULTS AND DISCUSSION

3.1 Volume Minimization of Conical Water Tanks

The following data was used. Young's Modulus of concrete E = 28.0 kN/m^2, Poisson's ratio $\nu = 1/6$, shear modulus of concrete G = 12.0 kN/m^2, weight density of water 9.81 kN/m^3 and weight density of concrete = 23.6 kN/m^3. In the finite element analysis 250 elements were used. It was observed that the bending stress did not govern the optimum design, as it was very much below the allowable value. Table I shows the results for a tank with R = 3 m having a semi-vertical angle of 30°, for various R/L ratio. It can be seen that with a single slope of the outerwall surface, appreciable saving in material is possible. With two slopes of the outer wall surface there is only marginal increase in the savings. The savings range from 20 to 33% for the cases represented herein. When a constraint of 50 mm on the minimum thickness was imposed, the savings reduced as shown in the table. This constraint is required for practical purposes. Analogous results were obtained for other angles ϕ.

3.2 Volume Minimization with Frequency Constraint

In the studies pertaining to frequency, 140 elements were used in the analysis. All examples treated in sections 3.2 and 3.3 had the following data. Young's Modulus E = 30 x 10^6 kN/m^2, mass density = 2410 kg/m^3, Poisson's ratio $\nu = 0.15$.

Figure 2 illustrates the variation of the frequency ω_A with volume of material for $\phi = 40°$, in the case of a shell having R = 1.25 m, R/L = 0.3. Though the range of frequency ω_A within which a volume minimization procedure could be carried out is small, the material saving resulting from two slopes instead of one in the design is very substantial. This saving is 71% at $\omega_A = 630$ rad/s. Other shells displayed analogous behaviour.

3.3 Frequency Maximization with Volume Constraint

Effects of the three parameters viz, R/L ratio, number of slopes of the outer wall and semi-vertical angle ϕ, on the frequency ω_A are investigated herein. For shells with R = 1.25 m, variation of the frequency ω_A with R/L ratio is shown in Figure 3 for $\phi = 20°$. It can be seen that ω_A increases with the number of slopes (at any R/L ratio). However, the incremental improvements become less significant as the number of slopes increases. The % increase in ω_A is highest at the lowest R/L ratio and reduce as R/L increases. At R/L = 0.1 the increase (in ω_A) is 94% for the shell with Ø = 20°. Analogous results were obtained with shells having other values of the radius R and Ø.

Since the treatment of non-axisymmetric vibration involves three dimensional frequency analysis, which is tedious, the same objective of frequency elevation is achieved in the present study by considering axisymmetric vibration. During the process of optimization as the frequency ω_A kept on increasing, the fundamental frequency ω_1 (which corresponds to a non-axisymmetric mode) also kept on increasing and both ω_A and ω_1 attain their peak values simultaneously (within the constraints of the axisymmetric shape and constant volume). Figures 4a, 4b and 4c show the initial uniform shell and the shapes of the optimized shells with one and two slopes of the outer wall respectively, together with the corresponding values of ω_A and ω_1. These shells have Ø = 20°, R = 1.25 m and R/L = 0.30 and the optimized shapes are typical for those obtained in (all) the numerical examples treated herein. The elevations in ω_A and ω_1 are 34% and 67% with one slope and 36% and 92% with two slopes. Similar results were obtained at other values of Ø. Thus it can be seen that substantial elevation in the fundamental frequency ω_1 is possible by the procedure described in the present study. As mentioned earlier, elevation in ω_1 will also result in elevation of all the other frequencies. Shells with other R and R/L values yielded analogous results.

4. CONCLUSIONS

Optimum design of conical shell structures subject to hydrostatic loads and vibration has been treated by using the finite element method and numerical optimization techniques. From the several numerical results, typical of which are presented in this paper, the following trends may be reported.

(1) By varying the thickness piecewise linearly appreciable material savings can be achieved in the case of conical shells under hydrostatic loading. A saving of up to 33% was possible in the cases reported and up to 40% was possible when Ø = 60°.

(2) For a prescribed value of the frequency ω_A, considerable saving in material was possible by having two slopes of the outerwall surface instead of one. This saving decreases with the R/L ratio and increases with the angle Ø. For cases reported herein up to 71% saving in material was possible.

(3) By altering the wall shape but keeping the (material) volume constant, it was possible to elevate the natural frequencies of conical shells by significant amounts. In the examples treated up to 100% increase in frequencies was possible.

(4) During the optimization process with respect to ω_A, the fundamental frequency ω_1 also kept on increasing and both ω_A and ω_1 attained their peak values simultaneously.

(5) Frequencies increased with R/L ratio and tend to converge at higher values of this ratio.

(6) Results in all the three cases studied improved with the number of slopes and moreover the optimum shapes were similar. This similarity in shapes will facilitate in the design of conical water tanks subjected to vibration.

TABLE I Results of Volume Minimization of a Conical Water Tank R = 3m, Ø = 30°

R/L	Type	Volume m^3	% Saving in Material
0.30	uniform	128.6	-
	1 slope	97.8	23.9
	2 slopes	86.4	32.8
	2 slopes*	88.5	31.2
0.40	uniform	56.5	-
	1 slope	41.4	25.6
	2 slopes	37.9	32.8
	2 slopes*	40.5	28.5
0.50	uniform	30.7	-
	1 slope	22.6	26.5
	2 slopes	20.5	33.2
	1 slope*	23.0	25.0
	2 slopes*	22.9	25.3
0.75	uniform	10.8	-
	1 slope	7.8	27.3
	2 slopes	7.2	32.9
	1 slope*	8.8	18.6
	2 slopes*	8.7	18.8
1.00	uniform	5.2	-
	1 slope	3.9	25.7
	2 slopes	3.6	31.3
	1 slope*	4.7	10.1
	2 slopes*	4.7	10.5

*with 50 mm minimum thickness.

REFERENCES

1. O.E. Lev (Ed.), Structural Optimization, American Society of Civil Engineers, New York, 1981.

2. V. Thevendran and D.P. Thambiratnam, Minimum Weight Design of Cylindrical Water Tanks, International Journal of Numerical Methods in Engineering, Vol. 23, No. 9, 1986, pp. 1679-1691.

3. D.P. Thambiratnam and V. Thevendran, Maximization of Natural Frequencies of Cylindrical Shells, Engineering Optimization,to appear in 1988.

4. BS 5337 The Structural Use of Concrete for Retaining Aqueous Liquids, (formerly CP 2007). British Standards Institution, London, 1976.

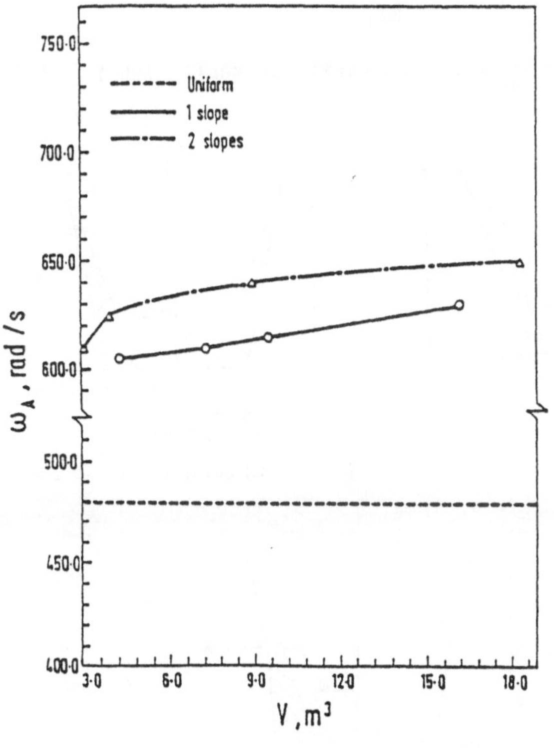

Figure 2 Variation of lowest axisymmetric frequency ω_A with volume V for a shell with R = 1.25 m, R/L = 0.30 and angle Ø = 40°, for one and two slopes of the outer wall surface.

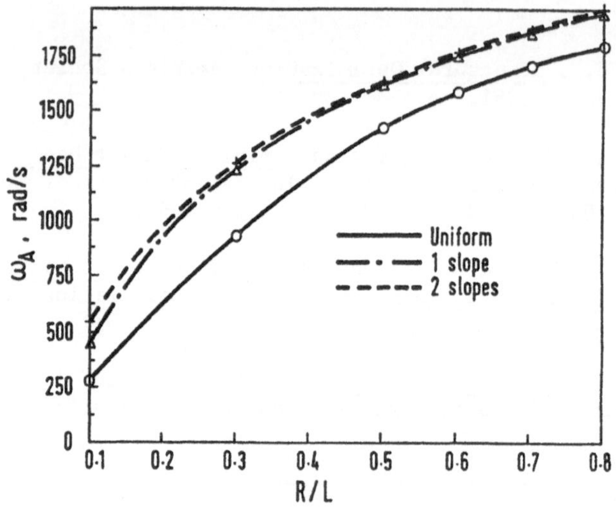

Figure 3 Variation of ω_A with R/L ratio for shell with $\emptyset = 20°$, R = 1.25 m.

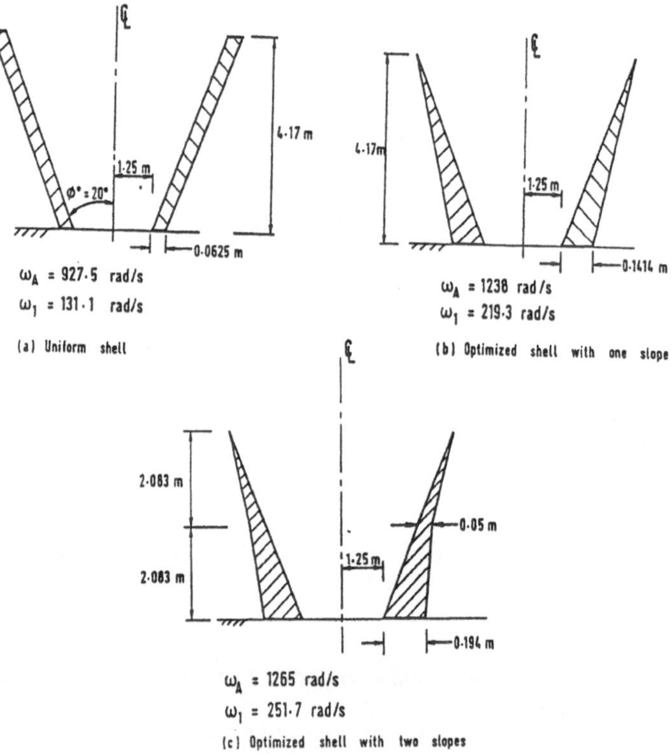

Figure 4 Shapes of uniform and optimized shells together with their frequencies.

OPTIMUM DESIGNS OF ROTATING SHAFT SYSTEMS FOR NONLINEAR DYNAMIC RESPONSES
(Optimum Shape Design and Optimum Operating Curve)

H. Yamakawa
Waseda University
Department of Mechanical Engineering
3-4-1 Okubo, Shinjuku-ku, Tokyo 160
Japan

ABSTRACT. Two recent studies of optimum designs of rotating shaft systems are shown here for nonlinear dynamic transient responses when the operating speeds pass through the critical ones. One is an optimum design problem to find a optimum shape of a rotating shaft which gives the minimum transient response where the shaft is supported by nonlinear bearings. The other is a optimum design problem to find an optimum operating curve for a rotating shaft system which yields the minimum transient response where the system is subjected to the operation with a limited power supply. To analyse the nonlinearities, the step-by-step integration techniques are used. Numerical results obtained by the gradient-based optimization methods are examined and discussed.

1. INTRODUCTION

Rapid developements of computer technology and many investigations on optimum designs enabled us to apply the existing optimum design methods to practical problems. However, further studies will be still necessary on certain classes of problems from a practical point of view. An optimum design of nonlinear dynamic problem is one of such problems because practical difficulties often lie in the increase of the computation time and the requirement of a large memory area of the computer during the nonlinear dynamic analysis and the optimization.

Various types of nonlinear dynamic problems are possibly found in rotating shaft systems which compose typical parts of machineries. For instance, nonlinear dynamic properties should often be considered in bearing systems and in the coupled motions of transverse and rotational or torsional displacements with the variation of the driving torque. Lately dynamic behavior of rotating shaft systems which are operating with limited power supplies have become special interest for many researchers.

Two of our recent studies on optimum designs of rotating shaft systems will be introduced and discussed in the following for their nonlinear dynamic transient responses when the systems have speeds to pass through the critical speeds. Those are the studies on the

G. I. N. Rozvany and B. L. Karihaloo (eds.), Structural Optimization, 363–370.

following problems :

(a) Optimum Shape Design
A problem to find an optimum shape of a rotating shaft to minimize the transient response where the shaft is supported by nonlinear bearings.

(b) Optimum Operating Curve
A problem to find an optimum operating curve for a rotating shaft system to minimize the transient response where the system is subjected to the operation with a limited power supply.

For these two types of problems, optimum design methods will be shown and numerical results obtained by use of the methods will be examined and discussed later.

2. METHODS OF NONLINEAR DYNAMIC ANALYSIS AND OPTIMUM DESIGNS

2.1. Nonlinear Dynamic Analysis of a Rotating Shaft System Sustained by Nonlinear Bearings

The objective rotating shaft is modeled here by finite number of uniform beam elements so as to find an approximated optimum shape with ease. The dynamic analysis and the optimum design of a system with a nonlinearity generally involve great difficulties in practical computation because two iteration procedures might be neccessary for the nonlinear analysis and the optimization which yield a considerable increase in computation time and require a large memory area in the computer. Efficient methods for such problems are desirable. The author has proposed a new general method for dynamic analysis by combining any of step-by-step integration techniques with the concept of ordinary transfer matrix method and named it the incremental transfer matrix method. In the method, the dynamic analysis is based on the following incremental transfer matrix of each element :

$$\begin{Bmatrix} q(t+\Delta t) \\ Q(t+\Delta t) \\ 1 \end{Bmatrix}^R = \begin{bmatrix} t_A & t_B & t_C \\ t_D & t_E & t_F \\ 0 & 0 & 1 \end{bmatrix} \begin{Bmatrix} q(t+\Delta t) \\ Q(t+\Delta t) \\ 1 \end{Bmatrix}^L \qquad (1)$$

where q : displacement vector , Q : force vector
$t_A \sim t_F$: element of incremental transfer matrix involves the response quantities at the time t.

A basic procedure is quite briefly shown in the later described Fig. 4 The Midpoint Runge-Kutta method is used there to improve the accuracy of nonlinear analysis. Further details will be referred to the literature (6).

2.2. Nonlinear Dynamic Analysis of a Rotating Shaft with a Limited Power Supply

2.2.1 Equations of Motion and Solutions. A model consisting of a massless flexible shaft with a rigid disk may promote a better understanding of the problem. For the model shown in Fig.1 the basic equations of motion are derived as follows :

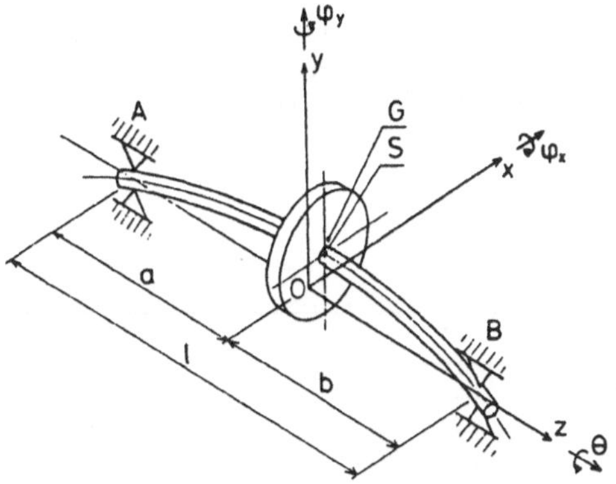

Fig.1 Basic rotor system

$$(I_p + m\varepsilon^2)\ddot{\theta} = T + m\varepsilon(\ddot{x}\sin\theta - \ddot{y}\cos\theta) - \frac{1}{2}I_p(\ddot{\varphi}_x\varphi_y - \varphi_x\ddot{\varphi}_y) \quad (2)$$

(ε : eccentric displacement)

$$m\ddot{x} + c\dot{x} + k_{11}x - k_{12}\varphi_y = m\varepsilon(\ddot{\theta}\sin\theta + \dot{\theta}^2\cos\theta) \quad (3)$$

$$m\ddot{y} + c\dot{y} + k_{12}y + k_{12}\varphi_x = m\varepsilon(\dot{\theta}^2\sin\theta - \ddot{\theta}\cos\theta) \quad (4)$$

$$I_d\ddot{\varphi}_x + I_p\dot{\theta}\dot{\varphi}_y + k_{12}y + k_{22}\varphi_x = -\frac{1}{2}I_p\ddot{\theta}\varphi_y \quad (5)$$

$$I_d\ddot{\varphi}_y - I_p\dot{\theta}\dot{\varphi}_x - k_{12}x + k_{22}\varphi_y = \frac{1}{2}I_p\ddot{\theta}\varphi_x \quad (6)$$

(I_d : moment of inertia
I_p : polar moment of inertia)

Equations (2) ~ (6) are simultaneous nonlinear equations and their numerical solutions are probably obtained directly by a step-by-step integration method together with piecewise linearizations. However, the procedure is very tedious and we can not always have accurate solutions. Hence the following new method is presented.

Applying the Newmark's β method to the evaluation of $x, y, \varphi_x, \varphi_y$ at the time $t+\Delta t$ after a short time increment Δt and rearranging them, we finally have the following single nonlinear differential equation only for the rotational angle $\theta(t+\Delta t)$. In Eq.(7) the coefficients $a_0 \sim a_7$, $b_0 \sim b_7$, $c_0 \sim c_7$, d_0, d_1 involve the response quantities at the time t. This equation can be solved by the Newton-Rapson method with high accuracy.

$$\begin{aligned} &a_7\dot{\theta}^7(t+\Delta t) + a_6\dot{\theta}^6(t+\Delta t) + a_5\dot{\theta}^5(t+\Delta t) \\ &+ a_4\dot{\theta}^4(t+\Delta t) + a_3\dot{\theta}^3(t+\Delta t) + a_2\dot{\theta}^2(t+\Delta t) \\ &+ a_1\dot{\theta}(t+\Delta t) + a_0 + \{b_6\dot{\theta}^6(t+\Delta t) + b_5\dot{\theta}^5(t \\ &+ \Delta t) + b_4\dot{\theta}^4(t+\Delta t) + b_3\dot{\theta}^3(t+\Delta t) \\ &+ b_2\dot{\theta}^2(t+\Delta t) + b_1\dot{\theta}(t+\Delta t) \\ &+ b_0\}\sin\{d_1\dot{\theta}(t+\Delta t) + d_0\} \\ &+ \{c_6\dot{\theta}^6(t+\Delta t) + c_5\dot{\theta}^5(t+\Delta t) \\ &+ c_4\dot{\theta}^4(t+\Delta t) + c_3\dot{\theta}^3(t+\Delta t) + c_2\dot{\theta}^2(t+\Delta t) \\ &+ c_1\dot{\theta}(t+\Delta t) + c_0\}\cos\{d_1\dot{\theta}(t+\Delta t) + d_0\} \\ &= 0 \end{aligned} \quad (7)$$

2.2.2. Dynamic Analysis under a Consideration of a Relation between the Driving Torque and the Torque given by the Operation Curve

Many rotating machineries have the chracteristics of decreasing torque as the rotational speeds increase. If the energy sources of such machineries do not have enough energy, in other words, they have limited power supplies, it may happen that the machineries cannot be operated beyond the critical speeds. Fig.2 illustrates the effect of the driving torque on the corresponding acceleration. Due to the restriction of torque characteristic of the machine, the operation can not be continued with a constant angular acceleration, in this case. That is, the angular acceleration of driving operation should follow to a_s for the interval B. Under a consideration of above mentioned restriction of the angular acceleration, a new method of dynamic analysis is presented whose procedure is shown in Fig.3.

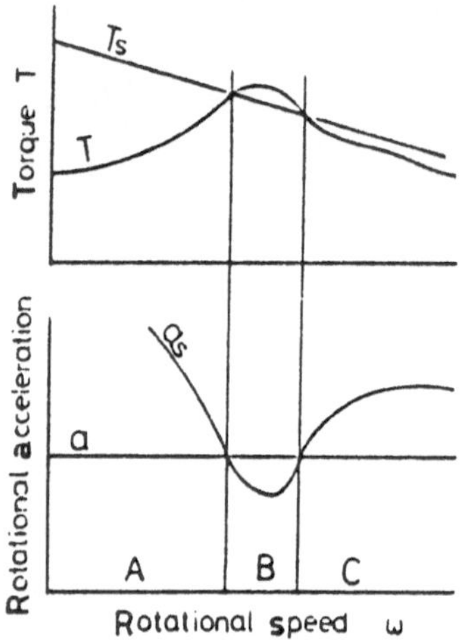

Fig.2 Effect of driving torque on angular acceleration

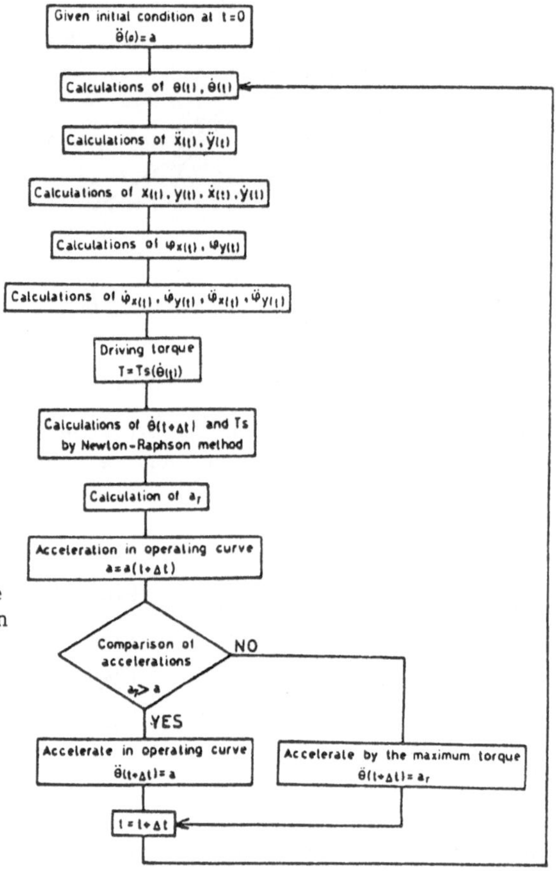

Fig.3 Dynamic analysis of rotor system with limited power supply

3. OPTIMUM DESIGN PROBLEMS AND OPTIMIZATION PROCEDURES

3.1. Optimum Shape Design of a Rotating Shaft System Supported by Nonlinear Bearings

Various types of optimum design problems may be interesting from a practical point of view but we only consider here a simple problem shown in Table I. This problem can be solved by the gradient projection method. Then the use of the incremental tranfer matrices and the incremental sensitivity matrices make it possible to have the gradients of the response quantities with respect to the design variables very conveniently and easily. The whole procedure is summarized in Fig.4

TABLE I Optimum Design Problem

Model	shaft
Loading	Unbalance force
Design variable	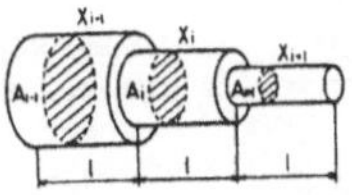 Mass ratio : $X_j = \frac{\text{Element mass}}{\text{Structural mass}} = \frac{\rho A_j l_j}{M^*}$
Condition of constraints	Constant total structural mass : $\sum_{j=1}^{n} X_j = 1$
Objective function	r.m.s. value of prescribed displacement $\bar{q}_i = \sqrt{\frac{1}{h}\sum_{k=0}^{h} q_{ik}^2}$ (h: Number of time partitions)
Optimum design problem	Minimization of $\bar{q}_i$
Gradient of objective function	$\frac{\partial \bar{q}_i}{\partial X_j} = \frac{\frac{1}{h}\sum_{k=0}^{h} q_{ik}\frac{\partial q_{ik}}{\partial X_j}}{\bar{q}_i}$

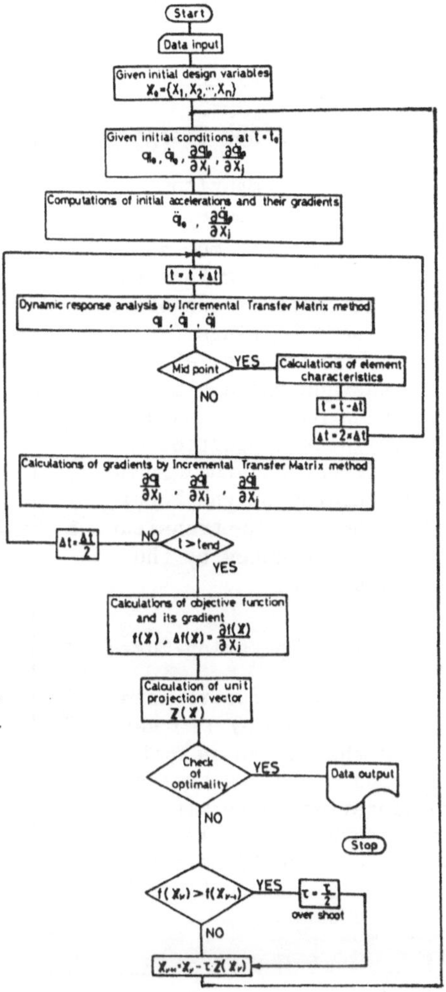

Fig.4 Procedure of optimum shape design for nonlinear dynamic response

3.2. Optimum Design of Operating Curve

From several calculations and experiments, the curve shown in Fig.5 can be considered as one of the most effective curves to reduce the transient response and the passing through the critical speed. Thus our main concern is that of finding the

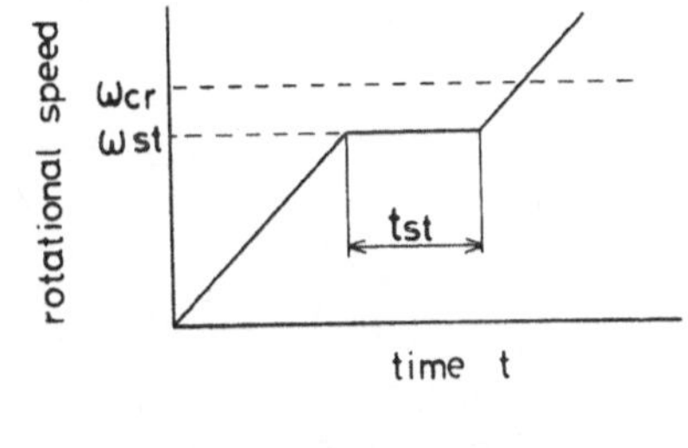

Fig.5 Effective operating curve

parameters, ω_{st} and t_{st} which specify the curve and we consider here a simple problem shown in Table II

4. NUMERICAL EXAMPLES

4.1. Optimum Shape Design

Fig.6 shows a comparison of the transient response (its envelope) of an optimum shaft with that of the uniform one. There the shaft is supported by nonlinear bearings (squeeze film dampers) at both ends and has a slight unbalanced mass at its center. The nonlinear properties of the stiffness and the damping are calculated by the lubrication theory. The transient response is reduced by the optimum design.

4.2. Optimum Operating Curve

A comparison of the transient responses(envelope) is made between the case of an optimum operating curve and that of the operating curve with a constant angular acceleration in Fig.7. The transient response is reduced by the optimum design.

Table II Optimum Design Problem

Design variable	Rotational speed ω_{st} ,	Time t_{st}
Condition of constraints	Rotational acceleration a ,	Characteristics of driving torque
Objective function	Max whirl displacement r_{max}	
Optimum design problem	Minimization of r_{max}	

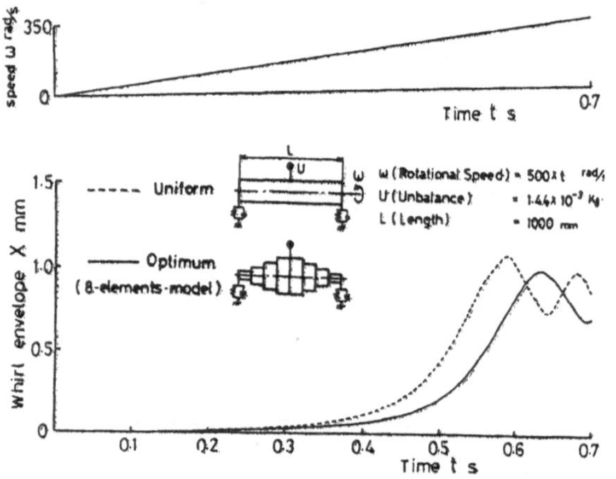

Fig.6 Result of optimum shape design

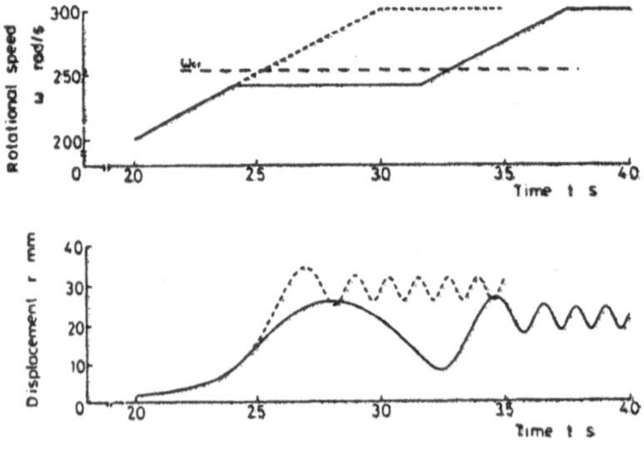

Fig.7 Result of optimum design of operating curve

Fig.8 shows the iso-displacement lines of the transient responses for various values of ω_{st} and t_{st} and some experimental results are also shown for comparison.

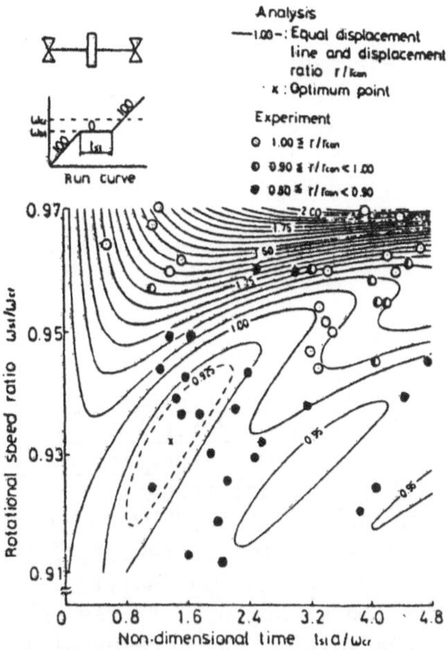

Fig.8 Iso-displacement lines and comparison with experimentalresults

5. CONCLUSIONS

Optimum Shape Design

(1) An optimum design method for nonlinear dynamic response is introduced in which the incremental transfer matrix method and the incremental sensitivity matrix method are utilized.
(2) The method can be applied to most optimum shape design problems with nonlinearities.
(3) The method can solve the difficulties in a practical computation. It can save computation time and the requirement of memory area.

Optimum Operating Curve

(1) A new method of dynamic transient analysis of a rotating shaft system passing through beyond the critical speed with a limited power supply.
(2) From several calculations and experiments it is seen that the operating curve which has a a constant rotational speed for certain interval just below the critical speed, is one of the effective curves for both the operation beyond the critical speed and the reduction of the transient response.

(3) The two parameters specifying the effective curve were determined optimally by the gradient based optimization technique and the results were examined and discussed.

Acknowledgements

The author would like to acknowledge the advice of Prof. A. Okumura of Waseda University. And thanks are due to the following people who helped the developement of computer programs and numerical calculations, some of which are appeared in this paper, during their studies at the master courses in Waseda University : N. Hirami, Y. Nishioka, Y.Suzuki

Refernces

(1) Tondl,A., Some Problems of Rotor Dynamics, (1969), Paval Dolan.
(2) Pfützner,R.G., Rotor Dynamik Eineführung, (1975), Springer-Verlag.
(3) Kononenko,V.O., Vibrating System with a Limited Power Supply, (1969), LondonIliffe.
(4) Iwatsubo,T. et al, "Vibrations through Critical Speeds of Axisymmetric Rotor with Limited Power", Trans. of JSME, 40, 335, (1974), P.1908.
(5) Yamakawa,H. and Ohnishi,T. , "Dynamic Response Analysis of Structures with Large Degrees of Freedom", Bull. of JSME,26, 211, (1983), p.109.
(6) Yamakawa,H., "An Incremental Transfer Matrix Method for Dynamic Analysis and Optimum Designs of Linear/Nonlinear Mechanical and Structural Systems", ASME Computers in Engineering- 1986, II ,P.313.

MINIMUM COMPLIANCE STIFFENER LAYOUT OF A PLATE

K. YAMAZAKI
Department of Mechanical Systems Engineering
Kanazawa University
Kodatsuno 2-40-20,
Kanazawa, 920 Japan

Abstract. An optimum design technique of stiffener layout to achieve the minimum compliance is developed. A thin plate with stiffeners is treated as anisotropic pseudo-continuum and discretized into finite elements. The minimum compliance design problem subjected to constant volume in which angles of stiffener arrangement and its density distributions in the orthogonal directions are varied, is solved by a recursive quadratic programming technique. Optimal stiffener layouts of a rectangular plate under some typical loading and supporting conditions are obtained numerically.

1. Introduction

Reinforcement of plates and shell structures by stiffeners is a useful and practical means for obtaining stronger and more rigid structures than, say solid plates of varying thickness having the same volume as the stiffened structures. While the optimum stiffener design problem of a thin plate is very important practically, there are very few works to predict the optimum stiffener shape and layout. Recently Cheng and Olhoff [(1),(2)] studied the minimum compliance design of annular plates with circumferential stiffeners by introducing the new concept of integral stiffeners, that is, the plate is equipped with an infinite number of integral stiffeners with varying height and infinitesimal width. This new plate model can express the theoretical limitation of unidirectionally stiffened plate structures, and they discussed the distribution of stiffeners. But the potential and limitations of bidirectional stiffeners to decrease the compliance and suitable directions of the stiffener arrangement have never been investigated.

In this paper a new plate model equipped with bidirectional integral stiffeners will be proposed to obtain the optimum layout and distribution of stiffeners which can achieve the minimum compliance under any given loading and supporting conditions. The model to minimize the plate compliance which can be used for searching the theoretical limitations will be developed from Cheng and Olhoff's integrally stiffened plate. The new design model of thin plate with stiffeners is discretized into finite

G. I. N. Rozvany and B. L. Karihaloo (eds.), Structural Optimization, 371–378.

elements with anisotropic material properties. Then the minimum compliance design problem subjected to volume constraint will be formulated by taking the directions and rigidities of the bidirectional stiffeners in each element as design variables, and solved by using a recursive quadratic programming technique. Pshenichny's linearization method will be adopted here together with design sensitivity analysis. To demonstrate the applicability of the bidirectional integral stiffener model, optimum stiffener layouts and density distributions of a rectangular plate under typical loading and supporting conditions will be determined numerically.

2. Minimum Compliance Design Formulation

2.1. INTEGRAL STIFFENER MODEL

Consider a thin plate equipped on the upper and lower surfaces by an infinite number of bidirectional stiffeners with continuously varying height and infinitesimal width- by the so-called integral stiffeners. At first the bidirectional integral stiffeners are assumed to cross orthogonally to one another and to be arranged along an orthogonal curvilinear net when the minimum compliance design is achieved, because the orthogonal arrangement of the stiffeners is the most effective in a biaxial moment field. Then the plate with integral stiffeners in the orthogonal directions can be treated as pseudo-orthotropic solid plate.

Let us begin to write the stress-strain relationship of this pseudo-continuum. The middle plane of the orthotropic solid plate will be taken as the xy- plane and the z- axis is taken perpendicular to the middle plane as shown in Fig. 1(a). The local Cartesian coordinates (α, β) are defined in the tangent directions to the curved integral stiffener. The angle between the x axis and the tangent to α- axis is denoted by φ. Consider an small element cut out from the plate by two pairs of planes normal to the αz and βz planes as shown in Fig. 1(b). The deflection of the middle plane being denoted by w, the stress components in the α and β directions may be obtained as follows. The thin plate theory gives the relationship between the stress components ($\sigma_{\alpha\alpha}{}^P$, $\sigma_{\beta\beta}{}^P$, $\sigma_{\alpha\beta}{}^P$) in the real solid part of the plate

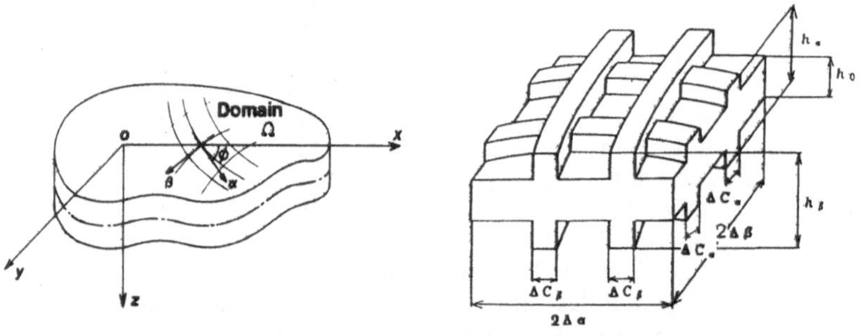

(a) Local coordinate system. (b) Detailed structure

Fig.1 Pseudo-continuum model of an integrally stiffened plate.

($2|z| \leq h_0$, where h_0 denotes the solid plate thickness) and the deflection w:

$$\sigma_{\alpha\alpha}{}^{p} = -\frac{Ez}{1-\nu^2}(w_{,\alpha\alpha} + \nu w_{,\beta\beta}),$$

$$\sigma_{\beta\beta}{}^{p} = -\frac{Ez}{1-\nu^2}(\nu w_{,\alpha\alpha} + w_{,\beta\beta}), \quad \sigma_{\alpha\beta}{}^{p} = -\frac{Ez}{1-\nu^2} w_{,\alpha\beta}, \quad \cdots\cdots\cdots\cdots\cdots \quad (1)$$

in which E and ν denote Young's modulus and Poisson's ratio of the plate, and subscripts ",$\alpha\alpha$", ",$\beta\beta$" and ",$\alpha\beta$" imply partial differentiation with respect to α and β. On the other hand, if the resistance of the orthogonally crossing integral stiffeners for the twisting deformation is neglected, one can use the beam theory and write the stress components ($\sigma_{\alpha\alpha}{}^{s}, \sigma_{\beta\beta}{}^{s}, \sigma_{\alpha\beta}{}^{s}$) in the integral stiffener part of the plate ($2|z| \leq h_0$) in terms of the deflection w as follows:

$$\sigma_{\alpha\alpha}{}^{s} = -Ezw_{,\alpha\alpha}, \quad \sigma_{\beta\beta}{}^{s} = -Ezw_{,\beta\beta}, \quad \sigma_{\alpha\beta}{}^{s} = 0. \quad \cdots\cdots\cdots\cdots\cdots\cdots \quad (2)$$

Next, let us consider the equilibrium of the element shown in Fig. 1(b). Denoting the heights of the integral stiffeners in the α- and β- directions by h_α and h_β, the normal stresses distributed over the lateral side of both parts of the solid plate and the stiffener in the α direction can be reduced to a couple, the magnitude of which per unit length is equal to the bending moment $M_{\alpha\alpha}$. In this way we obtain

$$\Delta\beta M_{\alpha\alpha} = -2\Delta\beta\int_0^{\frac{h_0}{2}} \sigma_{\alpha\alpha}{}^{p} z dz - 2\Delta C_\alpha \int_{\frac{h_0}{2}}^{\frac{h_\alpha}{2}} \sigma_{\alpha\alpha}{}^{s} z dz, \quad \cdots\cdots\cdots\cdots \quad (3)$$

where $\Delta\beta$ is the width in the α section and ΔC_α is the total stiffener width per $\Delta\beta$. By substituting from Eqs. (1) and (2) into Eq. (3) and noting that $\Delta\beta$ and ΔC_α are infinitesimal, the bending moment can be expressed via the deflection w as

$$M_{\alpha\alpha} = -D_0\left[1 + (1-\nu^2)\left\{\left(\frac{h_\alpha}{h_0}\right)^3 - 1\right\} b_\alpha\right] w_{,\alpha\alpha} - D_0 \nu w_{,\beta\beta}, \quad \cdots\cdots\cdots\cdots \quad (4)$$

in which $D_0 = Eh_0^3/12(1-\nu^2)$ and b_α is stiffener density defined as $b_\alpha = \Delta C_\alpha / \Delta\beta$. In the same way, one can define the bending moment $M_{\beta\beta}$ in the β-direction and the twisting moment $M_{\alpha\beta}$

$$M_{\beta\beta} = -D_0 \nu w_{,\alpha\alpha} - D_0\left[1 + (1-\nu^2)\left\{\left(\frac{h_\beta}{h_0}\right)^3 - 1\right\} b_\beta\right] w_{,\beta\beta}, \quad \cdots\cdots\cdots\cdots \quad (5)$$

$$M_{\alpha\beta} = -D_0(1-\nu) w_{,\alpha\beta}. \quad \cdots\cdots\cdots\cdots\cdots\cdots\cdots\cdots\cdots\cdots \quad (6)$$

In Eq. (5) $b_\beta = \Delta C_\beta / \Delta\alpha$ is stiffener density in the β-direction. By comparing these moment-deflection relationships with those of a solid plate of thickness h_0, one finds that the new stiffened plate model can be treated as pseudo-orthotropic plate, with the following stress-strain relationships:

$$\sigma_{\alpha\alpha} = \frac{E\{1 + (1-\nu^2)\rho_\alpha\}}{1-\nu^2}\varepsilon_{\alpha\alpha} + \frac{E\nu}{1-\nu^2}\varepsilon_{\beta\beta},$$

$$\sigma_{\beta\beta}=\frac{E\nu}{1-\nu^2}\varepsilon_{\alpha\alpha}+\frac{E\{1+(1-\nu^2)\rho_\beta\}}{1-\nu^2}\varepsilon_{\beta\beta},\quad \sigma_{\alpha\beta}=\frac{E}{2(1+\nu)}\varepsilon_{\alpha\beta},\quad\cdots\cdots\cdots (7)$$

in which ρ_α and ρ_β represent ratios of stiffener rigidity to that of the plate defined as

$$\rho_\alpha=\{(\frac{h_\alpha}{h_0})^3-1\}b_\alpha,\quad \rho_\beta=\{(\frac{h_\beta}{h_0})^3-1\}b_\beta.\quad\cdots\cdots\cdots (8)$$

The above derivation allows us to treat the minimum compliance design problem of a integrally stiffened plate as a material design problem of an anisotropic plate having a constant thickness h_0 and varying flexural rigidity.

2.2 MINIMUM COMPLIANCE DESIGN FORMULATION

If we consider the minimum compliance design problem of the pseudo-orthotropic plate having the stress-strain relationships Eq. (7), the objective function f with the angle φ of stiffener arrangement and the ratio of stiffener rigidities ρ_α, ρ_β as the design variables can be formulated as follows.

$$f(\varphi,\rho_\alpha,\rho_\beta)=\frac{1}{2}\int\int_\Omega pw\,dx\,dy\rightarrow \min,\quad\cdots\cdots\cdots (9)$$

subjected to

$$\rho_\alpha\geqq 0,\ \rho_\beta\geqq 0,\quad\cdots\cdots\cdots (10)$$

where the plate is subject to a distributed load p. In Eq. (9) the integration has to be carried out over the plate surface.

Equation (9) is equivalent to minimum strain energy of the loaded plate and implies maximization of the average flexural rigidity. Furthermore, the deflection w in Eq. (9) has to be stationary of the following total potential energy functional π

$$\pi=\frac{D_0}{2}\int\int_\Omega[\{1+(1-\nu^2)\rho_\alpha\}w_{,\alpha\alpha}^2+2\nu w_{,\alpha\alpha}w_{,\beta\beta}+$$
$$+\{1+(1-\nu^2)\rho_\beta\}w_{,\beta\beta}^2+2(1-\nu)w_{,\alpha\beta}^2]\,dx\,dy-\int\int_\Omega pw\,dx\,dy.\quad\cdots\cdots (11)$$

For existence of bounded values of ρ_α, ρ_β, we must constrain the upper limit of the total volume

$$g=\int\int_\Omega[h_0+(h_\alpha-h_0)b_\alpha+(h_\beta-h_0)b_\beta-$$
$$-\{\min(h_\alpha,h_\beta)-h_0\}b_\alpha b_\beta]\,d\alpha\,d\beta-V_0\leqq 0,\quad\cdots\cdots\cdots (12)$$

in which V_0 is prescribed total volume. Since it is difficult to write a relationship between the stiffener heights h_α, h_β and densities ρ_α, ρ_β when the minimum compliance is achieved, and since the stiffener height will be constrained by the buckling failure in a discrete realistic stiffener design, we will here deal with cases where the stiffener height or the density of width has a constant value.

3. Finite Element Model and Its Optimization

3.1. DISCRETIZATION INTO FINITE ELEMENTS

Let us discretize the new plate model into isoparametric elements with 8 nodes for seeking the optimum stiffener layout. If we introduce normalized curvilinear coordinates (ξ, η) defined on the middle plane of the plate and adopt a linear function $R_i(\xi, \eta)$ to interpolate the angle φ and the ratios ρ_α, ρ_β of stiffener rigidity at any point then

$$\varphi=\sum_i R_i\varphi_i,\ \rho_\alpha=\sum_i R_i\rho_{\alpha i},\ \rho_\beta=\sum_i R_i\rho_{\beta i}, \qquad (13)$$

in which φ_i, $\rho_{\alpha i}$ and $\rho_{\beta i}$ are the values at a corner node i.

If we denote the nodal displacement vector at node i as $\{\delta\}_i=(w_i, \theta_{xi}, \theta_{yi})^T$, in which θ_{xi} and θ_{yi} are the rotations about the x- and y-axes, and if a quadratic interpolation function $N_i(\xi, \eta)$ is adopted, we can write the displacement components u, v and w in the direction of the global, x, y and z-axes as

$$u=-\sum_i N_i\zeta\frac{h_0}{2}\theta_{yi},\quad v=\sum_i N_i\zeta\frac{h_0}{2}\theta_{xi},\quad w=\sum_i N_i w_i, \qquad (14)$$

where ζ is the normalized coordinate normal to the $\xi\eta$ plane(3),(4). If the total nodal displacement vector in the element is denoted by $\{\delta\}=(\delta_1,\cdots,\delta_8)^T$, the global coordinates of the strain vector $\{\varepsilon\}=(\varepsilon_x, \varepsilon_y, \gamma_{xy})^T$ in the element is given in the usual manner

$$\{\varepsilon\}=[B]\{\delta\}, \qquad (15)$$

where $[B]$ is the matrix of the displacement derivatives with respect to the global Cartesian coordinates. We are now interested in the strain components, which are denoted by $\{\hat{\varepsilon}\}=(\varepsilon_{\alpha\alpha}, \varepsilon_{\beta\beta}, \varepsilon_{\alpha\beta})^T$, in the α, β local coordinates, and these are obtained from Eq. (15) by coordinate transformation

$$\{\hat{\varepsilon}\}=[T][B]\{\delta\}, \qquad (16)$$

where

$$[T]=\begin{bmatrix} c^2 & s^2 & cs \\ s^2 & c^2 & -cs \\ -cs & cs & c^2-s^2 \end{bmatrix},\quad \begin{aligned} c&=\cos\varphi, \\ s&=\sin\varphi, \end{aligned} \qquad (17)$$

The corresponding stress components $\{\hat{\sigma}\}=(\sigma_{\alpha\alpha}, \sigma_{\beta\beta}, \sigma_{\alpha\beta})^T$ are given by

$$\{\hat{\sigma}\}=[D][T][B]\{\delta\}, \qquad (18)$$

in which $[D]$ is the stress-strain matrix derived from Eq. (7). Finally, we can evaluate the element stiffness matrix $[K]^e$ of the pseudo-continuum in the global coordinates as follows:

$$[K]^e=\int_{-1}^{1}\int_{-1}^{1}\int_{-1}^{1}[B]^T[T]^T[D][T][B]\,|J|\,d\xi d\eta d\zeta. \qquad (19)$$

in which $|J|$ represents the determinant of the Jacobian matrix.

3.2. OPTIMIZATION AND SENSITIVITY ANALYSIS

After the discretization into finite elements the values, φ_i, $\rho_{\alpha i}$ and $\rho_{\beta i}$ at every corner node i of each element are taken as the design variables, and the objective function Eq. (9) is minimized taking account of the constraints Eqs. (10) and (12).

We adopted Pshenichny's linearization method[5] as the practical optimization technique. This recursive quadratic programming technique is used repeatedly for improving the design point, at which the objective function is expanded to the second order and the constraints are linearized. Since this method needs design sensitivities of the objective and constraint functions, the sensitivity of objective function will be formulated in the following.

Dems and Mróz have derived the first variation of an arbitrary stress, strain and displacement functionals corresponding to variation of material parameters within a specified domain by using the solution for primary and adjoint systems[6]. We can use their formulation to the displacement functional of Eq. (9). If we denote the general design vector as $x_i = (\varphi_i, \rho_{\alpha i}, \rho_{\beta i})$ and the stress-strain relationship as $\sigma = D\varepsilon$, then the first variation of Eq. (9) can be written as

$$\delta f = -\frac{1}{2}\int_{\Omega} \varepsilon^{\mathrm{T}} \frac{\partial D}{\partial x_i} \varepsilon \delta x_i \mathrm{d}\Omega. \qquad (20)$$

In our case, since the strain vector ε and the D correspond to $\{\hat{\varepsilon}\}$ and $[T]^{\mathrm{T}}[D][T]$, respectively, the practical form of the first variation δf will be

$$\begin{aligned}\delta f = -\frac{1}{2}\int\int_{\Omega}\Big[\{\hat{\varepsilon}\}^{\mathrm{T}}\Big(\frac{\partial [T]^{\mathrm{T}}}{\partial \varphi_i}[D][T] + [T]^{\mathrm{T}}[D]\frac{\partial [T]}{\partial \varphi_i}\Big)\{\hat{\varepsilon}\}\delta\varphi_i - \\ -\{\hat{\varepsilon}\}^{\mathrm{T}}[T]^{\mathrm{T}}\frac{\partial [D]}{\partial \rho_{\alpha i}}[T]\{\hat{\varepsilon}\}\delta\rho_{\alpha i} - \\ -\{\hat{\varepsilon}\}^{\mathrm{T}}[T]^{\mathrm{T}}\frac{\partial [D]}{\partial \rho_{\beta i}}[T]\{\hat{\varepsilon}\}\delta\rho_{\beta i}\Big]\mathrm{d}x\mathrm{d}y,\end{aligned} \qquad (21)$$

because the matrix $[T]$ depends on only φ_i and the matrix $[D]$ on only $\rho_{\alpha i}$ and $\rho_{\beta i}$. From Eqs. (7) and (17) the components of $\partial [T]/\partial \varphi_i$, $\partial [D]/\partial \rho_{\alpha i}$ and $\partial [D]/\partial \rho_{\beta i}$ can be derived as

$$\frac{\partial [D]}{\partial \rho_{\alpha i}} = ER_i \begin{bmatrix} 1 & 0 & 0 \\ 0 & 0 & 0 \\ 0 & 0 & 0 \end{bmatrix}, \quad \frac{\partial [D]}{\partial \rho_{\beta i}} = ER_i \begin{bmatrix} 0 & 0 & 0 \\ 0 & 1 & 0 \\ 0 & 0 & 0 \end{bmatrix},$$

$$\frac{\partial [T]}{\partial \varphi_i} = \frac{\partial [T]}{\partial \varphi} R_i. \qquad (22)$$

4. Numerical Example

As a design example, let us consider a simply supported square plate of width $2a$ and thickness h_0 under a uniformly distributed load p, as shown

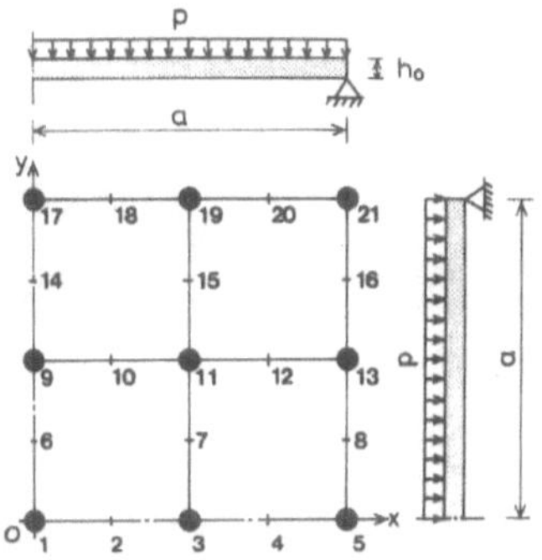

Fig. 2 Design model of simply supported plate under uniform load.

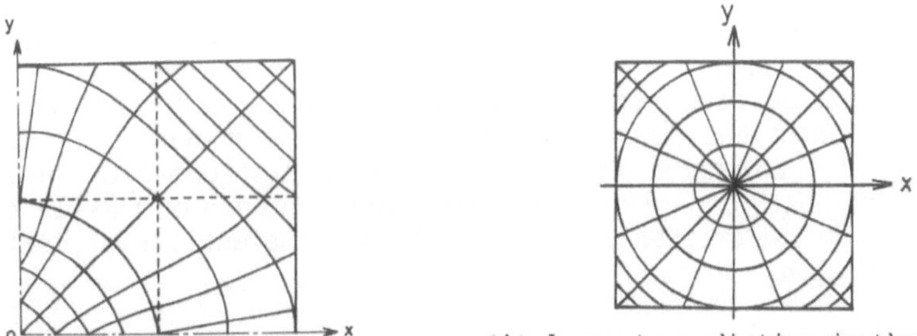

(a) Layout drawn numerically.

(b) Layout prediction in the full region.

Fig. 3 Optimum layout illustrations of integral stiffeners.

in Fig. 2. Because of symmetry a quarter region is discretized into 4 elements and the values φ_i, $\rho_{\alpha i}$ and $\rho_{\beta i}$ at each corner node are taken as the design variables. We prescribe the ratio $h_0/a = 0.01$, stiffener heights $h_\alpha = h_\beta = 2h_0$ and volume $V_0 = 1.19a^2h_0$, and adopt the initial values of $\varphi_i = 0$ and $\rho_{\alpha i} = \rho_{\beta i} = 0.7$.

After 113 repetitions of Pshenichny's linearization method the ratio f to f_0, which is the value of corresponding uniform plate with volume V_0, decreased from 1.27 to 0.811, that is, the flexural rigidity of the stiffened plate increased by 56.7% . The optimum layout of the integral stiffener is shown in Fig. 3(a), which is drawn by solving the following ordinary differential equation derived from $dy/dx = \tan\varphi$ for the optimum value of φ_i

$$\frac{d\xi}{d\eta} = -\frac{\sum_i \frac{\partial R_i}{\partial \xi}(y_i - x_i\tan\varphi)}{\sum_i \frac{\partial R_i}{\partial \eta}(y_i - x_i\tan\varphi)}. \qquad (23)$$

Many contours are calculated for arbitrary initial values φ_i, but the intervals between them don't relate to the stiffener density. From this figure we can notice that the optimum layout consists of stiffeners emanating from the center of the plate in the radial direction and of

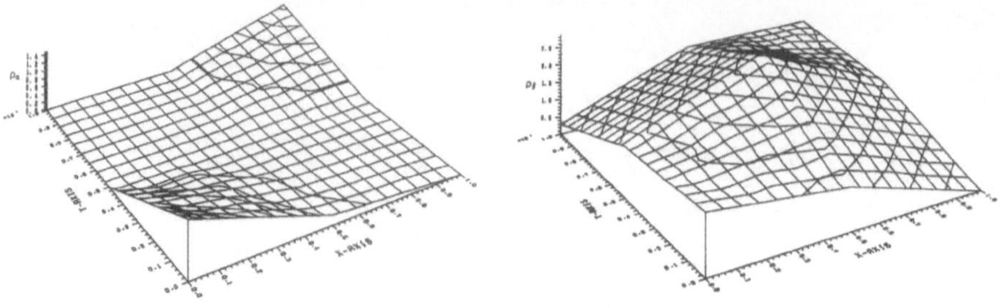

(a) Radial stiffeners. (b) Circumferential stiffeners.
Fig. 4 Distribution of stiffeners

concentric circumferential stiffeners. The optimum stiffener layout in the full region (drawn freehanded) by taking the characteristics of Fig. 3(a) and the condition of symmetry is shown in Fig. 3(b). Furthermore, the distribution of the rigidity ratios ρ_α and ρ_β is shown in Fig. 4 for a quarter region. From these figures the integral stiffeners in the radial direction distribute densely at the center and the edge of the plate, but the circumferential stiffener density achieves its maximum at center of the quarter region.

5. Conclusions

A new pseudo-continuum model of a plate with orthogonal stiffeners is proposed to obtain an optimum layout and distribution of stiffeners for minimum compliance. The stress-strain relationship of the continuum is derived and the minimum compliance problem subjected to volume constraint is formulated as a mathematical programming problem by using the finite element method. An example is solved numerically giving a peculiar layout of stiffeners under given support and loading conditions.

References

(1) Cheng, K.T., *Int. J. Solids & Struct.*, 17-8(1981), 795.
(2) Cheng, K.T. and Olhoff, N., *Int. J. Solids & Struct.*, 18-2(1982), 153.
(3) Zienkiewicz, O.C. et al., *Int. J. Num. Methods Engng.*, 3(1971), 275.
(4) Zienkiewicz, O.C. and Bruce, M.I., *Int. J. Num. Methods Engng.*, 2(1970), 419.
(5) Choi, K.K. et al., *Trans. ASME*, 105-1(1983), 91.
(6) Dems, K. and Mröz, Z., *Int. J. Solids & Struct.*, 19-8(1983), 679.

RECENT INVESTIGATIONS OF STRUCTURAL OPTIMIZATION BY ANALYTIC METHODS

Yeh Kai-yuan
Department of Mechanics
Lanzhou University
Lanzhou, Gansu 730001
People's Republic of China

1. INTRODUCTION

In this paper, we discuss two analytic methods suggested by the author for solving structural optimization problems, viz. the "stepped reduction method [1, 3-5] and a method based on the principle of complementary energy [2]. In the former, we discretize first the Young's modulus E(x) and the thickness of h(x) of structural elements. We use the initial parameter method and the Heaviside function for descriving results in analytic form and then we use a combination of homogeneous solutions to eliminate discontinuities of bending moment and shearing force at element boundaries. This means that we need to solve only simultaneous algebraic linear equations with a finite number of unknowns. The above methods will be illustrated on the following problems:

1. Optimal design of thin elastic solid annular plate under an arbitrary load (stepped reduction method) [6]; and
2. Uniform strength for statically indeterminate beams [2,7].

2. OPTIMAL DESIGN OF A THIN ELASTIC SOLID ANNULAR PLATE UNDER AN ARBITRARY DISTRIBUTED LOAD [6]

Consider an isotropic homogeneous thin elastic annular plate with an inner radius b, outer radius a, Young's modulus E and Poisson ratio γ. Let the center of the plate be the origin of a polar coordinate system, the inner radius of the plate $b=r_0$ and the outer radius $a=r_1$.

We only study axisymmetric thickness distributions, i.e. h(r) ($b\leq r\leq a$). Then the problem becomes, within a given volume (V_o) condition

$$\int_o^{2\pi}\int_b^a h(r)rdrd\theta = V_0 \tag{2.1}$$

and constraints on the thickness

$$h_{min}\leq h(r)\leq h_{max} , \tag{2.2}$$

G. I. N. Rozvany and B. L. Karihaloo (eds.), Structural Optimization, 379–386.

to find the optimal thickness distribution h(r) which minimizes the objective function (total compliance)

$$\Phi = \int_0^{2\pi}\int_b^a q(r,\theta)\, w(r,\theta) r dr d\theta \tag{2.3}$$

where $q(r,\theta)$ is arbitrarily distributed load acted on the plate and $w(r,\theta)$ is the deflection.

First, we expand the load into Fourier series:

$$q(r, \ ) = \sum_{m=o}^{\infty} q_m(r)\cos m\theta + \sum_{m=1}^{\infty} \tilde{q}_m(r)\sin m\theta \tag{2.4}$$

The corresponding deflection is

$$w(r, \ ) = \sum_{m=o}^{\infty} w_m(r)\cos m\theta + \sum_{m=1}^{\infty} \tilde{w}_m(r)\sin m\theta \tag{2.5}$$

Obviously, $w_m(r)$ and $\tilde{w}_m(r)$ depend on h(r) and can be expressed in an analytic form [9].

In order to use the stepped reduction method, we divide the whole plate into n+1 annular segments at constant thickness and h(r) is approximated by a segmentwise constant function. The radii of the boundary segments are $\xi_0=b$, ξ_1, $\xi_2, \ldots,$ $\xi_{n+1}=a$ (Fig. 1). This means that h(r) is replaced at n+1 variables h_0, $h_1, \ldots,$ h_n and thus $w_m(r)$ and $\tilde{w}_m$ reduce to a function of r, h_0, $h_1, \ldots,$ h_n. Letting

$$x = r/a, \ \beta_i=\xi_i/a \quad (i=0, 1,\ldots, n+1)$$

$$\delta^*=D_u/D_i=(h_u/h_i)^3 \quad (i=0, 1,\ldots, n)$$

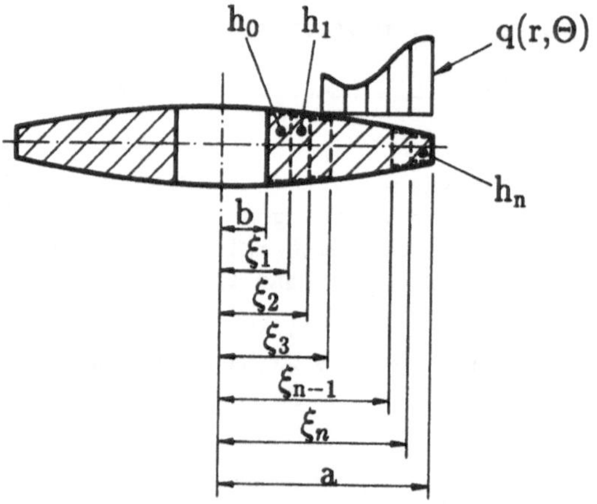

Figure 1 Discretization of the annular plate

where D_1 is stiffness of i-th section of the plate and D_u is stiffness of the uniform plate, h_u the thickness of an equivalent uniform plate with $V_0=h_u\pi(a^2-b^2)$, we can find the exact analytic expression for w_m and $\tilde{w}_m$,

$$w_m = w_m(x,\delta_0^*,\delta_1^*,\ldots,\delta_n^*) \tag{2.6}$$

$$w_m = w_m(x,\delta_0^*,\delta_1^*,\ldots,\delta_n^*) \tag{2.7}$$

Integration constants can be determined from the boundary conditions.

Substituting (2.6), (2.7) into (2.5), and then substituting the resulting expression and (2.4) into (2.3), we obtain

$$\Phi=\Phi(\delta_0^*,\delta_1^*,\ldots,\delta_n^*)$$

This means that the objective functional and condition (2.1) has been simplified into

$$\sum_{i=0}^{n}(\beta_{i+1}^2-\beta_i^2)\,(\delta_i^*)^{-1/3} = 1-\beta_0^2\ ,$$

$$\min \qquad \Phi_1= \Phi(\delta_0^*,\delta_1^*,\ldots,\delta_n^*) + \lambda(\sum_{i=0}^{n}(\beta_1^2+1-\beta_1^2)(\delta_i^*)^{-1/3}$$

subject to

$$\delta_{min}\le \delta_1^* \le \delta_{max}^* \tag{2.8}$$

Then the problem of optimal design is reduced to the problem of solving a set of nonlinear algebraic equations, i.e.

$$\frac{\partial\Phi_1}{\partial\delta_i^*} = 0\ (1=0,1,2,\ldots,n),\quad \frac{\partial\Phi_1}{\partial\lambda} = 0 \tag{2.9}$$

Numerically, (2.8) represents a problem of nonlinear programming problem. Uniqueness of the solution can be determined on the basis of (2.9).

As an illustration of the proposed method, we consider an annular plate clamped at the inner and outer edges and subjected to a load $q=C_m\cos m\theta$, with C_m=const, m=0, 1,2,...,N. In the following examples, we choose $\delta_{min}^*=0.027$, $\delta_{max}^*=3.375$, $\alpha=0.2=(b/a)$ and a, Poisson's ratio $\gamma=0.25$ which is equivalent to $h_u/h_{min}=1.5$, $h_u/h_{max}=0.3$ and $h_{max}/h_{min}=5$. This set of numerical values are the same as in [10].

First, we divide the plate into segments of equal width with m=0,3,4,7,8,10,16. For all cases, we only divide the whole plate into 5 elements. The results are shown in Fig. 2 in which Φ_o is the objective function for the plate of constant thickness and Φ_{min} is the optimal objective function value.

Fig. 3 shows solution for 40 elements and m=0, 4,8,16.

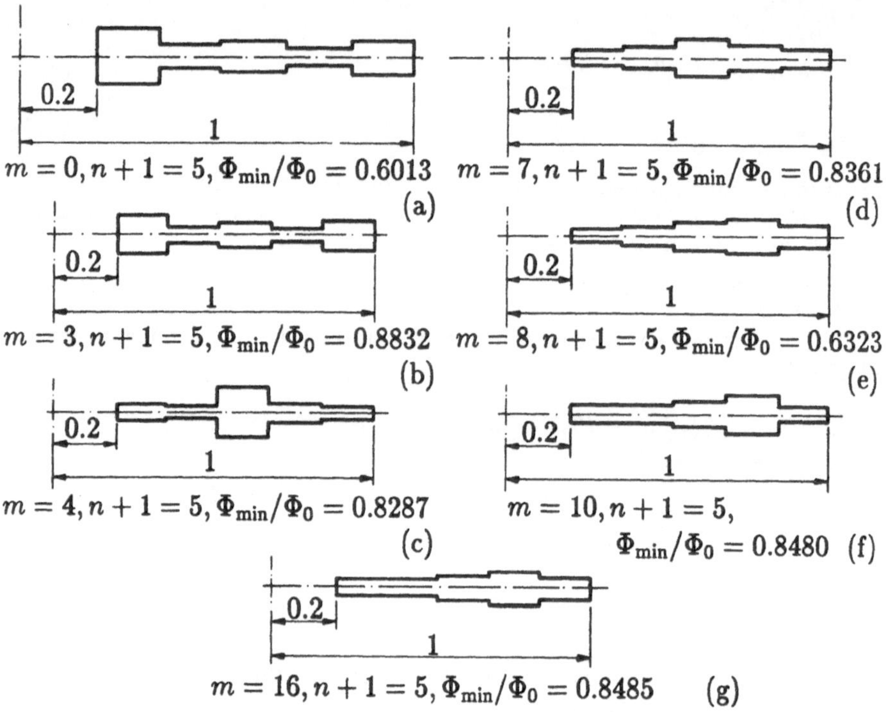
0.2
1
$m = 0, n + 1 = 5, \Phi_{min}/\Phi_0 = 0.6013$ (a)
$m = 3, n + 1 = 5, \Phi_{min}/\Phi_0 = 0.8832$ (b)
$m = 4, n + 1 = 5, \Phi_{min}/\Phi_0 = 0.8287$ (c)
$m = 7, n + 1 = 5, \Phi_{min}/\Phi_0 = 0.8361$ (d)
$m = 8, n + 1 = 5, \Phi_{min}/\Phi_0 = 0.6323$ (e)
$m = 10, n + 1 = 5, \Phi_{min}/\Phi_0 = 0.8480$ (f)
$m = 16, n + 1 = 5, \Phi_{min}/\Phi_0 = 0.8485$ (g)

Figure 2 The optimal variations of the cross section of an annular plate with five segments and both edges clamped.

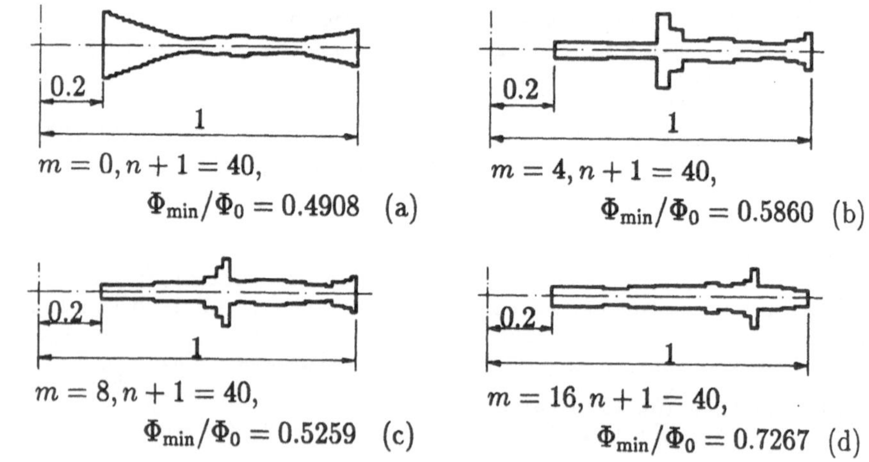
0.2
1
$m = 0, n + 1 = 40, \Phi_{min}/\Phi_0 = 0.4908$ (a)
$m = 4, n + 1 = 40, \Phi_{min}/\Phi_0 = 0.5860$ (b)
$m = 8, n + 1 = 40, \Phi_{min}/\Phi_0 = 0.5259$ (c)
$m = 16, n + 1 = 40, \Phi_{min}/\Phi_0 = 0.7267$ (d)

Figure 3 The optimal variations of the cross section of an annular plate with 40 segments and both edges clamped.

In order to compare our results with [11] and to decide if there exists an infinite number of discontinuities [10], we discuss the case with m=4 in greater detail, dividing the rib-like region of the optimal distribution into 400 elements and the rest into 40 elements. The result is shown in Fig. 4 which is contrary to the suggestions in [10].

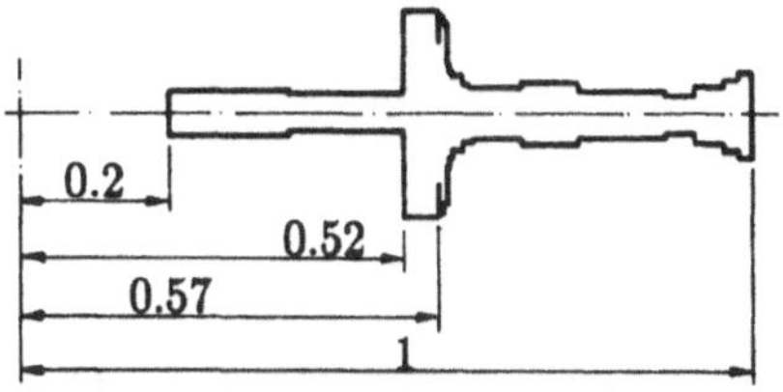

Figure 4 The optimal shape of the cross section of an annular plate with two clamped edges, 400 segments for the rib-like region and m=5, $\Phi_{min}/\Phi_0 = 0.5277$.

Comparing our results with [10], we can conclude that the author's method results in a faster convergence than the one in [10]. For m=0, [10] used 300 elements to obtain $\Phi_{min}/\Phi_0 = 0.463$, whilst the author obtained the ratio 0.498 with 40 elements. For m=5, [10] used 200 elements to get $\Phi_{min}/\Phi_0 = 0.525$, whilst the author derived the ratio 0.5277 with 40 elements.

When [10] used 150 elements to get $\Phi_{min}/\Phi_0 = 0.536$, ribs appeared in the solution. The same ribs did not appear in the author's solution even when he used 400 elements in the region of a potential stiffener. The author believes, therefore, that the conclusions of [10] and [11] should be reconsidered.

Using the author's method, Yu Huan-ran and the author studied the problem of optimal design for a minimum value or the maximum deflection over an annular plate [12]. Moreover, the authors, Yu-Huan-ran and Liu Xiang studied the problem of uniform strength annular plates [13]

3. UNIFORM STRENGTH DESIGN FOR STATICALLY INDETERMINATE BEAMS

Consider a beam having two or more supports and a length of L. The beam is to be designed to sustain an arbitrarily distributed load q(x), the concentrated forces P_i and couples T_j, as shown in Fig. 5.

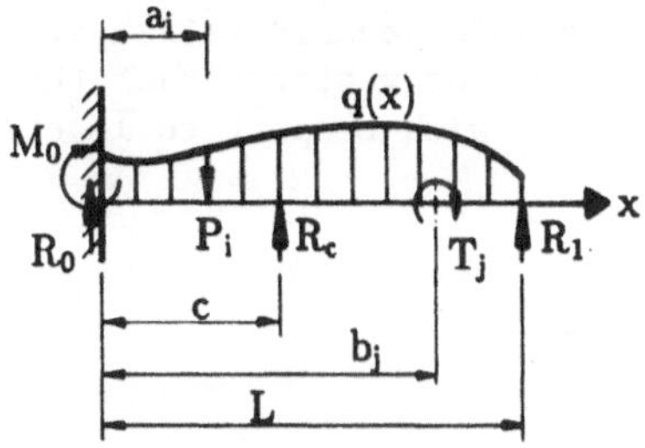

Figure 5 Uniform strength design

Let R_0, R_c and R_1 denote reaction forces at supports and M_0 the reaction moment at the left end of the beam (Fig. 5). Thus, the bending moment M(x) in the beam can be expressed as

$$M(x)=M_0+R_0x+R_c(x-c)\{x-c\}^0-\int_0^x(x-t)q(t)dt-\sum_i P_i(x-a)\{x-a_i\}^0+$$
$$+\sum_j T_j\{x-b_j\}^0. \qquad (3.1)$$

In Eq. (3.1), the Heaviside function $\{x-a\}^0$ is used. In order to make M(x) a statically admissible moment field, the moment at the right end of the beam must be zero:

$$M(1)=M_0+R_0 1+R_c(1-c) - \int_0^\ell(1-t)q(t)dt - \sum_i P_i(1-a_i)+\sum_j T_j=0 \qquad (3.2)$$

Moreover, the shearing force at the right end must equal the reaction ($R\nu$) of the support:

$$R_1 = \int_0^\ell q(t)dt + \sum P_i-R_0-R_c \qquad (3.3)$$

Therefore, out of the four reactions M_0, R_0, R_c and R_1, only two are independent variables. Taking M_0 and R_0 as being independent, we can obtain R_c from (3.2). The substitution of R_c into Eq. (3.1) then yields M(x) in terms of M_0 and R_0

$$M(x) = M_0[1-\frac{x-c}{1-c}\{x-c\}^0 + R_0[x-1\frac{x-c}{1-c}\{x-c\}^0] -$$
$$-\int_0^x(x-t)q(t)dt - \sum_i P_i(x-a_i)\{x-a_i\}^0 +$$
$$+\sum_j T_j\{x-b_j\}^0 + \frac{x-c}{1-c}[\sum_i P_i(1-a_i) +$$
$$+\int_0^\ell(1-t)q(t)dt - \sum_j T_j]\{x-c\}^0. \qquad (3.4)$$

Using the assumptions of linear elasticity and small displacements, the complementary energy of the beam is given by

$$U = \frac{1}{2}\int_0^\ell \frac{M^2(x)}{S(x)}\,dx \qquad (3.5)$$

If the supports of the beam under consideration are rigid, the theorem of the minimum complementary energy states that the partial derivatives of the complementary energy U with respect to independent variables M_0 and R_0 vanish, that is

$$\frac{\partial U}{\partial M_0} = \int_0^\ell \frac{M}{S}\frac{\partial M}{\partial M_0}\,dx = 0 \qquad (3.6)$$

$$\frac{\partial U}{\partial R_0} = \int_0^{\ell} \frac{M}{S} \frac{\partial M}{\partial M_0}\, dx = 0 \tag{3.7}$$

Two equations with unknowns M_0 and R_0 can be obtained by substituting M(x) in Eq. (3.4) into the integrands of Eqs. (3.6) and (3.7). Assume that the relation between section stiffness S(x) and section modulus W(x) can be written as

$$S(x) = \psi(x)W(x) \tag{3.8}$$

where $\psi(x)$ is a function dependent on the prescribed section form. If $\bar{\sigma}$ be allowable stress for bending, the uniform strength condition requires

$$W(x) = \frac{|M(x)|}{\bar{\sigma}} \tag{3.9}$$

Substituting Eq. (3.9) into (3.8) yields

$$S(x) = \psi(x) \frac{|M(x)|}{\bar{\sigma}} \tag{3.10}$$

Moreover, by substitution of Eq. (3.10) into Eqs. (3.6) and (3.7) we have

$$\int_0^{\ell} \frac{M}{\psi|M|} \frac{\partial M}{\partial M_0}\, dx = 0 \tag{3.11}$$

$$\int_0^{\ell} \frac{M}{\psi|M|} \frac{M}{\partial M_0}\, dx = 0 \tag{3.12}$$

The solution to Eqs. (3.11) and (3.12) gives the reactions M_0 and R_0 for a uniform strength beam. Then using Eqs. (3.4) and (3.10), we finally find the bending and stiffness distributions for uniform strength design.

It is clear that the foregoing discussion can directly be generalized to the case with n unknown reactions, in which there are (n-2) independent unknown variables and (n-2) equations in the form of expressions (3.11) or (3.12).

REFERENCES

[1] Yeh Kai-yuan, 'General solution on certain problems of elasticity with non-homogeneity and variable thickness', The Advances of Applied Mathematics and Mechanics in China, China Academic Publishers, Vol. 1, (1987), 240-272.

[2] X. Tang, Yeh Kai-yuan, 'Equi-strength design for statically indeterminate beams', Applied Mathematics and Mechanics (English Edition), 6,12 (1985), 1141-1148.

[3] Yeh Kai-yuan, Hsu Chin-yun, 'General solution on certain problems of elasticity with non-homogeneity and variable thickness I. Elastic and plastic stress analyses of high speed rotating disc with non-homogeneity and variable thickness in non-homogeneous steady temperature field', Journal of Lanzhou University, Special Number of Mechanics, No. 1, (1979), 60-74; Acta Mechanica Sinica, Special Number, (1981), 78-89; presented in the XVth ICTAM (Toronto), (1980).

[4] Yeh Kai-yuan, Lieu Ping, 'Steady heat conduction of a disc with non-homogeneity and variable thickness', Applied Mathematics and Mechanics (English Edition), 5,5 (1984), 1587-1593.

[5] Yeh Kai-yuan, Liu Ping, 'Equi-strength design of non-homogeneous variable thickness high speed rotating disk under steady temperature field', Applied Mathematics and Mechanics (English Edition), 7,9 (1986), 825-834.

[6] Yu Huanran, Yeh Kai-yuan, 'Optimal design of thin elastic plate under arbitrary load', presented in the XVIth ICTAM (Lyngby), (1984); Collect Works of Chinese Scholars, edited by the Chinese Association of Theoretical and Applied Mechanics, Dalian Institute of Technology Press, (1986), 348-356, (in Chinese).

[7] Yeh Kai-yuan, Yu Huanran, 'Static-indeterminate equi-strength beam beams', Journal of Lanzhou University, (Natural Science), Special Number of Mechanics, (1983), 1-9, (in Chinese).

[8] Holms, A.C. and Faldetta, R.D., 'Effects of temperature distribution and elastic properties of materials on gasturbinedisk stresses', NACA Report 864, (1947).

[9] Yeh Kai-yuan, Kue Jainhuo, 'General solution on certain problems of elasticity with non-homogeneity and variable thickness II. Bending of arbitrary axial-symmetrically non-homogeneous and variable thickness circular plates with holes at centers under arbitrary steady temperature field and arbitrarily distributed loads', Journal of Lanzhou University, Special Number of Mechanics, No. 1, (1979), 75-114, (in Chinese).

[10] Cheng Keng-tung, Olhoff, N., 'An investigation concerning optimal design of solid elastic plates', Int. J. Solids and Structures, 17,3 (1981), 305-323.

[11] Cheng Keng-tung, Olhoff, N., 'Regularized formulation for optimal design of axisymmetric plates', Int. J. Solids and Structures, 18,2 (1982), 153-169.

[12] Yu Huanran, Yeh Kai-yuan, 'Optimal design of minimax deflection of of an annular plate', Proceedings of International Conference of Optimization: Techniques and Applications,(Singapore), (1987), 1087-1095.

DETAILED MACHINE STRUCTURAL SHAPES GENERATED FROM SIMPLIFIED MODELS

M. Yoshimura
Department of Precision Engineering
Kyoto University
Sakyo-ku, Kyoto 606
Japan

ABSTRACT. In order to effectively obtain the optimum design for high precision machines requiring the evaluation of dynamic characteristics, two design optimization methods based on different simplification concepts have been proposed. In this paper a systematic methodology in which the two methods are synthesized is presented for obtaining more effectively the optimum detailed structural shapes of machine structural systems.

1. Introduction

Designers of machine products are always seeking a higher product performance and a lower product manufacturing cost. In industrial machines such as machine tools, evaluation of not only static characteristics but also dynamic characteristics of the machine structural systems is essential for enhancing performances such as higher accuracy and higher productivity [1]. Those characteristics depend on the machine structural shapes. The detailed machine structural shapes have been mainly determined based on the experiences of designers.

Machine structures basically consist of two components: the structural members and the joints which connect the structural members with each other. Determining the detailed shapes of these components is closely related to the following factors:

(1) to achieve the required product performance
(2) to realize practical functions
(3) to practically manufacture the designed machine elements or structural members

For example, practical structural members have complicated geometrical shapes having structural features such as ribbings and partitions for achieving the required product performance. Practical joints also have complicated contact shapes for realizing practical functions. Furthermore, determination of the structural members and contact shapes has a great influence on the manufacturing cost.

Obtaining the optimal design of machine structures without depending on designers' experiences is the ultimate objective of design processes. However, when design optimization is conducted for such machine

G. I. N. Rozvany and B. L. Karihaloo (eds.), Structural Optimization, 387–394.

structural systems, the following problems exist:

(i) In the design optimization of machine structures, the objective functions and constraints are usually nonlinear functions of design variables. In order to obtain the optimal solution, generally, a nonlinear mathematical programming method must be used. A solution obtained by a nonlinear mathematical programming method is basically only a local optimal solution. That is, it is very difficult to obtain the global optimal solution.

(ii) The decision making process of design is generally divided into two phases: fundamental design stage and detailed design stage. Usually, the design optimization of machine structures is scarcely applied to the fundamental design stage, but it is usually applied to the detailed design stage. However, for a design in which a practical design configuration is given, room for design change is small, and great improvement of the machine performance cannot be expected as a matter of course.

(iii) The relationships between design variables and dynamic characteristics are very complicated. If those complicated relations are not clarified before the formulation of the design optimization problem, the possibility of a poor convergence into a local optimum near the initial design is considered to be very high.

As a method for solving problems (i) and (ii), a multiphase design optimization method using simplified structural models has been proposed [2,3]. The process of the design optimization procedures is divided into three phases: simplification, optimization, and realization.

As a method for solving the problem (iii), a design optimization method of machine structural dynamics based on clarification of competitive-cooperative relationships between characteristics has been proposed [4]. In this method, the optimization problem is divided into three simpler (smaller) problems.

In this paper, the above mentioned methodologies based on two types of simplification are briefly explained. Then a systematic methodology in which the two methods are combined is proposed for obtaining more effectively the optimum detailed structural shapes of machine structural systems. Finally, the effectiveness of the proposed systematic method is demonstrated by applying it to a machine structural model.

2. Evaluative Characteristics of High Precision Machine Structural Systems

Fig.1 shows a simplified framework model of a milling machine and a single column vertical lathe. An example of the relative receptance frequency response between points A and B at the cutting point for such a machine tool is shown in Fig.2, where f_s is the static compliance (reciprocal of the static rigidity), and $R_{m \cdot max}$ indicates the maximum receptance value over the whole frequency range (the maximum value occurs at the natural frequency ω_m). The most troublesome vibrational phenomenon of machine-tool structures, that is, regenerative chatter, depends on the magnitude of this $R_{m \cdot max}$ [1]. In order to maximize the accuracy and productivity of machine tools, $R_{m \cdot max}$ must be minimized.

The transfer function of the receptance frequency response (D_G/F_E), concerning the exciting force F_E and the displacement D_G at a given frequency ω, is expressed approximately as the proportional vibrational damping system as follows:

$$R(\omega)=\frac{D_G}{F_E}(\omega)=\sum_{m=1}^{\infty}\left[\frac{f_{m(E,G)}}{1-\left(\frac{\omega}{\omega_m}\right)^2+2j\left(\frac{\omega}{\omega_m}\right)\zeta_m}\right] \quad (1)$$

where j designates the imaginary unit, and ω_m and ζ_m are the angular natural frequency and the damping ratio at the mth natural mode. $f_{m(E,G)}$ is the modal flexibility between the exciting force at point E and the displacement at point G [5].

3. Algorithmic procedures of design optimization using simplification concepts

3.1 Design optimization method using simplified structural models [2,3]

A design optimization method using simplified structural models is composed of three phases: simplification, optimization, and realization.

In the first phase, "simplification", mathematical models or simulation models which have structural properties equivalent to practical structures are constructed. The left hand side of Fig.3 shows examples of cross-sectional shapes for column members of practical machine structures. These are transformed into simplified box-type members having equivalent cross-sectional characteristic values (as shown on the right hand side of Fig.3). On the other hand, practical joints having complicated contact surface shapes are modeled by spring elements having equivalent rigidities and damper elements having equivalent energy dissipation ability.

In the second phase, "optimization", design optimization is conducted for the simplified model, using a mathematical programming method. Since the number of design variables is greatly reduced in this model compared to that of the practical structures, application of mathematical programming methods is easy, and possibility of a poor convergence into some local optimum solution can be reduced.

Finally, in the third phase, "realization", concrete configurations

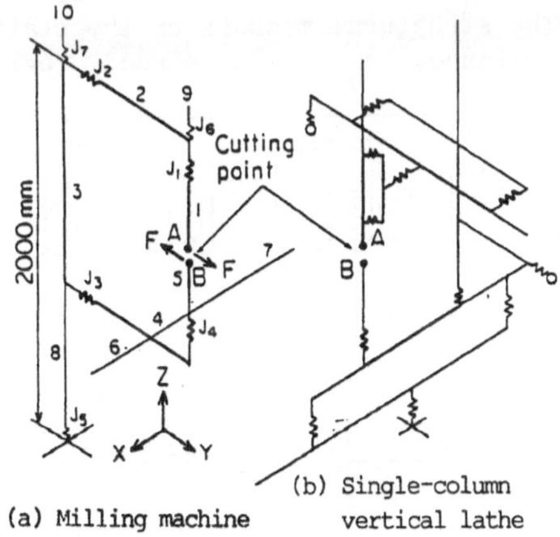

(a) Milling machine (b) Single-column vertical lathe

Fig.1 Structural models of machine tools

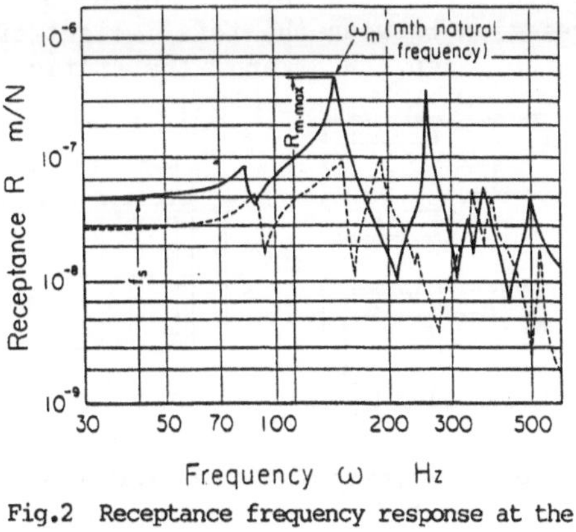

Fig.2 Receptance frequency response at the cutting point (the broken line indicates the result optimized in Chapeter 4)

(shapes) and dimensions of a practical machine structure are determined so that all requirements defined in the optimization phase are satisfied. If it is difficult to satisfy all the requirements because of having too many requirements, permissible ranges are defined for some of them by evaluating sensitivity of the objective function. In this realization phase, requirements for additional designs and functions are considered. For example, ribs and partition plates are attached to practical structural members in order to avoid local deformations of the members, and various elements are added from the standpoint of functional requirements. Fig.4 shows an example of practical cross-sectional shape patterns. When cross-sectional pattern (c) is selected, design variables ($X_1,X_2,\ldots,X_5$) are determined so that the optimized cross-sectional characteristic values are realized. In the realization stage of a joint design, a suitable joint configuration pattern is selected from among various joint shape patterns, considering functional requirements such as sliding and fixing. Those "realized" structural members and joints are synthesized into a complete machine structure having detailed shapes.

In this method, large design changes are made through a simplified model which allows much room for design choices and design changes. Hence, great improvements of characteristics can be expected.

3.2 Design optimization method based on division of the problem into simpler problems [4]

If the formulation of a design optimization problem is conducted without clarifying the relations among many evaluative factors and the solution is obtained by a mathematical programming method, the possibility of a poor convergence into a local optimum is very high. The basic concept of the design optimization method based on clarification of competitive-cooperative relationships is to construct a new design optimization algorithm after clarifying the competitive and cooperative relations between evaluative characteristics.

In the first phase, the design of the structural members supporting the static external force (for example, cutting force in the case of machine tools) is determined. Those members are here called "structural members on the static force loop". Fig.5 (in Chapter 4) shows the relation between the total weight of the structural members on the static force loop, W_s, and the static compliance, f_s. The shaded region

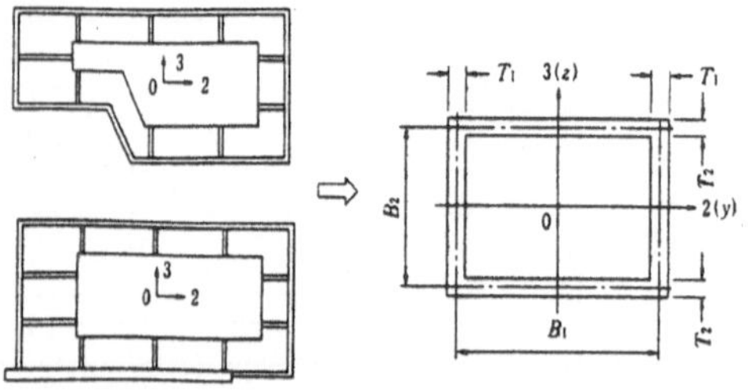

Examples of practical cross-section

Equivalent simple cross-section

Fig.3 Simplification of structural members

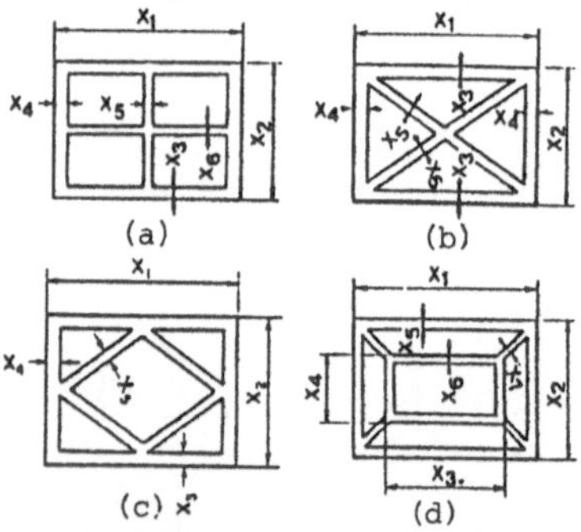

Fig.4 Examples of detailed cross-sectional patterns

corresponds to the design variable space feasible for W_s and f_s, and the heavy solid line indicates the Pareto optimum solution set of a multiobjective design optimization problem [6] minimizing both W_s and f_s. Each of the broken curves indicates the contour line of a constant maximum receptance value of the complete machine structural system. The feasible design solution at the point (on the Pareto optimum solution set) making contact with the contour line having the minimum receptance value corresponds to the design solution (on the static force loop) which minimizes $R_{m \cdot max}$. Based on such considerations, the design variables on the static force loop can be determined.

Next, in the second phase, the design of the structural members and joints outside the static force loop is determined so that the ratio f_m/f_s of the modal flexibility f_m at the mth natural mode having $R_{m \cdot max}$ to the static compliance f_s is minimized. This procedure is based on the analytical result that minimization of the ratio f_m/f_s brings about minimization of the maximum modal flexibility and maximization of the damping ratio both of which are required to minimize $R_{m \cdot max}$.

Finally, in the third phase, damping coefficients of all the joints in the complete machine structure are determined so that the maximum receptance value $R_{m \cdot max}$ is minimized.

By completing the forementioned procedures, a complicated design optimization problem is divided into three "simpler" problems and the optimum solution can easily be obtained.

3.3 A systematic methodology

The fundamental procedures of a systematic methodology in which two design optimization methods based on different kinds of simplification

Table 1 The fundamental procedures of a systematic methodology for design optimization based on two simplification concepts

	Phase 1 Simplification	***Phase 2*** Optimization	***Phase 3*** Realization
Phase I Modeling for a structure on the static force loop as a static deformation system	(i) Construction of simplified models of structural members on the static force loop	(iv) Obtaining the Pareto optimum solution set of a multiobjective problem minimizing W_s and f_s	(vii) Determination of detailed design of structural members on the static force loop
Phase II Modeling for a complete structure as an undamped vibrational system	(ii) Construction of simplified models of structural members and joints outside the static force loop	(v) Minimization of the ratio of the maximum modal flexibility $f_{m \cdot max}$ to the static compliance f_s	(viii) Determination of detailed designs of structural members outside the static force loop
Phase III Modeling for a complete structure as a nonproportionally damped vibrational system	(iii) Modeling of damping effects of all joints by damper elements	(vi) Minimization of the maximum receptance value $R_{m \cdot max}$ or maximization of the damping ratio	(ix) Determination of contact conditions of all joints

are combined are shown in Table 1. The horizontal and vertical flows correspond to the procedures described in Section 3.1 and the procedures described in Section 3.2, respectively.

Usually, it is not easy to obtain detailed designs of a complete machine structural system, especially when dynamic characteristics are evaluated. In the systematic procedures of Table 1, realization process (Phase 3) of detailed design is divided into three sub-processes: vii, viii, and ix. In sub-process vii(Phase I), detailed designs of the structural members and joints on the static force loop can be determined independently by evaluating only the static structural analysis of those parts.

For realization of detailed designs, the three factors: (1) realization of practical functional requirements, (2) achievement of required characteristics, and (3) economical manufacturing of the designed elements or structural members, should be considered. From consideration of factor (1), some of the structural shapes are given as shape patterns. Structural shapes of the other parts may also be given as shape patterns or are, in a more sophisticated manner, generated based on algorithmic procedures. As for factor (2), the requirements are expressed as constraints of a decision making problem of detailed designs. As for economical manufacturing of the designed machine elements or structural members, the manufacturing cost C generally becomes the objective function of the decision making problem where the function is minimized.

Eventually, the decision making problem for the realization of detailed designs is expressed:

Minimize C
subject to "the constraints that the characteristics required in the optimization phase (Phase 2 in Table 1) are satisfied or the degradation of the characteristics is within the permissible range". (2)

4. Example

Fig.1(a) shows a simplified milling machine model which is used to demonstrate the availability of the proposed procedures. The machine structural model is composed of 10 box-type members having square cross-sections and 7 joints expressed by spring marks. The static force F or the exciting force is applied between points A and B in Y-direction, and the relative displacement between points A and B is detected in the same direction. The structural members 1-5 and joints J_1-J_4 are the structural members and joints on the static force loop respectively, while the structural members 6-10 and the joints J_5-J_7 are, respectively, the structural members and joints outside the static force loop.

In process iv of Table 1, a multiobjective design optimization problem minimizing both the total weight of the structural members on the static force loop, W_s, and the static compliance, fs, is formulated. The multiobjective optimization problem can be solved using the Kuhn-Tucker necessary conditions for optimality. The Pareto optimal solution set is obtained, as shown in Fig.5, on the heavy curve in the functional space for two objectives minimizing both W_s and f_s. The shaded portion corresponds to the feasible design variable space. Point P_3 is considered to be the most preferable point for minimizing both W_s and f_s, that is, for minimizing the maximum receptance value $R_{m\ max}$. Hence, point P3 was selected as the solution of the design variables of the structural members on the static force loop.

In process vii of Table 1, detailed designs of structural members and joints on the static force loop are determined. In detailed designs of the column member (member 3 in Fig.1(a)), contact elements for bolted joints and slide ways are added to realize practical functional requirements. Ribbings and partitions are added to the box-type structural member in order to prevent local deformations. Those detailed designs must satisfy the rigidity required at the optimization phase for the simplified model. The requirement is expressed as constraints in the decision making problem of detailed designs.

Many feasible designs satisfying the constraints may exist. Among them, a design having the smallest manufacturing cost is selected as the solution of the decision making problem expressed in eq.(2). As for economical manufacturing of the designed member, two kinds of manufacturing costs: welding cost of ribbing and partition plate elements and machining cost of contact parts for bolted joints and slide-ways, are considered [3].

Here, only the design decisions for partitions and ribbings in the column member are explained. The simplest practical algorithms for the decision making are as follows:

Step 1. Select a design pattern having the smallest welding cost from the cross-sectional patterns (design patterns of longitudinal partitions and ribbings) of the structural member [as shown in Fig. 4]).

Step 2. Determine the dimensions of the cross-sectional pattern so that optimized cross-sectional values are satisfied.

Step 3. Construct the finite element model of the structural member using shell or plate elements and calculate the value of the total strain energy of the member, V_e.

Step 4. If the constraint in eq.(2) (the incremental quantity ΔV_e of

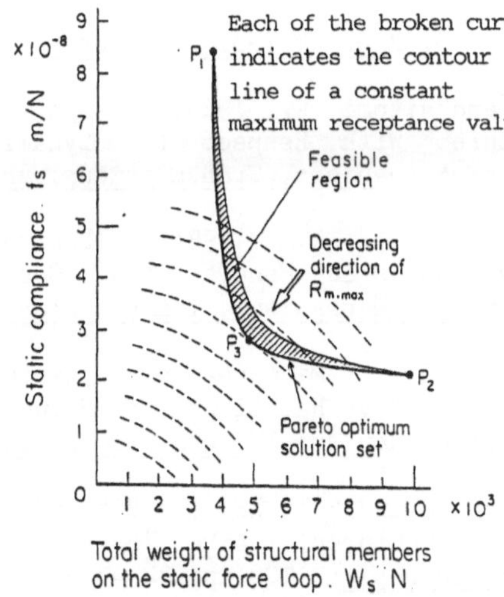

Fig.5 Functional space for W_s and f_s

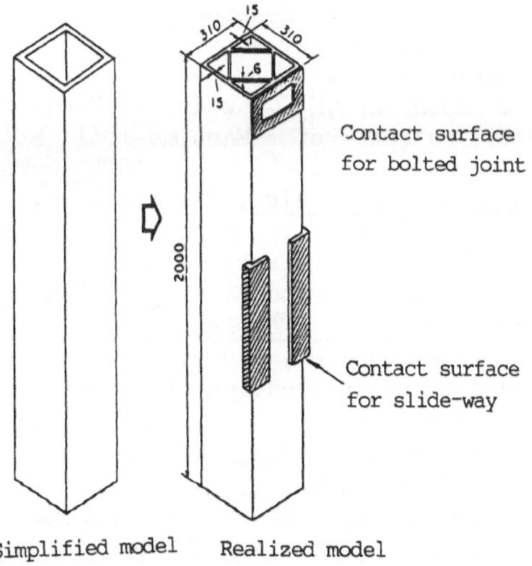

Fig.6 Detailed design process of column member

V_e from the total strain energy of the optimized simplified member, $\Delta V_e'$, must be less than the upper limit $\Delta V_e{}^u$) is satisfied, go to Step 5. Otherwise, select the next design pattern having the smallest welding cost, and go to Step 2.

Step 5. Adopt the design as the final detailed design.

The more sophisticated method generating detailed designs from the optimized simplified models is now being studied and will be presented in the forthcoming paper.

Fig.6 shows the detailed shapes of the column member obtained by the above-mentioned procedures (pattern (c) in Fig.4 was finally selected). Similar procedures are conducted for other members and joints (the detailed procedures are given in Ref.[2,3]), and the other processes of Table 1 are continued. Finally, the detailed design of the machine structural system having the smallest maximum receptance value is obtained.

5. Conclusions

In order to obtain optimum detailed designs of high precision machines requiring evaluation of dynamic characteristics, a design optimization method was proposed in which two types of simplification procedures: (1) use of simplified models and (2) division of the problem into simpler problems, are synthesized. The determination process of detailed designs from optimized simplified models is divided into three simpler processes. The proposed method enables the detailed optimum machine designs to be obtained practically and effectively.

Acknowledgments

The author wishes to express his deepest gratitude to Dr. Katsundo Hitomi, Professor at Kyoto University, for his support and encouragement during this study.

References

1 Koenigsberger,F., and Tlusty,J., 1970, Machine Tool Structures, 1, Pergamon Press.

2 Yoshimura,M., Hamada,T., Yura, K., and Hitomi, K., 1983, 'Design Optimization of Machine-Tool Structures With Respect to Dynamic Characteristics,' ASME Journal of Mechanisms, Transmissions, and Automation in Design,105, pp.88-96.

3 Yoshimura,M., Takeuchi,Y., and Hitomi, K., 1984, 'Design Optimization of Machine-Tool Structures Considering Manufacturing Cost, Accuracy and Productivity,' ASME Journal of Mechanisms, Transmissions, and Automation in Design, 106, pp.531-537.

4 Yoshimura,M., 1987, 'Design Optimization of Machine-Tool Dynamics Based on Clarification of Competitive-Cooperative Relationships Between Characteristics,' ASME Journal of Mechanisms, Transmissions, and Automation in Design, 109, pp.143-150.

5 Yoshimura,M., 1986, Evaluation of Forced and Self-Excited Vibrations at the Design Stage of Machine-Tool Structures, ASME Journal of Mechanisms, Transmissions, and Automation in Design, 108, pp.323-329.

6 Osyczka,A., 1984, Multicriterion Optimization in Engineering, Ellis Horwood.

ON SHAPE OPTIMIZATION OF SATELLITE TANKS

H.A. ESCHENAUER
University of Siegen, Research Laboratory for Applied Structural Optimization, 5900 Siegen, FRG

In order to reduce the transportation cost incurred when sending a satellite into orbit, the weight of the satellite should be as low as possible. Even small reductions in the weight of the individual components lead to a decrease in the costs. One of these components is the satellite fuel tank, which stores the fuel required for positioning purposes for the entire lifetime of the satellite.

The paper deals with the method of approach involved in finding the optimal design of a thin-walled satellite tank subject to constant internal pressure. The problem requires the fulfilling of the following demands:

- compliance with the specified design space
- the weight should be a minimum
- no exceeding of the permissible stresses
- no buckling
- no undercuts

The way to fulfil these demands is the selection of a suitable tank shape and wall thickness distribution. It yields an improved utilization of the material and a reduction in the stresses.

The investigations have shown that the bending effects as well as the influence of large deformations must be taken into account for the structural analysis to get sufficiently accurate results for deformations and stresses. After establishing a suitable structural analysis approach by transfer matrices which satisfies these requirements, the meridional shape and wall thickness distribution can be determined simultaneously with regard to all constraints by means of a special shape function (modified ellipses). The procedure is described in more detail.

This paper is part of a larger research project supported by MBB-ERNO Company, Bremen, FRG.

G. I. N. Rozvany and B. L. Karihaloo (eds.), Structural Optimization, 395.

DIVERGENCE INSTABILITY CONDITIONS IN THE OPTIMUM DESIGN OF NONLINEAR ELASTIC SYSTEMS UNDER FOLLOWER FORCES

A.N. KOUNADIS* & O. MAHRENHOLTZ**
*National Technical University of Athens, Greece
**Technical University of Hamburg-Harburg, 2100 Hamburg 90, FRG

Nonlinearly elastic discrete systems under nonconservative, compressive loading [1,2,3] of follower type, that may lose their stability through divergence, are considered. Using a general mathematical formulation a thorough parametric discussion of the critical, prebuckling and postbuckling, large displacement response, is comprehensively presented. The predominant effects on the nonlinear divergence instability of the material nonlinearity as well as of the loading parameters defining the degree of nonconservativeness, are completely revealed. Necessary and sufficient conditions for the existence of regions of devergence instability, are properly established. By means of these conditions the boundary between divergence (static) and flutter (dynamic) instability is found. The case of existence of a double critical point (coincidence of the first and second static buckling eigenmodes), obtained as a result of the linear stability analysis [4-6], is also discussed. At the aforementioned boundary, the (critical) buckling load corresponds to the maximum load–carrying capacity that can be determined by means of a nonlinear (static) stability analysis. Thus, a further insight into the role of certain parameters of paramount importance for the change of mechanism of instability from divergence to flutter, and vice–versa, is also gained.

A simple example of a nonlinear elastic model with two degrees of freedom under a follower load is used to illustrate the general considerations and findings presented herein.

References:

[1] Herrmann, G.; Bungay, R.W.: On the stability of elastic systems subjected to nonconservative forces. J. Appl. Mech., Vol. 31, pp. 435-440, 1964.

[2] Mahrenholtz, O.: Das Stabilitätsverhalten des durchströmten, freihängenden Rohres. Pflüger–Festschrift Hannover 1977.

[3] Bogacz, R.; Mahrenholtz, O.: On stability of a column under circulatory load. Arch. Mech., 38, 3, pp. 281-287, Warszawa, 1986.

[4] Kounadis, A.N.: On the static stability analysis of elastically restrained structures under follower forces. AIAA J., Vol. 18, pp. 473-476, 1980.

[5] Kounadis, A.N.: Divergence and flutter instability of elastically restrained structures under follower forces. Int. J. of Engng. Sci., Vol. 19, pp. 553-562, 1981.

[6] Kounadis, A.N.: The existence of regions of divergence instability for nonconservative systems under follower forces. Int. J. Solids and Structures, Vol. 19, No. 8, pp. 725-733, 1983.

G. I. N. Rozvany and B. L. Karihaloo (eds.), Structural Optimization, 396.

SOLUTION OF MAX–MIN PROBLEMS VIA BOUND FORMULATION AND MATHEMATICAL PROGRAMMING

NIELS OLHOFF
Department of Mechanical Engineering
Aalborg University, Aalborg, Denmark

Structural optimization problems pertaining to maximization of the minimum (or minimization of the maximum) of a set of weighted criteria for given cost, are considered. It is shown that maximization of the initially smaller of the (weighted) criteria and using this as a lower bound in an iterative mathematical programming formulation leads to a very efficient method for solution of problems with global as well as local objectives.

The method is exemplified on problems of maximizing multimodal structural eigenvalues and problems of multicriterion or multipurpose type, and experiences on using the dual method with mixed direct/reciprocal variables developed by Fleury and Braibant for the mathematical programming problem, are reported.

G. I. N. Rozvany and B. L. Karihaloo (eds.), Structural Optimization, 397.

NATURAL STRUCTURAL SHAPES OF MEMBRANE SHELLS

W. STADLER & V. KRISHNAN
San Francisco State University, San Francisco
California, USA

The concept of natural structural shapes is based on the simultaneous "minimization" of the mass and the strain energy of the loaded structure, a multicriteria optimization problem. Natural structural shapes are represented by "proper" Pareto optimal designs in the sense that the limiting minimum weight and minimum stored energy designs are omitted. The method here is used to obtain optimal shell designs within the membrane theory of shells. In order to make full use of standard control theoretic methods, the problems are restricted to axisymmetrically loaded shells of revolution, a formulation involving only one independent variable. The meridional radius of curvature is used as the design variable with the Cartesian coordinates of the midsurface and a modified meridional force per unit length of the midsurface as state variables. Necessary conditions for an Edgeworth–Pareto optimum are derived and only extremal solutions based on this condition are considered.

G. I. N. Rozvany and B. L. Karihaloo (eds.), Structural Optimization, 398.

OPTIMAL DESIGN OF STRUCTURES UNDER CREEP CONDITIONS

MICHAL ŻYCZKOWSKI
Politechnika Krakowska (Technical University of Cracow)
Kraków, Poland

Optimal design of structures, or rather just of simple structural elements working under creep conditions, belongs to the most recent branches of structural optimization: it was initiated by four papers published in the years 1967-8 (Reytman, Prager, Nemirovsky, Życzkowski). The most important differences with respect to elastic or plastic design are as follows: factor of time appearing in the constraints, a great variety of constitutive equations of creep or viscoplasticity, of creep rupture hypotheses, creep buckling theories, various definitions of creep stiffness etc. Moreover, the constraints related to stress relaxation are quite new. So, it is almost impossible to establish a sufficiently general theory and various types of problems must be treated separately by appropriate methods.

The paper gives a classification of problems and then a review of results obtained for bars, columns, arches, trusses, frames, plates and shells. Over 30 per cent of those results were obtained at the Technical University of Cracow.

Particular attention is paid to most recent results. They bring solutions to multimodal optimal design of perfect columns and arches for creep stability, design of imperfect columns and circular plates for creep buckling, ribbed pipelines for brittle creep rupture and creep stability, beams and membranes for viscous creep rupture, viscoplastic bars and beams for dynamic loadings. Optimal design of prestressed structural elements subject to creep and relaxation is also discussed, as well as perspectives of future research.

G. I. N. Rozvany and B. L. Karihaloo (eds.), Structural Optimization, 399.

SUBJECT INDEX

Acknowledgement

The first Editor wishes to acknowledge the contribution of his wife, Susann Rozvany in preparing the Subject Index, List of Contributors, Conference Programme and Contents as well as editing and processing some of the papers in this volume.